Flow Cytometry in Hematology

ANALYTICAL CYTOLOGY SERIES

Marvin A. Van Dilla, Phillip N. Dean, Ole D. Laerum, and Myron R. Melamed (eds)
Flow Cytometry: Instrumentation and Data Analysis, 1985
Joe W. Gray (ed.)
Flow Cytogenetics, 1989
Kohen (ed.)
Cell Structure and Function by Microspectrofluorometry, 1989
Ole D. Laerum and Robert Bjerknes (eds)
Flow Cytometry in Hematology, 1992

Flow Cytometry in Hematology

Edited by

Ole Didrik Laerum and Robert Bjerknes

University of Bergen
Haukeland Hospital
N-5021 Bergen, Norway

ACADEMIC PRESS
(Harcourt Brace Jovanovich, Publishers)

London San Diego New York
Boston Sydney Tokyo Toronto

This book is printed on acid free paper ♾

ACADEMIC PRESS LIMITED
24–28 Oval Road
LONDON NW1 7DX

United States Edition published by
ACADEMIC PRESS, INC.
San Diego, CA 92101

British Library Cataloguing in Publication is available

ISBN 0–12–432940–3

Typeset by Mathematical Composition Setters Ltd, Salisbury, England
and printed in Great Britain at The Bath Press, Avon

Contents

Colour plates 1–16 referring to Chapters 1.3 and 2.2 respectively appear between pages 116–117.

Contributors

MICHAEL ANDREEFF, University of Texas M. D. Anderson Cancer Center, 1515 Holcombe Boulevard, Box 61, Houston, Texas 77030, USA

KENNETH A. AULT, Maine Cytometry Research Institute, 190 Park Avenue, Portland, Maine 04102, USA

DAVID A. BASS, Department of Medicine, Bowman Gray School of Medicine, 300 South Hawthorne Road, Winston-Salem, North Carolina 27103, USA

ROBERT BJERKNES, Department of Pediatrics, University of Bergen, Haukeland Hospital, N-5021 Bergen, Norway

CURT I. CIVIN, Division of Pediatric Oncology, Johns Hopkins Oncology Center 3-121, Johns Hopkins University School of Medicine, 600 North Wolfe Street, Baltimore, Maryland 21205, USA

MARY J. FACKLER, Division of Pediatric Oncology, Johns Hopkins Oncology Center 3-121, Johns Hopkins University School of Medicine, 600 North Wolfe Street, Baltimore, Maryland 21205, USA

ROBERT S. HILLMAN, Maine Cytometry Research Institute, 190 Park Avenue, Portland, Maine 04102, USA

CHRIS J. JANSE, Laboratory for Parasitology, University of Leiden, Wassenaareweg 62, Leiden, The Netherlands

WALTER KNAPP, Tumor Immunology Laboratory, Institute of Immunology, Borschkegasse 8 A, A-1090 Vienna, Austria

OLE DIDRIK LAERUM, The Gade Institute, Department of Pathology, University of Bergen, Haukeland Hospital, N-5021 Bergen, Norway

ALAN LANDAY, Department of Immunology and Microbiology, Rush-Presbyterian-St. Luke's Medical Center, 1653 West Congress Parkway, Chicago, Illinois 69612, USA

MICHAEL R. LOKEN, Becton Dickinson Immunocytometry Systems, 2350 Qume Drive, San Jose, California 95131, USA

FRIDTJOF LUND-JOHANSEN, The Gade Institute, Department of Pathology, University of Bergen, Haukeland Hospital, N-5021 Bergen, Norway

JANE G. MITCHELL, Maine Cytometry Research Institute, 190 Park Avenue, Portland, Maine 04102, USA

BAREND MONS, Laboratory for Parasitology, University of Leiden, Wassenaareweg 62, Leiden, The Netherlands

ALEXANDER NAKEFF, Division of Hematology and Oncology, Wayne State University, P.O. Box 02188, Detroit, Michigan 4820-0188, USA

BETSY OHLSSON-WILHELM, Department of Medicine, University Hospital, The Milton S. Hershey Medical Center, P.O. Box 850, Hershey, Pennsylvania 17033, USA

C. RINDER, Maine Cytometry Research Institute, 190 Park Avenue, Portland, Maine 04102, USA

H. M. RINDER, Maine Cytometry Research Institute, 190 Park Avenue, Portland, Maine 04102, USA

MICHAEL SEEDS, Department of Medicine, Bowman Gray School of Medicine, 300 South Hawthorne Road, Winston-Salem, North Carolina 27103, USA

INE SEGERS-NOLTEN, Twente University, Enschede, The Netherlands
VIREN O. SHAH, Becton Dickinson Immunocytometry Systems, 2350 Qume Drive, San Jose, California 95131, USA
OLAV SLETVOLD, Department of Geriatrics, The Deaconess Hospital, Haraldsplass, N-5009 Bergen, Norway
RUNE SMAALAND, The Gade Institute, Department of Pathology, University of Bergen, Haukeland Hospital, N-5021 Bergen, Norway
HANS J. TANKE, Department of Cytochemistry and Cytometry, Sylvius Laboratories, University of Leiden, Wassenaarseweg 72, 2333 Al Leiden, The Netherlands
LEON W. M. M. TERSTAPPEN, Becton Dickinson Immunocytometry Systems, 2350 Qume Drive, San Jose, California 95131, USA
JAN W. M. VISSER, Radiobiological Institute TNO, 151 Lange Kleiweg, 2288 GJ Rijswijk, The Netherlands

Preface

Hemopoietic tissues and peripheral blood contain cell populations of great morphological, structural and functional complexity. Flow cytometric techniques have in recent years undergone a rapid development, and these assays have provided a powerful tool for the biological characterization of hematopoietic cells. Flow cytometry allows multiparameter analyses, and the high rates at which these measurements can be performed result in very reliable statistics. The inaccuracy implied by judgement of stained hemopoietic cells in smears can thus be replaced by precise measurements, also allowing subpopulations of cells to be identified and sorted for further characterization.

Some of the flow cytometric methods are already employed for routine use in hematology, e.g. leukocyte differential counting and immunophenotyping in malignant hematological diseases and AIDS. Most of the other techniques are today considered mainly experimental tools. We hold the opinion that the rapid expansion in this research field, combined with the general applicability of flow cytometric methods, give a great potential for future use in experimental as well as clinical hematology. We therefore felt the need for a monograph which surveyed the various applications of flow cytometry in hematology. This topic also reflects our own field of research throughout several years, and it is our sincere hope that this volume may serve as a useful guide.

OLE DIDRIK LAERUM AND ROBERT BJERKNES
Bergen
May, 1991

1

Normal Hemopoiesis

1.1

Introduction and General Outline of Hemopoiesis

OLE DIDRIK LAERUM

University of Bergen
Department of Pathology, The Gade Institute
Haukeland Hospital, Bergen, Norway

I. HISTORICAL BACKGROUND OF FLOW CYTOMETRY

Although the forerunners of the automated cytometric methods gave important contributions from the 1930s to the 1950s, the main methodological development occurred in the late 1960s. From this period automated devices became available where single cells could be quantitated in suspension at a high rate. This started with the observation that single cells passing a narrow laser beam would spread the light proportional to their size. In addition, absorption of light from stained cells could be automatically measured. This was followed by the observation that the staining of different cell components with a fluorochrome could enable the measurement of fluorescent signals as they passed the light source.

A brief history of the field was given by Shapiro in 1984 (*19*) with main emphasis on the developments which occurred in the United States. In parallel with this Göhde and collaborators (*10*) developed a microscope-based flow cytometric device with a degree of resolution that even today has hardly been competed.

Rather early these methods were used for measurement of hemopoietic cells, including studies on DNA distributions in leukemia cells (*6*). It was shown that aneuploidy was a frequent finding in leukemia (for review see *7*). In addition, the cell cycle distribution of normal and malignant hemopoietic cells could be measured, and this was later used for treatment monitoring in leukemia patients (*1,5,6,20*). In the beginning this received little attention. With the introduction of the acridine orange staining of leukemia cells, both RNA and DNA

FLOW CYTOMETRY IN HEMATOLOGY
ISBN 0–12–432940–3

could be measured simultaneously, and subpopulations could be discriminated (for review see *2*). This was both used for diagnostic purposes and for studying effects of therapy (see also Chapter 3.5).

In the 1970s one of the research directions in cancer chemotherapy was to explore the possibilities for synchronizing the proliferating cells. The idea was to use small doses of cytostatic agents, such as hydroxyurea, in order to synchronize all proliferating cells in the same cell cycle phase, e.g. in the DNA synthetic phase. By releasing the cells afterwards, normal cells such as the bone marrow cells were expected to proceed more rapidly through the cell cycle than the highly pathological, malignant cell population. By giving phase-specific cytotoxic agents, one could therefore achieve selective kill of the tumor cells. Although this was not successful, it greatly increased the interest for monitoring the cell cycle both in normal cells and the tumor population as a background for cancer chemotherapy studies. For the first time flow cytometry offered a technique which was rapid enough for this purpose and where small cell samples could be studied sequentially. In fact, several early studies showed that cytostatic treatment induced altered cell cycle distribution in the peripheral blood and bone marrow of patients with acute leukemia (*1,5,6,20*).

Another important objective was the study of aneuploid cells in malignant conditions of hemopoietic tissues. These were identified both in peripheral blood, in lymph nodes and in the bone marrow and were early considered as characteristic markers of the disease. Since the resolution of the DNA measurements was rather variable, mainly gross deviations from the normal cellular DNA content could be identified (*2,14*).

Rather early, on the basis of two-parameter measurements of acridine orange-stained leukemia cells, both DNA distribution, aneuploidy and RNA content could be studied simultaneously and related to chemotherapy (Chapter 3.5). The early use of flow cytometry in hematology, however, received relatively little attention.

There were also several objections against the use of flow cytometric analysis as a supplement to the common hematological methods at that time. First, the bone marrow was considered as too complex a cell population for being characterized by single-parameter flow cytometry, e.g. DNA distribution. The few available studies on two-parameter measurements were likewise not considered specific enough. Second, therapy monitoring by using peripheral blood was not considered representative of the large mass of leukemia cells which resided in the bone marrow and other tissues of the body. The danger of biased measurements on a small non-representative cell sample was therefore considered as high.

The diagnostics both on peripheral blood and bone marrow were traditionally based on morphological recognition in cell smears. Pure morphology was also the basis for recognition of the different cell lineages and their maturation stages. This was the strongest objection against the use of flow cytometry. Therefore, microspectrophotometry was more used, where single cells could be measured after they had been identified by the eye. In man, the knowledge of cell kinetics in the hemopoietic system was largely based on a combination of morphology, cytochemistry and to some extent autoradiography, and the same largely applied to experimental animals as well (see, e.g. *12*).

For cells that were so few that they could hardly be identified on smears, one was dependent on functional tests. This particularly applied to the studies of stem and progenitor cells. Quantitation of the stem cells was dependent on the spleen colony technique with transplantation of a bone marrow cell suspension to lethally irradiated host mice and subsequent counting of spleen colonies (*16*). Characterization of progenitor cells was mainly based on colony formation in semisolid agar or methylcellulose medium (*17*) and to some extent other techniques, such as *in vivo* culture in intraperitoneal diffusion chambers (*4,21*).

II. SOME FEATURES OF RECENT DEVELOPMENTS OF FLOW CYTOMETRY RELATED TO HEMATOLOGY

In recent years an expanding research activity has totally changed our picture of hematopoietic tissues. Based on multiple marker studies they seem to consist of even more complex populations than earlier believed. The development of monoclonal antibodies has led to the recognition of multiple functional markers of hematopoietic cells. By use of fluorescent ligands these markers can be directly visualized. An extensive combination of immunocytochemistry with other methods, such as cell and tissue culture, molecular biology studies, etc., has led far away from the classical cytological classification of the cells. Today cytology only makes a frame for biological characterization of the complex hematopoietic cell populations.

The recent developments in the study of various cytokines and growth factor regulation of the various cell lines and maturation stages have also led to a more "biological" view on hemopoiesis. This both applies to their binding to various target cells, their combined actions on various subpopulations and characterization of their receptors and gene sequences (see, e.g. *18*).

Improved culture methods have greatly facilitated the propagation of restricted cell populations for prolonged periods. In addition, new separation techniques, both combining sedimentation and centrifugation methods as well as removal of unwanted cells by cytotoxic antibodies or antibody-coated magnetic beads, have undergone a rapid development. Not least, the use of flow sorting in combination with monoclonal antibodies has been a great advancement.

On this background, the various techniques of flow cytometry have become important tools not only in experimental hematology, but also for extensive clinical studies. One basis for this has been the further development of flow sorting methods. The available machines are more stable and simpler to use than the earlier ones, and this enables their daily and routine use for morphological and functional tests on sorted subpopulations.

The rapid development in computer technology has enabled simple procedures for discrimination analysis of multiple parameters. This again can be used for setting windows whereby even very small subpopulations can specifically be quantitated and sorted. The use of list mode function in the computers has also facilitated multiparameter analyses of cells, especially of small samples, whereby multiple parameters measured on the same cells can be stored. Later, the different parameters can be analysed in relation to each other and subpopulations further characterized (for a review, see *22*).

From the very beginning one main advantage of studying blood and bone marrow cells was that they could easily be obtained as single cell suspensions. Therefore these cell types have not been so dependent on improved preparation techniques as in flow cytometric studies of other tissues. In addition, the rapid development in the use of various fluorescent dyes has facilitated the development of multiparameter studies. Both fluorochromes with differing light emission wavelength and excited by the same laser at one wavelength, as well as excitation at different wavelengths with several lasers, have been used.

Further, when different fields of research are expanding individually, and their potentials are then combined, this gives an almost exponential increase in the number of possible uses.

III. OUTLINE OF HEMOPOIESIS

The bone marrow is one of the most complex organ systems in the body. Distributed as red marrow all over the skeleton, but mainly in the pelvis, sternum and the vertebrae, an adult

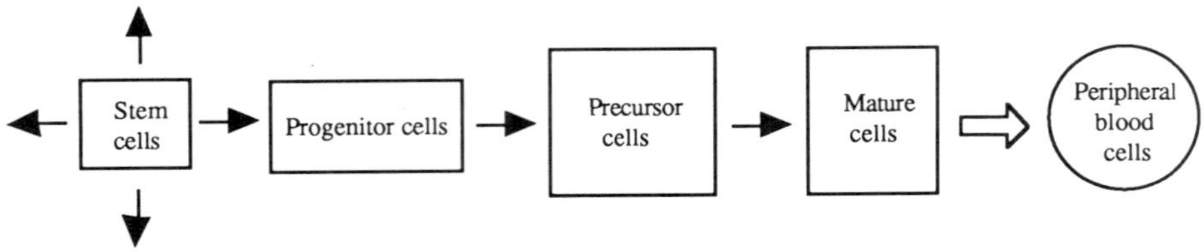

Fig. 1. Schematic diagram of the different compartments of hemopoiesis. A multipotent stem cell, usually 1 per 2000 bone marrow cells, can give rise to all subclasses of hemopoietic cells. This occurs through a state of progenitor cells committed for one lineage, followed by several stages of maturation in the precursor cell pool, until the end stage as non-proliferative mature cells is reached; i.e. through combined cell division and maturation a few immature cells make many mature cells.

person of 70 kg body weight has roughly 1.5 kg hemopoietic tissue. This comes in addition to 5–6 l of blood, of which more than 2 kg consist of red blood cells.

The main principles of the production of blood cells are the following:

(1) A few stem cells make a large number of progeny cells. This goes on continuously, and a single stem cell may in a week or two have up to 10^6 descendants. The stem cells also maintain their own numbers by additional cell division (Fig. 1).

(2) Immature cells make mature blood cells which are able to perform specialized functions in the body.

(3) A limited number of cell divisions go in parallel with maturation, and the eventual mature cells lose their ability to divide.

(4) Mature cells are stored for a while in the bone marrow until they enter the bloodstream to exert their specialized functions in the blood or in tissues.

(5) The multipotent stem cells have the ability to direct their development into all subclasses of hemopoietic cell types, including erythrocytes, granulocytes (neutrophilic, eosinophilic and basophilic), mast cells, monocytes and macrophages, megakaryocytes and plateletes, as well as the different types of lymphocytes, including B, T, O and NK cells.

The period from a stem cell starting to divide until its progeny reaches the stage of a non-proliferative mature cell usually takes 2 weeks in man, and 1 week in mice. As a rough estimate this implies that the whole bone marrow is replaced at least twice every month. For a man of 70 kg body weight this would amount to a production of 3 kg of hemopoietic tissue a month and one-half his body weight every year. In the lifetime of man this will amount to a total production of *three tons of blood cells*.

In consequence, hemopoiesis should be considered as a continuous stream of cell divisions, the expression of the genetic programs for different maturation lines, cell maturation, cell migration and cell loss. Once the cells enter the bloodstream, they will not return to the bone marrow. Erythrocytes in average live 120 days in man, but far less time in animals with shorter lifespan. In man about one-half of the neutrophilic granulocytes leave the bloodstream every 7 hours and migrate into the tissues where they are lost in the course of a few days. Monocytes also migrate into the tissues where they divide once or twice and develop into macrophages. A high proportion of these reside in the lungs, but there are also numerous macrophages in other sites such as the abdominal cavity.

The high rate at which mature blood cells are lost in the body makes it mandatory for the bone marrow to maintain a high rate of cell division. The main cell loss occurs in the granulocytes, which in man constitute about 60% of the leukocytes in peripheral blood, the rest being lymphocytes. In the mouse 20–30% of the leukocytes are granulocytes, and the rest lymphocytes. The various classes of lymphocytes are more longlived and can recirculate between blood and the lymphoid system for many months or longer.

Production of new blood cells in the bone marrow does not occur at a constant rate. First, the rate is continuously adjusted proportional to the cell loss in order to keep the equilibrium. Second, the rate of cell production is highly dependent on the age of the individual, with young persons tending to have more active hemopoiesis than the older ones. In addition, variations can be seen with the season and with the time of day (Chapter 1.5). It is therefore justified to state that hemopoiesis occurs in waves (for more details about the kinetics of hemopoiesis, readers are referred to more general surveys, e.g. *3,8,9,15*).

As mentioned earlier, recent hematological research has given entirely new insights into the biological properties of the different hemopoietic cell classes. This has led to an increased need for quantitative methods by which biological properties of individual cells could be quantitated and classified. Analytical cytological methods, and mainly flow cytometry, have been important tools for this research and have greatly contributed to these new insights. This has led to the development of a *biologically* and not only a *morphologically* based hematology. The main areas are the following:

— Antigenic development of hemopoietic cells, both related to the maturation compartments and the maturation stages (*13*; Chapters 1.2 and 1.3).
— Proliferative activity of the different compartments and maturation stages, as well as the division rate of the whole tissue (Chapter 1.5).
— Differing sizes and optical properties of the various classes of hemopoietic cells. This makes it possible to use the scattering of light at different angles as a valuable tool for discriminating subpopulations (Chapters 1.2 and 2.2).
— Functional tests at the single cell level. This is mainly related to the maturing and fully mature blood cells, for example phagocytosis, oxidative metabolism, intracellular pH (Chapters 2.3 and 2.4).
— Regulatory aspects. By labeling of fluorochromes both to regulatory molecules (fluorochrome-labeled ligands), their receptors, the kinetics of binding and internalization can be quantitated on individual target cells. Since this is a more general approach related to hormone and growth factor research and not only hematology, this is not dealt with in the present volume. Readers are therefore referred to general literature.
— Knowledge of how the above-mentioned biological properties are combined at the single cell level. Hence, multiparameter measurements can be used for automatic discrimination between subpopulations of hemopoietic cells and leukocytes, both for functional analysis (Chapters 2.3 and 2.4), and their use for automated leukocyte differential counting (Chapter 2.2).
— New insight into the nature of different pathological conditions, by which measurements of deviating biological properties can be used for diagnostic and prognostic purposes (Chapters 3.1–3.5).

IV. IMPLICATIONS AND FUTURE PROSPECTS

Flow cytometry should not be regarded as a separate area of research, but merely as an important tool for multiparameter single cell measurements in complex tissues and with a high velocity. The use of flow sorting not only on fixed cells, but also on living cells under sterile conditions, enables a close connection between analytical cytology and morphology as well as cell culture and further biological analysis. Although the research and practical applications of FCM in hematology today have an overweight of analysis on antigenic properties, the rapid development in molecular biology should be kept in mind. In general there is a rapid expansion in the development of fluorescent DNA probes and hybridization techniques for obtaining genetic fingerprints. This is an actual expansion area not only in the study of single cells, but also isolated chromosomes (*11*).

In the coming years a more extensive clinical use of flow cytometry in hematology is expected. This is both for diagnostic use in daily routine and research, i.e. for further biological characterization of the cells with establishment of prognosis of a given disease, and also in therapy monitoring. Both identification of malignant cells and establishing the degree of side effects of cytostatic therapy on normal hemopoiesis are possible.

It is also expected that the available cheap automated instruments combined with more standardized preparation methods and improved procedures for data analysis will further enhance the practical value of flow cytometric measurements for daily clinical needs.

REFERENCES

1. Aardal, N. P., Talstad, I., and Laerum, O. D. (1979). Sequential flow cytometric analysis of cellular DNA-content in peripheral blood during treatment for acute leukemia. *Scand. J. Haematol.* **22**, 25–32.

2. Andreeff, M., Darzynkiewics, Z., Sharpless, L. T. K., Clarkson, B. D., and Melamed, M. R. (1980). Discrimination of human leukemia subtypes by flow cytometric analysis of cellular DNA and RNA. *Blood* **55**, 282.

3. Benestad, H. B., and Laerum, O. D. (1989). The neutrophilic granulocyte. *In*: "Current Topic in Pathology. Vol. 79: Cell Kinetics of the Inflammatory Reaction" (Iversen, O. H., ed.), pp. 7–36. Springer-Verlag, Berlin.

4. Böyum, A., Boecker, W., Carsten, A. L., and Cronkite, E. P. (1972). Proliferation of human bone marrow cells in diffusion chambers implanted into normal or irradiated mice. *Blood* **40**, 163–173.

5. Büchner, T., and Hiddeman, W. (1978). Kinetic response to antileukemic drugs. *In*: "Puls-Cytophotometry III" (Lutz, D. ed.), pp. 637–641. European Press Medikon, Ghent.

6. Büchner, Th., Dittrich, W., and Göhde, W. (1971). Die Impulscytophotometrie in der hämatologischen Cytologie. *Klin Wschr.* **49**, 1090–1092.

7. Büchner, Th., Bloomfiel, C. D., Hiddemann, W., Hossfeld, D. K., and Schumann, J., eds (1985). Tumor Aneuploidy. Springer-Verlag, Heildelberg.

8. Dancey, J. T., Deubelbeiss, K. A., Harker, L. A., and Finch, C. A. (1976). Neutrophil kinetics in man. *J. Clin. Invest.* **58**, 705–715.

9. Dresch, C., Troccoli, G., and Mary, J. Y. (1986). Growth fraction of myelocytes in normal human granulopoiesis. *Cell Tissue Kinet.* **19**, 11–22.

10. Göhde, W. (1973). Automation of cytofluorometry by use of the impuls-microphotometer. *In*: "Fluorescence Techniques in Cell Biology" (A. A. Thaer and M. Sernetz, eds), pp. 81–88. Springer-Verlag, Berlin.

11. Gray, J. W., ed. (1990). "Flow Cytogenetics." Analytical Cytology Series. Academic Press, London.

12. Killman, S.-A., Cronkite, E. P., Fliedner, T. M., and Bond, V. P. (1962). Mitotic indicies of human bone marrow cells. I. Number and cytologic distribution of mitoses. *Blood* **19**, 743–750.

13. Knapp, W., ed. (1989). "Leukocyte Typing IV." Oxford University Press, Oxford.

14. Laerum, O. D., and Farsund, T. (1981). Clinical application of flow cytometry: A review. *Cytometry* **2**, 1–13.

15. Lajtha, L. G. (1983). Stem cell concepts. *In* "Stem Cells. Their Identification and Characterisation" (Potten, C. S., ed.), pp. 1–11. Churchill Livingstone, Edinburgh.

16. Lord, B. I. (1983). Hemopoietic stem cells. *In* "Stem Cells. Their Identification and Characterisation" (Potten, C. S., ed.), pp. 118–154. Churchill Livingstone, Edinburgh.

17. Metcalf, D. (1977). "Hemopoietic Colonies." Springer-Verlag, Berlin.

18. Nicola, N. A. (1989). Hemopoietic cell growth factors and their receptors. *Annu. Rev. Biochem.* **58**, 45–77.

19. Shapiro, H. M. (1988). "Practical Flow Cytometry," 2nd edn. Alan R. Liss, New York.

20. Smets, L. A., Mulder, E., de Waal, F. C., Cleton, F. J., and Blok, J. (1976). Early responses to chemotherapy detected by pulse cytophotometry. *Br. J. Cancer* **34**, 153–161.

21. Steinberg, H. N., and Robinson, S. H. (1985). Laboratory methods in experimental hematology. I. The diffusion-chamber technique in experimental hematology. *Exp. Hematol.* **13**, 16–22.

22. Van Dilla, M. A., Dean, P. N., Laerum, O. D., and Melamed, M. R., eds (1985). "Flow Cytometry: Instrumentation and Data Analysis." Analytical Cytology Series. Academic Press, London.

1.2

Isolation and Biological Characterization of Bone Marrow Stem Cells and Progenitor Cells

JAN W. M. VISSER

Institute of Applied Radiobiology and Immunology TNO
P.O. Box 5815
2280 HV Rijswijk, The Netherlands

I. INTRODUCTION

All blood cells, most bone marrow cells and many specialized macrophages and lymphocytes in all organs are derived from a common ancestor, the pluripotent hemopoietic stem cell (PHSC). This cell type is defined by its capability to replicate extensively and to differentiate into progenitor cells which are committed to one or more hemopoietic lineages. In the clinic, these capacities of the PHSC are employed to reconstitute the hemopoietic system in patients who receive bone marrow transplantation after therapy. In preclinical tests, gene therapy through transplantation of DNA-manipulated stem cells has proven possible in the mouse and monkey. The human hematoimmunological system and its reactions upon infections can be studied in specially conditioned mice by transplantation of human stem cells and some of their regulatory environment. Because of these

FLOW CYTOMETRY IN HEMATOLOGY
ISBN 0-12-432940-3

applications, and also because of the intriguing regulation of their proliferation and differentiation, PHSCs are the subject of many and extensive investigations. This chapter describes the involvement of flow cytometry in these studies.

During embryogenesis, the PHSCs are first detectable in the yolk sac, and later in the fetal liver. In adult life, PHSCs reside mainly in the bone marrow, probably at specific sites or niches, where they are kept in a quiescent state, allowed to self-renew, or differentiate. A cascade of differentiated offspring also fills the bone cavities, so that less than 1% of the bone marrow cells are pluripotent stem cells. Because of this low incidence, and because the PHSCs closely resemble their daughter cells the committed progenitors, they cannot be studied by direct microscopical examination. Therefore, a number of indirect techniques has been developed to analyze the characteristics of the PHSC, and to elucidate the regulation of their proliferation and differentiation. One group of such techniques is based on the capacity of PHSCs to generate extensive offspring, making use of a variety of culture systems. Other studies employ cell separation techniques in an analytical fashion using a well-defined physical or cytochemical property, such as size, density or specific membrane antigens as a basis for the separation. These studies combine the separation with cell culture to demonstrate the presence of PHSC in certain fractions. In other studies cell separation is employed in a preparative way with the intention to purify the stem cells and to study them by direct methods. All separation protocols for the isolation of stem cells up to now consist of a combination of techniques, of which flow cytometry is a most important one.

A. Strategies for Sorting Stem Cells

The applications for purified PHSCs generally require high numbers of bone marrow cells which at present could not be processed only by fluorescence-activated cell sorting. Because PHSCs are cells of low buoyant density (*14,65,98,148*), one of the first steps in their isolation often consists of equilibrium density centrifugation. Between 5 and 30% of the bone marrow cells are recovered from the density gradient fractions which contain the stem cells. In some laboratories counterflow centrifugal elutriation (CCE) is used as the first step (*116,165,170*). Because elutriation separates cells primarily on the basis of differences in size, and because PHSCs have a similar diameter to most lymphocytes and many other bone marrow cells (*3,43–46,137,153,155,171*), one has to modify the CCE technique to obtain fractions with a useful pre-enrichment. The second step often consists of the removal of mature bone marrow cells by binding them to each other, to flasks or to beads using antibodies. Using cocktails of antibodies specifically directed against erythrocytes, lymphocytes, monocytes, granulocytes and the committed precursors of these cell types, 90–98% of the bone marrow cells are removed (*23,24,66,130,162*). This so-called negative selection is a good strategy if there is no antibody or lectin available which binds only a few of the bone marrow cells including the PHSCs, as is the case for mouse and rat. However, monkey and human PHSCs have an antigen, CD34, which occurs on only 1–4% of all bone marrow cells (*35,36,163*). Therefore, the isolation of these stem cells is performed more efficiently by positive selection, binding the PHSC to the flasks or beads. Several methods are available at present to release the bound cells (*8,37*). In a number of protocols for PHSC isolation, especially in those for human PHSCs, an adherence step is included, with the intention of removing adherent cells, predominantly monocytes. Recently, however, Bearpark and Gordon (*13*) reported that an important subpopulation of stem cells is also removed by the adherence step, both with mouse and with human bone marrow. This implies further, that each step of the total procedure should be performed under conditions which prevent adherence.

The last step in the purification of stem cells

consists of fluorescence-activated cell sorting, which now is facilitated by the 30- to 100-fold pre-enrichment. The flow cytometer provides complementary discriminating parameters for sorting, such as forward and perpendicular light scatter, supravital staining with DNA or mitochondria specific dyes. In addition, using fluoresceinated second layer antibodies the flow cytometer may serve to control the quality of the immunological pre-enrichment procedures, and to correct and compensate errors and irregularities which may occur during the first separation steps. Furthermore, the sorter can be used to accurately count the sorted cells, and to deposit specified numbers of stem cells into culture dishes to clone them, or on nitrocellulose filters for the analysis of gene expression.

B. Stem Cell Assays

a. *The CFU-S Assay*

The hemopoietic stem cells are detected by their capacity to produce blood and bone marrow cells *in vitro* or *in vivo*. For many years mouse and rat PHSCs were detected and enumerated by the spleen colony forming unit (CFU-S) assay (*141*), a clonal *in vivo* test for which colonies of mature hemopoietic cells on the spleens of lethally irradiated mice were assumed to originate from transplanted stem cells. Careful examination of the growth of these spleen colonies and differences in colony formation by different fractions of sorted cells revealed that only part of the spleen colonies, more specifically even only part of the colonies visible at 12–16 days after transplantation (day 12 CFU-S), are due to pluripotent stem cells (*5,63,67,83,84,101,112–115,157*). The majority of the spleen colonies in the mouse, all of the day 8 CFU-S and many of the day 12 ones, arise from committed stem cells with a limited self-renewal capacity. Therefore, alternative stem cell assays have received more attention in recent years. Such alternative methods had already been studied for a long period in analysis of dog, monkey and human stem cells, for which no CFU-S assay was available. The isolation of PHSCs by different methods gave different results in the various assays. On the one hand, it could be concluded that the PHSCs are heterogeneous. On the other hand, the specificity of each test for detecting and counting PHSC subpopulations differing in their self-renewal potential and pluripotency could be evaluated using flow-sorted fractions of purified stem cells. There are a number of candidate PHSC assays to be included in these evaluations and comparative studies.

b. In Vitro *Tests*

Formation of hemopoietic colonies in semi-solid media such as agar or methylcellulose, in the presence of growth factors has been proposed for this purpose. The CFU-Mix or CFU-GEMM (colony forming unit-granulocyte erythrocyte monocyte megakaryocyte) assays count colonies which have to have been caused by multipotent stem cells because of their multilineage cell content (*49,62,70*). The pluripotency of the cells which give rise to these colonies has been questioned, however, especially with respect to their capacity for self-renewal. Reconstitution of the hemopoietic system has occurred *in vivo* by sorted bone marrow grafts which did not contain CFC-GEMM, cells capable of generating CFU-GEMM (*73,172*). *In vitro* treatment of bone marrow cells with 4-hydroperoxycyclophosphamide (4-HC) kills all CFU-GEMM, but it spares the PHSCs which give long-term CFU-GM production *in vitro* (*57,118,122,126,128,168*). In addition, with respect to CFU-GEMM, the plating efficiency of the assay is not known, so that absolute quantitation of CFC-GEMM or PHSCs with this test is as yet not possible. CFU-GELMM, also containing lymphocytes, have also been proposed to enumerate stem cells, however until now not in combination with cell sorting. The blast cell colony assay, which counts small colonies of undifferentiated hemopoietic cells stimulated by a combination of the growth factors interleukin 3 (IL-3) and interleukin 6 (IL-6) or G-CSF (granulocyte-colony stimulating factor), is a good candidate as a test system for PHSCs (*68,78,134,135*). Similarly, so-called HPP-CFC-1 (high proliferative potential-colony forming cells) which are grown in agar in the presence of IL-3, IL-6 and M-CSF (monocyte-colony stimulating factor) and form very large colonies, are likely to be at least in part similar to PHSCs (*20–22,90*). Both the blast cell colony formation and the HPP-CFU-C assay have either a very low plating efficiency or they detect cells which are very rare. Purification of stem cells and subsequent culture will reveal which of these two possibilities is the case, and which of the tests really measures PHSCs. The CFU-A assay (*119*) has also been reported to measure early hemopoietic cells using a combination of growth factors. However, its results correlate very well with those of the day 8 CFU-S assay for which it is clear that it does not detect pluripotent stem cells.

A number of new stem cell tests have been developed from long-term bone marrow culture techniques (*143*). Hemopoiesis can be maintained *in vitro* for several months in and on layers of stromal cells which seem to provide an adequate microenvironment for self-renewal and differentiation of PHSC, which adhere to the stroma. In Dexter-type cultures the production of non-adherent progenitor cells of granulocytes and monocytes (CFU-GM) can be quantitated weekly upon the changing of the medium (*3,41,137,168*). Similarly, in Whitlock–Witte cultures the ongoing production of lymphocyte precursors may serve as a measure for the presence of PHSC (*106*). Several studies have been reported to modify the stromal layer cultures into a clonal assay (*58,120*) by inoculation of bone marrow cells on the layer, which were allowed to adhere; subsequently the non-adherent cells were washed away, and a semi-solid medium was added to prevent migration of newly formed cells. This adherent blast cell colony assay is also a likely candidate for a PHSC test, which has to be compared with the other assays using highly enriched and purified fractions of stem cells and progenitor cells. Another promising clonal test from the long-term bone marrow cultures is the cobblestone area forming cell (CAFC) assay, which counts only blast cell colonies growing in between stromal cells. CAFCs, which give colonies later than 4 weeks after the initiation of the culture, are thought to be PHSCs. Earlier colonies, and colonies growing on top of the stromal layer are derived from committed progenitor cells. The assay has recently been modified using small-scale cultures and limiting dilution analysis to give quantitative information about the cell type generating the cobblestone areas (*117*).

c. In Vivo *Tests*

PHSCs can be demonstrated *in vivo* by their reconstitution of the hemopoietic system after transplantation into lethally irradiated or genetically anemic recipients. A difference again must be made with respect to long-term and short-term reconstitution and hemopoietic chimerism. Established *in vivo* tests for PHSCs are the 30-day survival assay, which compares the numbers of bone marrow cells required for the protection of lethally irradiated mice, and the marrow repopulating ability (MRA) assay, which determines the number of CFU-M (monocyte colony forming units) per femur 14 days after irradiation and transplantation (*20,67*). Recently, the validity of these tests for determining PHSC has become suspect, now that fractions of cells can be obtained with a relatively low radioprotective capability and a high MRA, few CFU-S and good long-term repopulating capacity (*22,72,113–115,158*). It appears that the earliest stem cells with the best long-term repopulating capacity and poor spleen colony formation require too much time to produce platelets and macrophages in the host and that, therefore, their radioprotective capability is somewhat smaller than that of the cells represented by most of the day 12 CFU-S. This conclusion affects the assumption that the PHSC content of different grafts can be compared by examining the reappearance of reticulocytes and leukocytes in the peripheral blood after transplantation, as is being done for monkey PHSCs, and in clinical studies retrospectively for human PHSCs. Short- and long-term hemopoietic reconstitution and chimerism can also be demonstrated using a variety of markers (*4*,*92*). If male stem cells are transplanted into female recipients, Y chromosome *in situ* hybridization can now be applied to examine the chimerism in all differentiation lineages. Adequate probes for such studies are available for humans and for mice (*39*,*129*,*158*). Figure 1 indicates that the Y chromosome detection in interphase nuclei can be performed by flow cytometry facilitating accurate quantitation of the chimeric state.

Another useful test for long-term repopulation by PHSCs is the cure of genetically anemic W/W^{v} mice, which can be monitored by examination of the size of the erythrocytes for example, by measuring their forward light scatter or Coulter volume (*27*). Similarly, the cure of alpha-thalassemic mice by transplantation of PHSCs can be determined by monitoring the size of their erythrocytes (Fig. 2) (*161*). With both of these assays it must be realized that only erythroid chimerism is determined, which is not necessarily a measure for PHSCs.

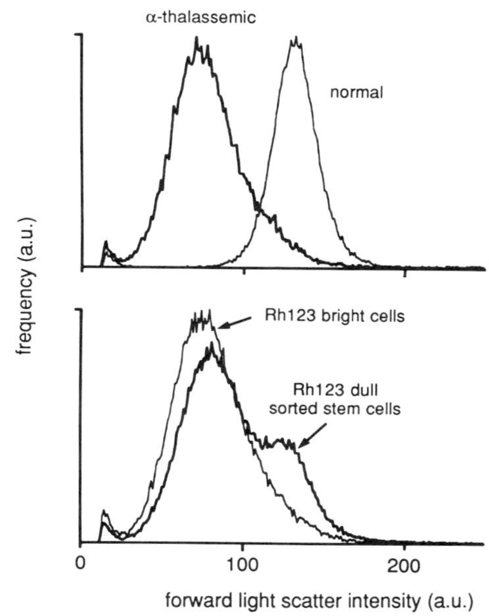

Fig. 2. Histograms of the forward light scatter intensities of red blood cells (upper panel: thick line) from alpha-thalassemic mice and from normal mice (thin line). Lower panel: from mice 10 weeks after 3 Gy irradiation and transplantation with 3000 sorted rhodamine 123 bright (thin line) or dull (thick line) stem cells (*150*).

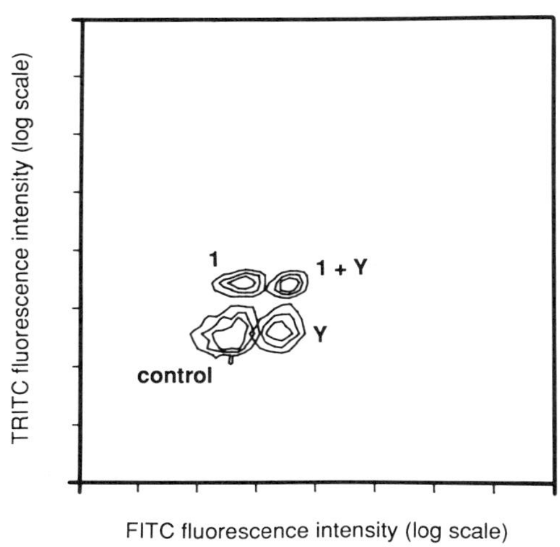

Fig. 1. Bivariate distributions of red (TRITC) and green (FITC) fluorescence intensities of human lymphocyte nuclei in G_0/G_1, which are hybridized *in situ* in suspension with a chromosome Y-specific (FITC) and a chromosome 1 (TRITC)-specific DNA probe, as well as of nuclei hybridized with no or only one probe (*39*).

Colony formation on the thymus after lethal irradiation and intrathymic transplantation has recently been argued to be a measure of PHSCs (*130*). Although the correlation between the results of this test and especially those of the CFU-S assay are good, also with highly purified mouse stem cells, the nature of the colonies and their origin has to be better established to accept it for the quantitation of PHSCs. Repopulation of the thymus with T cells by transplantation of purified PHSCs which also efficiently give day 12 spleen colonies, clearly demonstrates the pluripotent character of the sorted stem cells, as was earlier demonstrated by Mulder *et al.* (*100*,*104*).

ROCEDURES FOR FLOW SORTING MURINE HEMOPOIETIC STEM CELLS

A. Adult Mouse Bone Marrow-derived Stem Cells

The purification of PHSCs from mouse bone marrow by flow cytometry was preceded by a series of analytical separation studies elucidating properties of the CFU-S (*152*). It had been established that CFU-S are of low density and average size (*14,91,98,153,171*). Their forward and perpendicular light scatter characteristics had been carefully studied (Figs 3 and 4) (*43,45,46,154*). CFU-S were H-2K positive (*19,44,47,123,147*), Qa positive (*21,51,63,64,76*), I-E positive (*51*), anti-brain positive (*42,56,97*), anti-sperm positive (*19*), Pgp-1 positive (*11,12*), GM-1.2 negative (*11*) and Thy-1 negative (*140,147*) according to the earlier tests; Thy-1 low and not negative according to the later, more detailed investigations (*9,18,29,95,106,125,139*).

Separations by free flow electrophoresis of neuraminidase-treated bone marrow had indicated that the CFU-S belong to the cells with the highest density of membrane sialic acid groups, and that as a consequence wheat germ lectin (WGA) should be an adequate label for sorting stem cells (*28,149,156*), Hoechst 33342 had been observed to make a distinction between day 8 and day 12 CFU-S, and between HPP-CFC and other *in vitro* tests (*5,111*) probably because the earliest PHSCs are deeply quiescent with highly condensed chromatin. The absence of proliferation in most of the earliest PHSCs helps them to survive treatment with 5-fluorouracil (5-FU) (*67*), which in a number of isolation protocols is used to remove the committed progenitor cells from the bone marrow *in vivo* prior to the PHSC isolation. We have observed that three times more day 12 CFU-S are needed from 5-FU-treated mice than from untreated ones to protect mice after lethal irradiation in the 30-day survival assay (*150*). Therefore, either the day 12 CFU-S after 5-FU treatment are not representing PHSCs, or they are, but then the 30 day survival assay is not adequate to detect pluripotent stem cells. Because MRA cells also survive 5-FU treatment, pre-enrichment *in vivo* using 5-FU seems to be a useful first step before PHSC isolation, if attention is given to conflicting results in the different test systems. In this respect it should be noted that the observed differences may also be due to

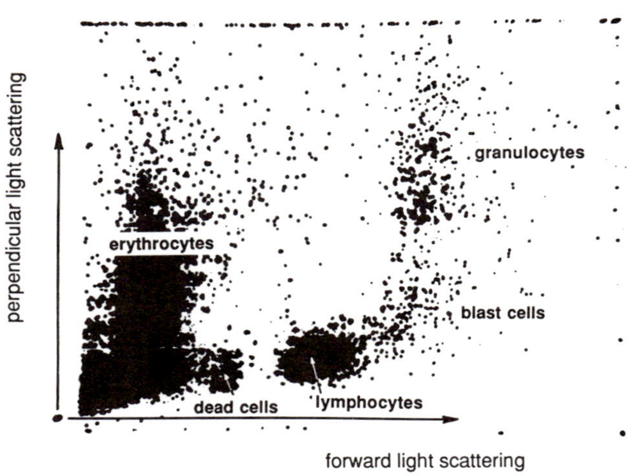

Fig. 3. Bivariate distribution of the forward and perpendicular light scatter intensities of mouse bone marrow cells (*43,45,154,155*).

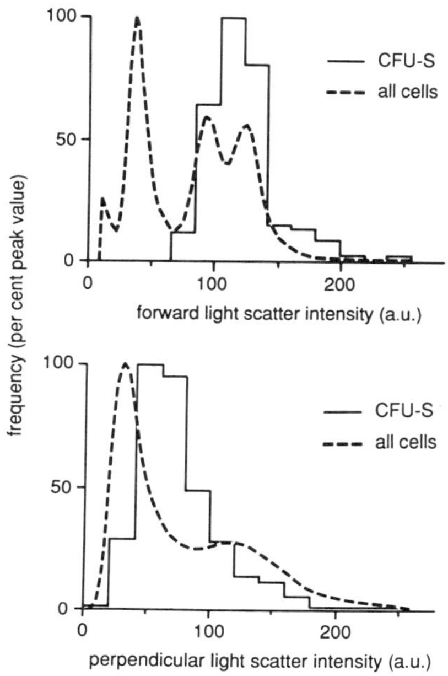

Fig. 4. Histograms of the forward and perpendicular light scatter intensities of mouse bone marrow cells and of CFU-S (*43,45,154,155*).

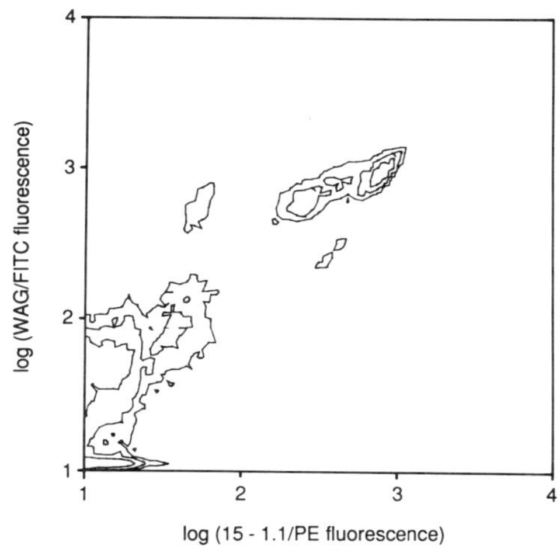

Fig. 5. Bivariate distribution of the WGA/FITC and 15-1.1/PE fluorescence intensities of low density mouse bone marrow cells. The small WGA positive, 15-1.1 negative cluster contains all CFU-S (*151*).

differences in the number of days between treatment and the collection of bone marrow. The earlier after 5-FU treatment, the more pluripotent the spleen colony forming cells seem to be (*100*). However, a period of about 2 days seems to be necessary to reduce the number of unwanted cells in the bone marrow sufficiently.

The results of these analytical separation studies were combined in a variety of protocols to purify the PHSCs. We employed the flow cytometer to isolate CFU-S using WGA labeling and light scatter parameters (*149*). Sorting with this lectin in combination with anti-H-2K (*157*) yielded up to 10 day 12 CFU-S and 1 to 2 day 8 CFU-S per 100 injected cells in a number of experiments (*93,94,124*), suggesting 100% purity. Later, we replaced the anti-H-2K by the antibody 15-1.1 (*151,152, 159,160*), which is directed against monocytes, granulocytes and their committed precursors, and which does not bind to stem cells (*60,158–160*). Figure 5 shows a bivariate distribution of the red and green fluorescence intensities of mouse bone marrow cells labeled with WGA/FITC and 15-1.1/PE. The cluster of WGA positive, 15-1.1 negative cells contains all PHSCs. The monoclonal antibody 15-1.1 is preferred over anti-H-2K ones, because it is more generally applicable, also with non-H-2K mouse strains, and secondly because it yields unlabeled stem cells. The WGA label can be removed from the stem cells by a short incubation with a competitive sugar, N-acetyl-D-glucosamine (*149,157*). The presence of WGA on the cells, however, does not seem to affect their function (*81*). The presence of anti-H-2K on stem cells severely diminished their take even in syngeneic transplantations, unless biotinylated antibody was used with avidin/FITC as a second layer (*10,157*). The large avidin molecules then seem to shield the Fc part of the antibody from recognition by opsonizing cells.

An alternative procedure with WGA was discovered by Lord and coworkers, who ran the cells through the sorter twice, improving the enrichment factor for day 12 CFU-S by a factor of two (*81,132*). Because the same sorting windows were used, this improvement cannot be attributed to a better selection of wanted cells. The repeated sorting probably causes the loss of unwanted cells such as monocytes by adherence, thereby enriching for CFU-S. Our choice of antibody 15-1.1 serves the same purpose of removing monocytes, probably more specifically than by adherence which may also remove stem cells (*13*). Spooncer and coworkers (*132*) observed that the sorted stem cells did not maintain hemopoiesis in long-term bone marrow cultures. This indicates that only part of the CFU-S isolated by the double WGA-sort are PHSCs. Miyama-Inaba *et al.* (*96*) and Sugiara *et al.* (*136*) combined the WGA labeling and flow sorting with 5-FU pretreatment of the mice, and with methods to remove T cells. Because WGA discriminates between stem cells and lymphocytes, the latter combination seems to be redundant. As was discussed above, pretreatment with 5-FU may yield high numbers of CFU-S types which do not represent PHSC. Baines and coworkers (*6*) combined density centrifugation and WGA labeling and sorting to enrich for high proliferative potential colony forming cells (HPP-CFC) from mouse bone marrow. The enrichment for this cell type was not as high as that for CFU-S obtained earlier using the same protocol (*149*). More recently, however, it has been shown that the HPP-CFC are heterogeneous, and that at least two types can be distinguished (*20,21,90*). Therefore, the possibility of employing the HPP-CFC assay to detect and quantitate pluripotent stem cells has to be further studied. A combination of WGA sorting and elutriation was studied by Ploemacher and coworkers (*112,113*). Although their enrichments for CFU-S were somewhat lower than those obtained by other methods, they observed that the cells giving 30-day survival could be separated from those giving spleen colonies. This prompted them to compare other PHSC assays with flow sorted fractions using labels and stains which were known to distinguish subpopulations of stem cells (*114,115*). One of these, rhodamine 123 (Rh123) had been reported by Bertoncello *et al.* (*20*) to separate MRA and PRA (platelet repopulating ability) cells from most of the CFU-S. We had demonstrated (*101*) that Rh123 dull bone marrow cells contained day 12 spleen colony forming cells, but no day 8 CFU-S. Rh123 bright bone marrow cells contained day 8 and day 12 CFU-S, but no MRA. Rh123 staining and sorting now forms the last step in our PHSC isolation procedure (*60,150,158*). Using Y chromosome *in situ* hybridization, we were able to show recently (*158*) that WGA positive, 15-1.1 negative, Rh123 dull cells which were 200-fold enriched for MRA gave long-term reconstitution of hemopoiesis *in vivo*, whereas WGA positive, 15-1.1 negative, Rh123 bright cells which gave high numbers of spleen colonies gave no long-term reconstitution. The sorted Rh123 dull cells seem to be similar to pre-CFU-S, the type of PHSC which survive 5-FU treatment (*20,67*). Figure 6 shows the histograms of the Rh123 fluorescence of bone marrow cells which had been sorted to isolate PHSC. Rh123 clearly distinguishes two types of stem cells, dull ones and bright ones with no intermediate ones, indicating that the transition from one state to the other is relatively rapid. Cells isolated by a double WGA sort had more Rh123 bright cells than those of the other protocols (Fig. 6) indicating that this procedure yields many cells of a type that will not give long-term reconstitution, and therefore will not maintain hemopoiesis in long-term cultures (*132*). Our Rh123 dull sorted cells (WGA positive, 15-1.1 negative; Fig. 6B) were demonstrated recently to give long-term cobblestone area formation in bone marrow stroma cultures, whereas the sorted Rh123 bright cells could do so only for some weeks (*158*). The WGA positive, anti-H-2K positive sorted cells (*157*) also showed two populations with Rh123 (Fig. 6C); the majority of these were Rh123 dull, and therefore of the PHSC type providing long-term reconstitution and having MRA. It may be speculated that these Rh123 dull cells do not use their mitochondria, and are in fact deeply quiescent stem cells of the type which would probably also have relatively condensed chromatin.

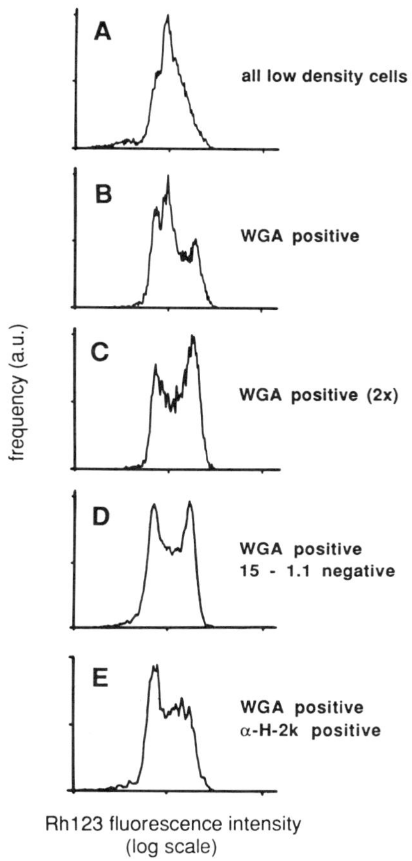

Fig. 6. Histograms of the rhodamine 123 fluorescence intensities of low density mouse bone marrow cells (A–E) sorted according to: (B) Visser and Bol (*149*), (C) Bart-Baumeister *et al.* (*8*), (D) Visser and de Vries (*150,151,159*), (E) Visser *et al.* (*157*).

In accordance with this a subpopulation of CFU-S which could not be easily stained with Hoechst 33342, and which gave high numbers of HPP-CFU-C, has been described (*5*). Pallavicini *et al.* (*111*) have demonstrated that two-laser flow sorting using light scatter and Hoechst 33342 fluorescence measurements could yield a 100-fold enrichment for CFU-S, which then are probably representing PHSC. McAlister *et al.* (*85*) recently confirmed this observation, and they also observed that the addition of WGA labeling to this combination does not result in a higher enrichment. A combination of Thy-1 and H2-K labeling with light scatter measurements was successfully employed recently to highly (100×) enrich CFU-S and long-term repopulating cells from the bone marrow of 5FU-treated mice (*139*). Combinations of negative selection to remove all mature and committed cells and positive selection to isolate subpopulations of CFU-S have also been reported to yield PHSC suspensions of high purity. Watt *et al.* (*162*) and Hoang *et al.* (*66*) selected a variety of monoclonal antibodies which were binding mature blood and bone marrow cells, but no CFU-S. Muller-Sieburg *et al.* (*106*) applied a cocktail of such antibodies to isolate pre-B cells by flow cytometry, and obtained a high enrichment for CFU-S, if they also made use of a Thy-1 label. Their so-called Thy-1^{lo}Lin^{-} cells were further segregated using an antibody called Sca-1, and an additional negative selection step was included in the combination by Spangrude *et al.* (*130*). Although the Thy-1^{lo}Lin^{-}Sca-1^{+} cells were almost 750-fold enriched for CFU-S, they gave similar numbers of day 12 spleen colonies as the WGA/H-2K or WGA/15-1.1 methods because the incidence of CFU-S in the starting material of Spangrude *et al.* was much lower than that in the mouse strains used in other laboratories (*40*). The Thy-1^{lo}Lin^{-}Sca-1^{+} cells can be divided into two equal fractions using rhodamine 123 (Spangrude and Bertoncello, personal communication), indicating that about half of them would not provide long-term hemopoietic reconstitution.

Bertoncello *et al* (*23*) also recently reported a considerable improvement in the efficiency of the PHSC isolation protocols by inclusion of negative selection using immunomagnetic beads and the antibodies 7/4 and B220. They furthermore pretreated the donor mice with 5-FU to obtain a significant pre-enrichment of the PHSCs, which they select with an antibody against Qa-m7 (*21*). In our laboratory immunomagnetic bead separation has also been useful for large-scale PHSC sorting. Enrichments of up to 60-fold could be obtained using 15-1.1 and 45D8 (*12*), and up to 10^6 stem cells could be sorted in one experiment. However, three difficulties were revealed by those experiments. First, many donor mice are needed to obtain many PHSC, and consequently the preparation of the bone marrow suspensions is time consuming. Second, high amounts of antibodies and beads are required, which can become costly. Third, we noticed that an antibody 6A2, which was known not to be present on CFU-S, whereas it labeled many, mainly erythroid, bone marrow cells, removed MRA. Consequently, the antibodies to be used for negative selection have to be tested separately for all the PHSC assay systems under study. We are now testing 6A2 to further evaluate its use for sorting pre-CFU-S. It seems that the antigen is conserved on stem cells and progenitor cells which differentiate into the erythroid lineage, and is lost in the other lineages.

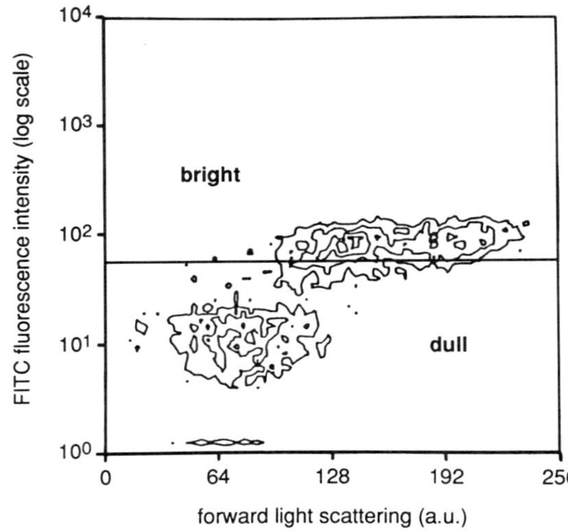

Fig. 7. Bivariate distribution of the PWM/FITC fluorescence of fetal liver cells which were pre-enriched for stem cells using density gradient centrifugation and negative selection with immunomagnetic beads and antibodies 15-1.1 and 10-2.2 (*159*). Courtesy of A. Corso.

B. Fetal Liver Stem Cells

The PHSCs from fetal liver are in several aspects different from the adult bone marrow ones. Analysis of their characteristics by equilibrium density centrifugation indicated that in the fetal liver between day 12 and day 14 of gestation the CFU-S change their buoyant density and possibly their size (*98*). Because the hemopoietic organ expands during that period, the PHSCs are thought to be proliferating, in contrast to the PHSCs in adult bone marrow. Membrane properties are different. H-2K could hardly be detected on day 12 CFU-S (*19*), nor could Qa (*76*). The antibody AA4 was successfully employed by McKearn *et al.* (*89*) to enrich 50- to 100-fold for CFU-GEMM, CFU-GM and BFU-E from fetal liver. Pokeweed mitogen (PWN) was found to be useful to enrich for CFU-S and committed progenitor cells (*71,108,109*). No purification of fetal liver PHSC has been reported as yet. Watt *et al.* (*162*) described the use of YW13.1.1 to remove the erythroid cells which constitute 90–95% of the fetal liver, whereas that antibody did not bind to CFU-S. We recently combined negative selection with a similar antibody, 10.2.2. (*159*), using immunomagnetic beads, with density gradient centrifugation and PWM flow sorting and obtained a 90-fold enrichment of CFU-S from day 14 fetal liver (A. Corso, unpublished results). This enrichment was obtained in both the PWM dull and the PWM bright fraction (Fig. 7). It was observed, however, that the PWM dull fraction contained the cells providing radioprotection in the 30-day survival assay. In the experiments with the highest enrichment (190- to 240-fold) the formation of spleen colonies by the PWM dull fraction was delayed and predominantly day 12 CFU-S were obtained. These observations are in agreement with the conclusions of Johnson *et al.* (*71*), that the CFU-S from the PWM dull fraction represent PHSCs, whereas those from the PWM bright cells probably arise from committed progenitor cells.

III. ISOLATION OF RAT BONE MARROW STEM CELLS

Pluripotent hemopoietic stem cells of the rat can be detected by their spleen colony formation and hemopoietic reconstitution after transplantation both in rat and in mice. However, most of the membrane properties of rat PHSCs which were discovered until now are different from those of mouse PHSCs. Rat PHSCs are amongst the most positive Thy-1 cells (*55,140*), and WGA dim (*86*).The majority of Thy-1 positive cells, the lymphocytes, can be removed from the bone marrow by treatment with cortisone. Goldschneider *et al.* (*55*) achieved an enrichment of 320-fold for CFU-S from the bone marrow of hydrocortisone-treated rat by sorting the most positive Thy-1 cells. The enrichment factor was equal to 151 if compared to the CFU-S numbers in the bone marrow of treated rat. Similar results were obtained by Castagnola *et al.* (*33*). This suggests that purification of normal rat marrow CFU-S results in the removal of weakly Thy-1-positive suppressor cells for colony formation (*55*). McCarthy *et al.* (*88*) obtained similar enrichments without using cortisone pretreatment of the donor rat, and they found no evidence for such suppressor cells. They made use of a combination of antibodies (*87,88*): (1) an anti-Thy-1 one called Ox7, for which the CFU-S are positive; (2) Ox22 against high molecular weight common leukocyte antigen, which is not on CFU-S; (3) monoclonal W3/13 against an antigen on a sialoglycoprotein, which divides the $Ox7^{++}Ox22^{-}$ cells into two subpopulations. A 350-fold enrichment for radioprotective cells was obtained by sorting the $Ox7^{++}Ox22^{-}$ cells. Prothymocytes, however, were also 282-fold enriched in this fraction (*61*). This indicates that further purification of the rat PHSC should be possible.

IV. ANTIGENS ON DOG, MONKEY AND HUMAN STEM CELLS

A. CD34 AND CD33

The purification of human PHSCs is facilitated by the presence of the CD34 antigens on these cells. Antibodies MY-10, HPCA-1 (*34,35,133*), Bi-3C5 (*75,142*), 12-8 (*2*), ICH3 and QBN-10 (*163*), 11G7 (H. J. Bühring, personal communication) and 5.1 (*77*) can be clustered in three groups recognizing at least three distinct epitopes of the CD 34 molecule (*77*). Between 1 and 4% of human bone marrow cells are positive for these antibodies, amongst which are nearly all colony forming progenitors, the BFU-E, the day 7 and day 14 CFU-GM (*7,75,163*), the CFU-GEMM or CFU-Mix (*75,163*), the blast cell colony forming cell (*30,78*) as well as the precursors of colony-forming cells in long-term marrow cultures (*2*). Berenson *et al.* (*17*) demonstrated that lethally irradiated baboons can be hematologically reconstituted using CD34 positive baboon bone marrow cells sorted by immunoadsorption using the antibody 12.8. Bart-Baumeister *et al.* (8) recently demonstrated that this can also be performed using ICH3 and rhesus monkeys with sorted bone marrow cells. This clearly demonstrates that PHSC are CD34 positive. It also shows that the homing of the PHSCs is not affected by labeling with the antibody, in spite of the proposed function of CD34 as an adherence molecule (M. F. Greaves, personal communication). Because only 12.8 and ICH3 have been shown to react with non-human primate bone marrow cells, whereas MY10 and other antibodies were negative, this discrepancy may be due to species and epitope differences.

Combinations of anti-CD34 and anti-CD33 have been successfully applied to further unravel the hemopoietic stem cell compartment. The monoclonal antibodies L4F3, L1B2 (*1*), and MY-9 (*59*) identify CD33. Andrews *et al.* (*1*) showed that CD33 is expressed on CFU-GEMM, CFU-GM and BFU-E, but not on their precursors. The PHSCs are, therefore, the $CD34^{+}CD33^{-}$ ones. Such cells now have been shown to form colonies and to generate CFU-GM in long-term cultures (*3*). $CD34^{+}CD33^{-}$ cells with low forward light scatter (FLS) were found to generate greater numbers of CFU-GM after several weeks in long-term cultures than cells with high ("blast cell") FLS. This was also reported by

Sutherland *et al.* (*137*). Bühring *et al.* observed that the $CD34^+CD33^-$ cells were more enriched for day 14 CFU-GM than cells from the double positive fraction, and they also concluded that during myelopoiesis CD34 antigens are gradually lost while CD33 becomes expressed.

A number of investigators have simultaneously analyzed CD34 and HLA-DR antigens on human PHSCs. This combination will be discussed in the next section.

B. HLA-DR Antigens

Most clonogenic hemopoietic cells have been found to express Ia-like or HLA-DR class II antigens. BFU-E and CFU-GM (*24,38,48,50, 69,74,99,121,145,146,166,167*) were shown to express HLA-DR both by complement-dependent cytotoxicity assays and by flow cytometry. It was concluded that all CD34 positive bone marrow cells express class II (*34*). Fitchen *et al.* (*52*) demonstrated that CFU-GEMM are HLA-DR, as well as HLA-A,B,C, positive. However, blast cell colony forming cells were reported to be DR negative (*30*). Lu *et al.* (*82*) and Ottmann *et al.* (*110*) described differences between the progenitor cells with respect to the amount of anti-DR antibody bound; the earlier the cell type the duller the fluorescence intensity. Some controversy exists with respect to the HLA-DR expression on the precursors of CFU-GM in long-term cultures. Some investigators observed expression (*48*), whereas others showed that the precursors were DR negative in long-term cultures which produced DR positive CFU-GM. Sutherland *et al.* (*137*) recently demonstrated that the precursors can be at least 800-fold enriched by selecting CD34 positive cells which are also HLA-DR low with a low forward as well as perpendicular light scatter intensity. These data seem to indicate that the PHSC are HLA-DR dull or low.

On the other hand, it has been shown that lethally irradiated dogs cannot be hematologically reconstituted after depletion of Ia-like antigen expressing cells from the graft (*138*). Transplantation of isolated Ia-positive dog bone marrow cells resulted in durable engraftment and hematologic and immunologic reconstitution (*16*). Also in rhesus monkeys the reconstituting stem cells are DR-positive (*53,54*). Figure 8 shows bivariate distributions of the CD34 (ICH3) and HLA-DR antigen labeling of rhesus monkey bone marrow cells. Transplantation of sorted cells from the different clusters into lethally irradiated recipients indicated that the repopulating stem cells are HLA-DR dull (*164*). In this species, CFU-GEMM was shown also to express DR antigens (*169*). It may be speculated that the choice of sorting windows and setting of thresholds to discriminate positive and negative cells contributes to the controversy. It may also be assumed that the degree of expression is different in other species, probably also depending on determinants.

With regard to other class II antigens, it may be mentioned that human bone marrow long-term repopulating stem cells were reported to be DQ negative, whereas they were found to be DR positive in the same study (*48*). Data on the expression of HLA-DQ by BFU-E, CFU-GM, and CFU-GEMM, however, are controversial

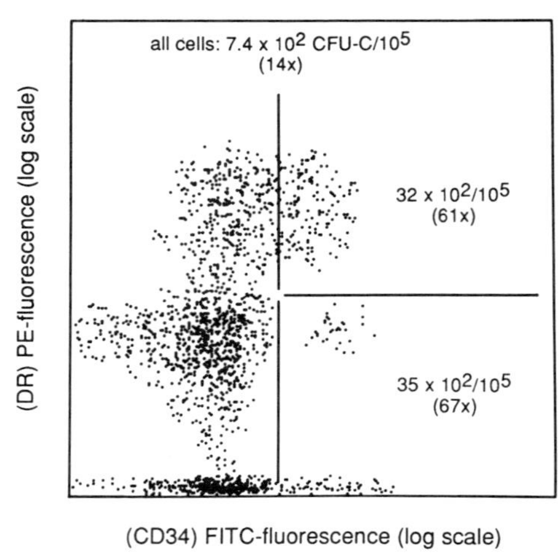

Fig. 8. Bivariate distribution of the (CD34) ICH3/FITC and HLA-DR/PE fluorescence intensities of low density rhesus monkey bone marrow cells (*164*).

(*32,80,110,127,131,144*). A small portion of these cells was reported to be DP-positive (*32,80,110,144*). The class II antigens obviously are not co-expressed on the stem cells and progenitor cells.

Class I antigens were found on all PHSC in mouse (*19,44,47,51,63,64,103,147,157*) and human (*52*) bone marrow. Because these antigens occur on many other cells too, and because they differ between individuals, they are not very practical in a general purification protocol. In several studies using specific strains of mice, the H-2K antigen, however, was successfully used to purify PHSC and to determine further characteristics (*93,94,102, 104,105,124,157*).

V. PURIFICATION OF HUMAN HEMOPOIETIC STEM CELLS

Because the stem cells are defined as having an extensive capacity for proliferation and differentiation, and because of the above described findings concerning modulation of CD34, CD33 and HLA-DR expression, the long-term production of CFU-GM *in vitro* appears to be an appropriate test system for PHSCs. The assay is not yet quantitative in an absolute sense, but it can be used to compare sorted fractions. In addition, it can be employed both for mouse and human PHSCs. Using this assay, the human PHSCs have been shown to be $CD34^+CD33^-$ and HLA-DR dull with low forward and perpendicular light scatter (*3,31,137*). $CD34^+$, HLA-DR dull cells were also reported to produce blast cell containing colonies (*30*), making that assay a strong candidate for quantitating PHSC.

Combination of equilibrium density centrifugation and flow cytometry resulted in an 800-fold enrichment of ($CD34^+$, HLA-DR dull, low FLS) human PHSCs (*137*). A similar enrichment is probably obtained using CD33 in combination with CD34 labeling, however, no figures for this are given in those studies (*3,31*). Bodger *et al.* (*25,26*) demonstrated that the monoclonal antibody RFB-1 in combination with other separation techniques can be employed to enrich more than 200-fold for CFU-GEMM from human cord blood, a rich source for this cell type. Bodger (*26*) also showed that the plating efficiency of CFU-GEMM is four- to five-fold increased by removal of $CD15^+$ granulocytic cells prior to culture. Further improvements of the purification protocols can be expected with regard to the amount of cells isolated. This can perhaps be achieved by pretreatment of bone marrow cells with 5-FU, or 4-HC which kills many bone marrow cells and spares the PHSCs (*20,67,118, 126,128,168*). Panning or immunomagnetic bead techniques will be efficient to deplete the bone marrow from most mature cells, and from many early cells too if anti-CD33 is included in the negative selection cocktail. The flow cytometer then will be useful to evaluate the performance of the large-scale separation steps, to correct small errors occurring during these steps, and to further purify the PHSCs using light scatter and fluorescence characteristics. Because of the quiescent state of the PHSC, it may be speculated that not only mouse PHSCs, but also those of other species are rhodamine 123 dull, and not easily stained with Hoechst 33342. These two intracellular dyes then should be useful in the last sorting steps using flow cytometry.

VI. APPLICATIONS OF SORTED STEM CELLS

Apart from the obvious clinical relevance of applying sorted stem cells in composed grafts for bone marrow transplantation, a variety of studies benefit from the availability of pure PHSCs. Some of these are described in more detail below. The availability of purified stem cells is of use to improve and design gene therapy protocols. For clinical purposes it may be mentioned in addition, that the purification of PHSC is expected to result in a PHSC-specific staining, which will have diagnostical applications. It will allow examination of the number and characteristics of PHSCs in patients without the need to sort or to culture the cells.

A. Culture of Sorted Stem Cells

The culture of PHSCs to examine the regulation of their proliferation and differentiation is facilitated by the availability of recombinant growth factors and serum free culture medium. However, both growth factors and inhibitors are produced by the offspring of the stem cells, by mature blood cells. Purification of PHSCs prior to culture eliminates this contribution to the cultures, which is otherwise difficult to control. High plating efficiencies have been reported for sorted fractions of stem cells (*11,107,108,157*). However, the long-term repopulating PHSCs were found to be difficult to culture *in vitro* (*107,158,159*). This is probably due to their deep quiescent state, and to the lack of an appropriate growth factor or microenvironment to induce these cells to proliferate in the presence of added growth factors. The sensitivity of PHSCs to these factors can be studied directly by DNA analysis of stimulated purified PHSC. As shown in Fig. 9 the sorted cells are in the G0 or G1 state of the cell cycle; they do not proliferate extensively during the first 24 hours of culture with IL-3 and, subsequently, they start to divide and to produce offspring (*157*). Similar studies directly indicated that the PHSCs respond to erythropoietin in synergy with IL-3 (*94*). The response of the sorted stem cells may be similarly studied in a direct fashion employing dyes which probe calcium release, and pH changes, as well as antibodies detecting the modulation of growth factor receptors and adherence molecules. Caux *et al.* (*34*) recently employed FACS sorting of cultured sorted stem cells to probe changes in the expression of HLA-DR and CD34 during culture in the presence of IL-3. They concluded that DR expression correlates with overall proliferative capacity of hemopoietic cells. This is in agreement with the findings described above that the PHSCs are quiescent and HLA-DR dull or negative.

The combination of culture and sorting has revealed subpopulations of stem cells in the past, and has improved the detection and analysis of PHSCs. It may be expected that this combination will ultimately result in a reliable and quantitative stem cell assay.

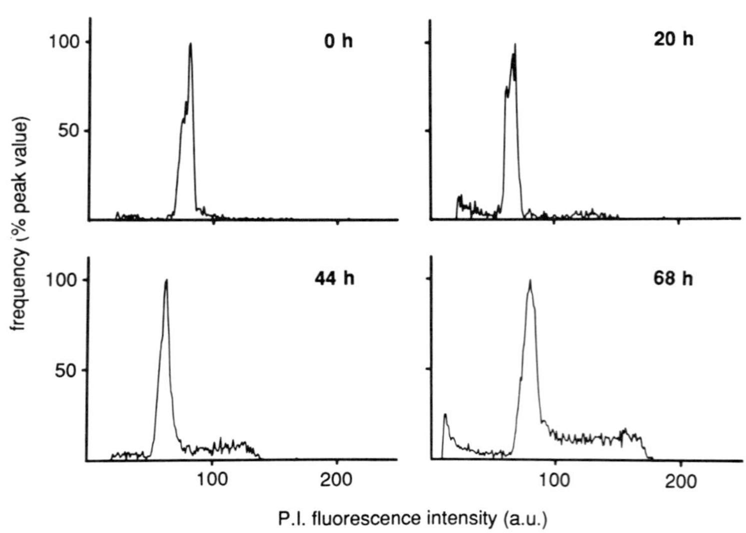

Fig. 9. Histograms of the propidium iodide fluorescence intensities of sorted mouse bone marrow cells before (0 h) and after culture in the presence of IL-3 (*157*).

B. *In Vivo* Tracking

One of the questions regarding the regulation of hemopoiesis concerns the homing of the PHSCs after transplantation. Purified stem cells can be labeled with tracking dyes and subsequently i.v. injected into lethally irradiated and otherwise conditioned hosts. Analysis of the tissues of the recipient at different time points thereafter reveals the homing and divisions of the stem cells, and the migration of their daughter cells. Fractions of mouse bone marrow cells which were highly purified for CFU-S could be shown to home in the spleen (10%), the bone marrow (20%) and the liver (20%). The homing in the thymus was less than 0.13% (*100,102*). In contrast to this, highly purified CFU-S were reported to home in the thymus and to efficiently produce colonies there if the cells were inoculated directly in the thymus (*130*). Earlier observations using tracking dyes had indicated that pluripotent stem cells home in the thymus (*79*). However, flow cytometric analysis of the cells which home in the thymus clearly indicated that the light scatter characteristics of those cells is different from that of PHSCs (*102*). In combination with other data it could be concluded that the first donor-derived prothymocytes enter the thymus at 2 days after transplantation (*105*).

C. Examination of Gene Expression in Stem Cells

Several laboratories recently purified stem cells to isolate mRNA and to establish cDNA libraries. Using PCR techniques such libraries can be obtained from very few, even from single cells (*15*). In order to detect the genes which become expressed upon induction of quiescent stem cells, as well as those which are blocked, not only mRNA of purified stem cells, but also that of purified and subsequently cultured cells has to be amplified. The cDNA bank of the responding cells can be subtracted from the bank of the sorted cells before culture. In addition, it can be subtracted from a bank made from stimulated lymphocytes, so that the abundant house keeping genes are also largely removed. Different clones are now being analyzed in a number of laboratories, and interesting genes which may be specific for stem cell regulation are detected this way. The material is also used to check whether known genes of interest are being differentially expressed in quiescent and proliferating stem cells. For such studies on the regulation of hemopoiesis at the molecular level it is not essential to start with pluripotent cells. A purified population of for example CFU-GM (*165*) may also serve as starting material for a subtraction procedure. It will be clear that the combination of PCR techniques and stem cell purification opens a variety of possibilities to study hemopoietic regulation at the molecular level.

REFERENCES

1 Andrews, R. G., Takahashi, M., Segal, G. M., Powell, J. S., Bernstein, I. D., and Singer, J. W. (1986a). The L4F3 antigen is expressed by unipotent and multipotent colony-forming cells but not by their precursors. *Blood* **68**, 1030–1035.

2. Andrews, R. G., Singer, J. W., and Bernstein, I. D. (1986b). Monoclonal antibody 12-8 recognizes a 115-kd molecule present on both unipotent and multipotent hematopoietic colony-forming cells and their precursors. *Blood* **67**, 842–845.

3. Andrews, R. G., Singer, J. W., and Bernstein, I. D. (1989). Precursors of colony-forming cells in humans can be distinguished from colony-forming properties. *J. Exp. Med.* **169**, 1721–1731.

4. Ansell, J. D., and Micklem, H. S. (1986). Genetic markers for following cell populations. *In* "Handbook of Experimental Immunology" (D. M. Weir, L. A. Herzenberg, C. C. Blackwell, and L. A. Herzenberg, eds), chapter 56. Blackwell, Oxford.

5. Baines, P., and Visser, J. W. M. (1983). Analysis and separation of murine bone marrow stem cells by H33342 fluorescence-activated cell sorting. *Exp. Hematol.* **11**, 701–708.

6. Baines, P., Lajmanovich, A., and Hollard, D. (1984). Enrichment of high proliferation potential colony forming cells from mouse marrow by selecting low-density cells expressing receptors for wheat germ agglutinin. *Leuk. Res.* **8**, 853–861.

7. Baines, P., Mayani, H., Bains, M., Fisher, J., Hoy, T., and Jacobs, A. (1988). Enrichment of

CD34(My10)-positive myeloid and erythroid progenitors from human marrow and their growth in cultures supplemented with recombinant human granulocyte-macrophage colony-stimulating factor. *Exp. Hematol.* **16**, 785–789.

8. Bart-Baumeister, J. A. K., Wielenga, J. J., Levinsky, R. J., and Wagemaker, G. (1989). Positive selection of Rhesus monkey hemopoietic stem cells by immunomagnetic beads separation. *Exp. Hematol.* **17**, 6, abstr 54.
9. Basch, R. S., and Berman, J. W. (1982). Thy-1 determinates are present on many murine hematopoietic cells other than T cells. *Eur. J. Immunol.* **12**, 359–364.
10. Bauman, J. G. J., Mulder, A. H., and Engh, G. J. van den (1985). Effect of surface labeling on spleen colony formation of the indirect immunofluorescence and the biotin-avidin methods. *Exp. Hematol.* **13**, 760–767.
11. Bauman, J. G. J., Wagemaker, G., and Visser, J. W. M. (1986). A fractionation procedure of mouse bone marrow cells yielding exclusively pluripotent stem cells and committed progenitors. *J. Cell. Physiol.* **128**, 133–142.
12. Bauman, J. G. J., Vries, P. de, Pronk, B., and Visser, J. W. M. (1988). Purification of murine hemopoietic stem cells and committed progenitors by fluorescence activated cell sorting using wheat germ agglutinin and monoclonal antibodies. *Acta Histochem.* Suppl.-band XXXVI, 241–253.
13. Bearpark, A. D., and Gordon, M. Y. (1989). Adhesive properties distinguish sub-populations of haemopoietic stem cells with different spleen colony-forming and marrow repopulating capacities. *Bone Marrow Transplant.* **4**, 625–628.
14. Bekkum, D. W. van, Noord, M. J. van, Maat, B., and Dicke, K. A. (1971). Attempts at identification of hemopoietic stem cell in mouse. *Blood* **28**, 547–558.
15. Belyavsky, A., Vinogradova, T., and Rajewsky, K. (1989). PCR-based cDNA library construction: general cDNA libraries at the level of a few cells. *Nucleic Acid Res.* **17**, 2919–2932.
16. Berenson, R. J., Bensinger, W. I., Kalamasz, D., Schuening, F., Deeg, H. J., Graham, T., and Storb, R. (1987). Engraftment of dogs with Ia-positive marrow cells isolated by avidin-biotin immunoadsorption. *Blood* **69**, 1363–1367.
17. Berenson, R. J., Andrews, R. G., Bensinger, W. I., Kalamaz, D., Knitter, G., Buckner, C. D., and Bernstein, I. D. (1988). Antigen $CD34^+$ marrow cells engraft lethally irradiated baboons. *J. Clin. Invest.* **81**, 951–955.
18. Berman, J. W., and Basch, R. S. (1985). Thy-1 antigen expression by murine hematopoietic precursor cells. *Exp. Hematol.* **13**, 1152–1156.
19. Berridge, M. V., Ralph, S. J., Seng-Tan, A., and Jeffrey, K. (1984). Changes in cell surface antigens during stem cell ontogeny. *Exp. Hematol.* **12**, 121–129.
20. Bertoncello, I., Hodgson, G. S., and Bradley, T. R. (1985). Multiparameter analysis of transplantable hemopoietic stem cells. I. The separation and enrichment of stem cells homing to marrow and spleen on the basis of rhodamine-123 fluorescence. *Exp. Hematol.* **13**, 999–1006.
21. Bertoncello, I., Bartelmez, S. H., Bradley, T. R., Stanley, E. R., Harris, R. A., Sandrin, M. S., Kriegler, A. B., McNiece, I. K., Hunter, S. D., and Hodgson, G. S. (1986). Isolation and analysis of primitive hemopoietic progenitor cells on the basis of differential expression of Qa-m7 antigen. *J. Immunol.* **136**, 3219–3224.
22. Bertoncello, I., Hodgson, G. S., and Bradley, T. R. (1988). Multiparameter analysis of transplantable hematopoietic stem cells. II. Stem cells of long-term bone marrow-reconstituted recipients. *Exp. Hematol.* **16**, 245–249.
23. Bertoncello, I., Bradley, R., and Hodgson, G. S. (1989). The concentration and resolution of primitive hemopoietic cells from normal mouse bone marrow by negative selection using monoclonal antibodies and Dynabead monodisperse magnetic microspheres. *Exp. Hematol.* **17**, 171–176.
24. Beverley, P. C. L., Linch, D., and Delia, D. (1980). Isolation of human haematopoietic progenitor cells using monoclonal antibodies. *Nature* **287**, 332–333.
25. Bodger, M. P. (1987). Isolation of hemopoietic progenitor cells from human umbilical cord blood. *Exp. Hematol.* **15**, 869–876.
26. Bodger, M. P., Hann, I. M., Maclean, R. F., and Beard, M. E. J. (1984). Enrichment of pluripotent hemopoietic progenitor cells from human bone marrow. *Blood.* **64**, 774–779.
27. Boggs, D. R., Boggs, S. S., Saxe, D. F., Gress, L. A., and Canfield, D. R. (1982). Hematopoietic stem cells with high proliferative potential. Assay of their concentration in marrow by the frequency and duration of cure of W/W^v mice. *J. Clin. Invest.* **70**, 242–253.
28. Bol, S. J. L., Vliet, M. van, and Slingerland, R. (1981). Electrophoretic mobility properties of murine hemopoietic cells from mouse bone marrow. *Exp. Hematol.* **9**, 431–443.
29. Boswell, S. H., Wade, P. M. Jr., and Quesenberry, P. J. (1984). Thy-1 antigen cells. *J. Immunol.* **133**, 2940–2949.
30. Brandt, J., Baird, N., Lu, L., Srour, E., and Hoffman, R. (1988). Characterization of a human hematopoietic progenitor cell capable of forming blast cell containing colonies in vitro. *J. Clin. Invest.* **82**, 1017–1027.
31. Bühring, H. J., Asenbauer, B., Katrilaka, K., Hummel, G., and Busch, F. W. (1989). Sequential

expression of CD34 and CD33 antigens on myeloid colony-forming cells. *Eur. J. Haematol.* 42, 143–149.

32. Busch, F. W., Langer, M., Pawelec, G., Ziegler, A., Wernet, P., Buhring, H. J., Meyer, P., and Muller, C. (1987). HLA-class II antigens on human hematopoietic progenitors. *Blut* **54**, 179–188.

33. Castagnola, C., Visser, J. W. M., Boersma, W., and Bekkum, D. W. van (1981). Purification of rat pluripotent hemopoietic stem cells. *Stem Cells* **1**, 250–260.

34. Caux, C., Favre, C., Saeland, S., Duvert, V., Mannoni, P., Durand, I., Aubry, J.-P., and Vries, J. E. de (1989). Sequential loss of CD34 and Class II MHC antigens on purified cord blood hematopoietic progenitors cultured with IL-3: characterization of CD34-, HLA-DR + cells. *Blood* **74**, 1287–1294.

35. Civin, C. I., and Loken, M. R. (1987). Cell surface antigens on human marrow cells: dissection of hematopoietic development using monoclonal antibodies and multiparameter flow cytometry. *Int. J. Cell Cloning* **5**, 267–288.

36. Civin, C. I., Strauss, L. C., Brovall, C., Fackler, M. J., Schwartz, J. F., and Shaper, J. H. (1984). Antigenic analysis of hematopoiesis. III. A hematopoietic progenitor cell surface antigen defined by a monoclonal antibody raised against KG-Ia cells. *J. Immunol.* **133**, 157–165.

37. Civin, C. I., Strauss, L. C., Fackler, M. J., Trischman, T. M., Wiley, J. M., and Loken, M. R. (1990). Positive stem cell selection—basic science. *Prog. Clin. Biol. Res.* **333**, 387–401.

38. Cline, M. J., and Billing, R. (1977). Antigens expressed by human B-lymphocytes and myeloid stem cells. *J. Exp. Med.* **146**, 1143–1145.

39. Dekken, H. van, Arkesteijn, G. J. A., Visser, J. W. M., and Bauman, J. G. J. (1990). Flow cytometric quantification of human chromosome specific repetitive DNA sequences by single and bicolor fluorescent *in situ* hybridization to lymphocyte interphase nuclei. *Cytometry* **11**, 153–164.

40. Dexter, T. M., and Lord, B. I. (1989). Reply to letter by Weisman *et al. Immunology Today*, **10**, 185–186.

41. Dexter, T. M., Allen, T. D., and Lajtha, L. G. (1977). Conditions controlling the proliferation of haemopoietic stem cells *in vitro. J. Cell. Physiol.* **91**, 335–344.

42. Engh, G. J. van den, and Golub, E. S. (1974). Antigenic differences between hemopoietic stem cells and myeloid progenitors. *J. Exp. Med.* **139**, 1621–1627.

43. Engh, G. J. van den, and Visser, J. W. M. (1979), Light scattering properties of pluripotent and committed stem cells. *Acta Haematol.* **62**, 289–298.

44. Engh, G. J. van den, Russell, J., and Cicco, D. de (1978). Surface antigens of hemopoietic stem cells: The expression of BAS, Thy-1, and H-2 antigens on CFU-s. *In* "Experimental Hematology Today 1978" S. J. Baum and G. D. Ledney, eds), pp. 9–15. Springer Verlag, New York.

45. Engh, G. J. van den, Visser, J., and Trask, B. (1979). Identification of CFU-s by scatter measurements on a light activated cell sorter *In* "Experimental Hematology Today 1979" (S. J. Baum and G. D. Ledney, eds), pp. 19–26. Springer Verlag, New York.

46. Engh, G. J. van den, Visser, J. W. M., Bol, S. J. L., and Trask, B. (1980). Concentration of hemopoietic stem cells using a light-activated cell sorter. *Blood Cells* **6**, 609–623.

47. Engh, G. J. van den, Trask, B., and Visser, J. W. M. (1981). Surface antigens of pluripotent and committed haemopoietic stem cells. *In* "Modern Trends in Human Leukemia IV; Haematology and Blood Transfusion, Vol. 26" (R. Neth, R. C. Gallo, T. Graf, K. Mannweiler, and K. Winkler, eds), pp. 305–308. Springer, Berlin.

48. Falkenburg, J. H. F., Fibbe, W. E., Goselink, H. M., Rood, J. J. van, and Jansen, J. (1985). Human hemapoietic progenitor cells in long-term cultures express HLA-DR antigens and lack HLA-DQ antigens. *J. Exp. Med.* **162**, 1359–1369.

49. Fauser, A. A., and Messner, H. A. (1979). Identification of megakaryocytes, macrophages and eosinophils in colonies of human bone marrow containing neutrophilic granulocytes and erythroblasts. *Blood* **53**, 1023–1027.

50. Fitchen, J. H., Ferrone, S., Quaranta, V., Molinaro, G. A., and Cline, M. J. (1980). Monoclonal antibodies to HLA-A,B and Ia-like antigens inhibit colony formation by human myeloid progenitor cells. *J. Immunol.* **125**, 2004–2008.

51. Fitchen, J. H., Foon, K. A., and Cline, N. J. (1981). The antigenic characteristics of hemopoietic stem cells. *New Engl. J. Med.* **305**, 17–25.

52. Fitchen, J. H., Le Fevre, C., Ferrone, S., and Cline, M. J. (1982). Expression of Ia-like and HLA-A,B antigens on human multipotential hematopoietic progenitor cells. *Blood*, **59**, 188–190.

53. Gerritsen, W. R. (1989). Bone marrow transplantation with T lymphocyte depleted and stem cells enriched bone marrow grafts. Experiments in Rhesus monkeys. M.D. Thesis, Erasmus University, Rotterdam.

54. Gerritsen, W.-R, Wagemaker, G., Jonker, M., Kenter, M. J. H., Wielenga, J. J., Hale, G., Waldmann, H., and Bekkum, D. W. van (1988). The repopulation capacity of bone marrow grafts following pretreatment with monoclonal antibodies against T lymphocytes in rhesus monkeys. *Transplantation* **45**, 301–307.

55. Goldschneider, I., Metcalf, D., Battey, F., and Mandel, T. (1980). Analysis of rat hemopoietic cells

on the fluorescence-activated cell sorter. I. Isolation of pluripotent hemopoietic stem cells and granulocyte-macrophage progenitor cells. *J. Exp. Med.* **152**, 419–437.

56. Golub, E. S. (1972). Brain-associated stem cell antigen; an antigen shared by brain and hemopoietic stem cells. *J. Exp. Med.* **136**, 369–374.

57. Gordon, M. Y., Goldman, J. M., and Gordon-Smith, E. C. (1985a). 4-Hydroperoxycyclophosphamide inhibits proliferation by human granulocyte-macrophage colony-forming cells (GM-CFC) but spares more primitive progenitor cells. *Leuk. Res.* **9**, 1017–1021.

58. Gordon, M. Y., Hibbin, J. A., Kearny, L. U., Gordon-Smith, E. C., and Goldman, J. M. (1985b). Colony formation by primitive haemopoietic progenitors in cocultures of bone marrow cells and stromal cells. *Br. J. Haematol.* **60**, 129–136.

59. Griffin, J. D., Linch, D., Sabbath, K., Larcom, P., and Schlossman, S. F. (1984). A monoclonal antibody reactive with normal and leukemic myeloid progenitor cells. *Leuk. Res.* **8**, 521–534.

60. Hagenaars, C. E., Kraan, A. A. M. van der, Kawilarang-de Haas, E. W. M., Visser, J. W. M., and Nijweide, P. J. (1989). Osteoclast formation from cloned pluripotent hemopoietic stem cells. *Bone Mineral*, **6**, 179–189.

61. Hale, M. L., Greiner, D. L., and McCarthy, K. F. (1987). Characterization of rat prothymocyte with monoclonal antibodies recognizing rat lymphocyte membrane antigenic determinants. *Cell Immunol.* **107**, 188–200.

62. Hara, H., and Ogawa, M. (1978). Murine hemopoietic colonies in culture containing normoblasts, macrophages, and megakaryocytes. *Am. J. Hematol.* **4**, 23–34.

63. Harris, R. A., Hogarth, P. M., Wadeson, L. J., Collins, P., McKenzie, I. F. C., and Penington, D. G. (1984). An antigenic difference between cells forming early and late haematopoietic spleen colonies (CFU-S). *Nature* **307**, 638–641.

64. Harris, R. A., Sandrin, M. S., Sutton, V. R., Hogarth, P. M., McKenzie, I. F. C., and Pennington, D. G. (1985). Variable expression of Qa-m7, Qa-m8, and Qa-m9 antigenic determinants on primitive hemopoietic precursor cells. *J. Cell. Physiol.* **123**, 451–458.

65. Haskill, J. S., McNeill, T. A., and Moore, M. A. S. (1970). Density distribution analysis of *in vivo* and *in vitro* colony forming cells in bone marrow. *J. Cell. Physiol.* **75**, 167–180.

66. Hoang, T., Gilmore, D., Metcalf, D., Cobbold, S., Watt, S., Clark, M., Furth, M., and Waldmann, H. (1983). Separation of hemopoietic cells from adult mouse marrow by use of monoclonal antibodies. *Blood*, **61**, 580–588.

67. Hodgson, G. S., and Bradley, T. R. (1979). Properties of haematopoietic stem cells surviving 5-fluorouracil treatment: evidence for a pre-CFU-S cell? *Nature* **281**, 381–382.

68. Ikebuchi, K., Wong, C. G., Clark, S. C., Ihle, J. N., Hirai, Y., and Ogawa, M. (1987). Interleukin 6 enhancement of interleukin 3-dependent proliferation of multipotential hemopoietic progenitors. *Proc. Natl. Acad. Sci. USA* **84**, 9035–9039.

69. Janossy, G., Francis, G. E., Capellaro, D., Gladstone, A. H., and Greaves, M. F. (1978). Cell sorter analysis of leukemia-associated antigens on human myeloid precursors. *Nature* **276**, 176–178.

70. Johnson, G. R., and Metcalf, D. (1977). Pure and mixed erythroid colony formation in vitro stimulated by spleen conditioned medium with no detectable erythropoietin. *Proc. Natl. Acad. Sci. USA* **74**, 3879–3882.

71. Johnson, G. R., and Nicola, N. A. (1984). Characterization of two populations of CFU-S fractionated from mouse fetal liver by fluorescence-activated cell sorting. *J. Cell. Physiol.* **118**, 45–52.

72. Jones, R. J., Celano, P., Sharkis, S. J., Sensenbrenner, L. L. (1989). Two phases of engraftment established by serial bone marrow transplantation in mice. *Blood* **73**, 397–401.

73. Kaizer, H., Stuart, R. K., Brookmeyer, R., *et al.* (1985). Autologous bone marrow transplantation in acute leukemia: a phase I study of *in vitro* treatment of marrow with 4-hydroperoxycyclophosphamide to purge tumor cells. *Blood* **65**, 1504–1510.

74. Kaplan, J., Inoue, S., and Ottenbriet, M. J. (1978). Myeloid colony-forming cells express human B lymphocyte antigens. *Nature* **271**, 458–459.

75. Katz, F. E., Tindle, R., Sutherland, D. R., and Greaves, M. F. (1985). Identification of a membrane glycoprotein associated with haemopoietic progenitor cells. *Leuk. Res.* **9**, 191–198.

76. Kincade, P. W., Flaherty, L., Lee, G., Watanabe, T., and Michaelson, J. (1980). Qa antigen expression on functional lymphoid, myeloid, and stem cells in adult mice. *J. Immunol.* **124**, 2879–2885.

77. Lansdorp, P. M., Dougherty, G. J., and Humphries, R. K. (1989). CD34 epitopes *In* "Leucocyte Typing IV" (W. Knapp, ed.), pp. 826–827. Oxford University Press, Oxford.

78. Leary, A. G., and Ogawa, M. (1987). Blast cell colony assay for umbilical cord blood and adult bone marrow progenitors. *Blood* **69**, 953–956.

79. Lepault, F., and Weissman, I. L. (1981). An in vivo assay for thymus homing bone marrow cells. *Nature* **293**, 151–154.

80. Linch, D. C., Nadler, L. M., Luther, E. A., and Lipton, J. M. (1984). Discordant expression of human Ia-like antigens on hematopoietic progenitor cells. *J. Immunol* **132**, 2324–2329.

81. Lord, B. I., and Spooncer, E. (1986). Isolation of haemopoietic spleen colony forming cells. *Lymph. Res.* **5**, 59–72.

82. Lu, L., Walker, D., Broxmeyer, H. E., Hoffman,

R., Hu, W., and Walker, E. (1987). Characterization of adult human marrow hematopoietic progenitors highly enriched by two-color cell sorting with My-10 and major histocompatibility class II monoclonal antibodies. *J. Immunol.* **139**, 1823–1829.

83. Magli, M. C., Iscove, N. N., and Odartchenko, N. (1982). Transient nature of early haematopoietic spleen colonies. *Nature* **295**, 527–529.

84. Martens, A. C. M., Bekkum, D. W. van, and Hagenbeek, A. (1986). Heterogeneity within the spleen colony-forming cell population in rat bone marrow. *Exp. Hematol.* **14**, 714–718.

85. McAlister, I., Wolf, N. S., Pietrzyk, M. E., Rabinovitch, P. S., and Priestley, G. (1990). Transplantation of hematopoietic stem cells obtained by a combined dye method fractionation of murine bone marrow. *Blood* **75**, 1240–1246.

86. McCarthy, K. F., and Hale, M. L. (1988). Flow cytometry techniques in radiation biology. *Toxicol. Letters* **43**, 219–233.

87. McCarthy, K. F., Hale, M. L., and Fehnel, P. L. (1985). Rat colony-forming unit spleen is OX7 positive, W3/13 positive, OX1 positive, and OX22 negative. *Exp. Hematol.* **13**., 847–854.

88. McCarthy, K. F., Hale, M. L., and Fehnel, P. L. (1987). Purification and analysis of rat hematopoietic stem cells by flow cytometry. *Cytometry* **8**, 296–305.

89. McKearn, J. P., McCubrey, J., and Fagg, B. (1985). Enrichment of hematopoietic precursor cells and cloning of multipotential B-lymphocyte precursors. *Proc. Natl. Acad. Sci. USA* **82**, 7414–7418.

90. McNiece, I. K., Bradley, T. R., Kriegler, A. B., and Hodgson, G. S. (1986). Subpopulations of mouse bone marrow high-proliferative-potential colony-forming cells. *Exp. Hematol.* **14**, 856–860.

91. Metcalf, D., Moore, M. A. S., and Shortman, K. (1971). Adherence column and buoyant density separation of bone marrow stem cells and more differentiated cells. *J. Cell. Physiol.* **78**, 441–450.

92. Micklem, H. S., Lennon, J. E., Ansell, J. D., and Gray, R. A. (1987). Numbers and dispersion of repopulating hematopoietic cell clones in radiation chimeras as functions of injected cell dose. *Exp. Hematol.* **15**, 251–257.

93. Migliaccio, A. R., and Visser, J. W. M. (1986). Proliferation of purified murine hemopoietic stem cells in serum-free cultures stimulated with purified stem-cell-activating factor. *Exp. Hematol.* **14**, 1043–1048.

94. Migliaccio, G., Migliaccio, A. R., and Visser, J. W. M. (1988). Synergism between erythropoietin and interleukin-3 in the induction of hematopoietic stem cell proliferation and erythroid burst colony formation. *Blood* **72**, 944–951.

95. Miller, B. A., Lipton, J. M., Lynch, D. C., Burakoff, S. J., and Nathan, D. G. (1985). Thy-1 is a differentiation antigen that characterizes immature murine erythroid and myeloid hematopoietic progenitors. *J. Cell. Physiol.* **123**, 25–32.

96. Miyaba-Inaba, M., Ogata, H., Toki, J., Kuma, S., Sugiura, K., Yasumizu, R., and Ikehara, S. (1987). Isolation of murine hemopoietic stem cells in the Go phase. *Biochem. Biophys. Res. Comm.* **147**, 687–694.

97. Monette, F. C. (1979). Antibodies against pluripotent stem cells: their use in studying stem cells function. *Blood Cells* **5**, 75–191.

98. Moore, M. A. S., McNeill, T. A., and Haskill, J. S. (1970). Density distribution analysis of *in vivo* and *in vitro* colony forming cells in developing fetal liver. *J. Cell. Physiol.* **75**, 181–192.

99. Moore, M. A. S., Broxmeyer, H. E., Sheridan, A. P. C., Meyers, P. A., Jacobsen, N., and Winchester, R. J. (1980). Continuous human bone marrow culture: Ia antigen characterization of probable pluripotential stem cells. *Blood* **55**, 682–690.

100. Mulder, A. H. (1986). Differentiation of hemopoietic stem cells in the thymus. Ph.D. Thesis, Erasmus University, Rotterdam.

101. Mulder, A. H., and Visser, J. W. M. (1987). Separation and functional analysis of bone marrow cells separated by rhodamine-123 fluorescence. *Exp. Hematol.* **15**, 99–104.

102. Mulder, A. H., and Visser, J. W. M. (1988). The entry of the prothymocyte into the thymus after lethal irradiation and bone marrow transplantation. I. Seeding of bone marrow cells into the thymus. *Thymus* **11**, 15–27.

103. Mulder, A. H., Bauman, J. G. J., Visser, J. W. M., Boersma, W. J. A., and Engh, G. J. van den (1984). Separation of spleen colony-forming units and prothymocytes by use of a monoclonal antibody detecting an H-2K determinant. *Cell. Immunol.* **88**, 401–410.

104. Mulder, A. H., Visser, J. W. M., and Engh, G. J. van den (1985). Thymus regeneration by bone marrow cell suspensions differing in the potential to form early and late spleen colonies. *Exp. Hematol.* **13**, 768–775.

105. Mulder, A. H., Visser, J. W. M., Zoetelief, J., and Bekkum, D. W. van (1988). The entry of the prothymocyte into the thymus after lethal irradiation and bone marrow transplantation. II. Time of entry. *Thymus* **11**, 29–41.

106. Muller-Sieburg, C. E., Whitlock, C. A., and Weissman, I. L. (1986). Isolation of two early B lymphocyte progenitors from mouse marrow: a committed pre-pre-B cell and a clonogenic Thy-1^{lo} hematopoietic stem cell. *Cell* **44**, 653–662.

107. Muller-Sieburg, C. E., Townsend, K., Weissman, I. L., and Rennick, D. (1988). Proliferation and differentiation of highly enriched mouse hematopoietic stem cells and progenitor cells in response to defined growth factors. *J. Exp. Med.* **167**, 1825–1840.

108. Nicola, N. A., Burgess, A. W., Staber, F. G., Johnson, G. R., Metcalf, D., and Battye, F. L. (1980). Differential expression of lectin receptors during hemopoietic differentiation: enrichment for granulocyte-macrophage progenitor cells. *J. Cell. Physiol.* **103**, 217–237.

109. Nicola, N. A., Metcalf, D., Von Melcher, H., and Burgess, A. W. (1981). Isolation of murine fetal hemopoietic progenitor cells and selective fractionation of various erythroid precursors. *Blood* **58**, 376–386.

110. Ottman, O. G., Nocka, K. H., Moore, M. A. S., and Pelus, L. M. (1988). Differential expression of class II MHC antigens in subpopulations of human hemopoietic progenitor cells. *Leukemia* **2**, 677–686.

111. Pallavicini, M. G., Summers, L. J., Dean, P. N., and Gray, J. W. (1985). Enrichment of hemopoietic clonogenic cells by multivariate analyses and sorting. *Exp. Hematol.* **13**, 1173–1181.

112. Ploemacher, R. E., and Brons, N. H. C. (1988a). Cells with marrow and spleen repopulating ability and forming spleen colonies on day 16, 12, and 8 are sequentially ordered on the basis of increasing rhodamine 123 retention. *J. Cell. Physiol.* **136**, 531–536.

113. Ploemacher, R. E., and Brons, N. H. C. (1988b). Isolation of hemopoietic stem cell subsets from murine bone marrow: II. Evidence for an early precursor of day-12 CFU-S and cells associated with radioprotective ability. *Exp. Hematol.* **16**, 27–32.

114. Ploemacher, R. E. and Brons, N. H. C. (1988c). Isolation of hemopoietic stem cell subsets from murine bone marrow: I. Radioprotective ability of purified cell suspensions differing in the proportion of day-7 and day-12 CFU-S. *Exp. Hematol.* **16**, 21–26.

115. Ploemacher, R. E., and Brons, N. H. C. (1989). Separation of CFU-S from primitive cells responsible for reconstitution of the bone marrow hemopoietic stem cell compartment following irradiation: evidence for a pre-CFU-S cell. *Exp. Hematol.* **17**, 263–266.

116. Ploemacher, R. E., Brons, R. H. C., and Leenen, P. J. M. (1987). Bulk enrichment of transplantable hemopoietic stem cell subsets from lipopolysaccharide-stimulated murine spleen. *Exp. Hematol.* **15**, 154–162.

117. Ploemacher, R. E., Sluijs, J. P. van der, Voerman, J. S. A., and Brons, N. H. C. (1989). An *in vitro* limiting dilution assay of long-term repopulating hemopoietic stem cells in the mouse. *Blood* **74**, 2755–2763.

118. Porcellini, A., Manna, A., Talevi, N., Sparaventi, G., Marchetti-Rossi, M.-T, Baronciani, D., and De Biagi, M. (1984). Effect of two cyclophosphamide derivatives on hemopoietic progenitor cells and pluripotential stem cells. *Exp. Hematol.* **12**, 863.

119. Pragnell, I. B., Wright, E. G., Lorimore, S. A., Adam, J., Rosendaal, M., Delamarter, J. F., Freshney, M., Eckmann, L., Sproul, A., and Wilkie, N. (1988). The effect of stem cell proliferation regulators demonstrated with an *in vitro* assay. *Blood* **72**, 196–201.

120. Reincke, U., Rosenblatt, M., and Hellman, S. (1984). An *in vitro* clonal assay of adherent cells (ASC) in mouse bone marrow. *J. Cell. Physiol.* **121**, 275.

121. Robinson, J., Sieff, C., Delia, D., Edwards, P. A. W., and Greaves, M. (1981). Expression of cell surface HLA-DR, HLA-ABC and glycophorin during erythroid differentiation. *Nature* **289**, 68–71.

122. Rowley, S. D., Colvin, O. M., and Stuart, R. K. (1985). Human multilineage progenitor cell sensitivity to 4-hydroperoxycyclophosphamide. *Exp. Hematol.* **13**, 295–298.

123. Russell, J. L., and Engh, G. J. van den (1979). The expression of histocompatibility-2 antigens on haemopoietic stem cells. *Tissue Antigens* **13**, 45–52.

124. Scheven, B. A. A., Visser, J. W. M., and Nijweide, P. J. (1986). *In vitro* osteoclast generation from different bone marrow fractions, including a highly enriched haematopoietic stem cell population. *Nature* **321**, 79–81.

125. Schrader, J. W., Battye, F., and Scollay, R. (1982). Expression of Thy-1 antigen is not limited to T cells in cultures of mouse hemopoietic cells. *Proc. Natl. Acad. Sci. USA* **79**, 4161–4165.

126. Sharkis, S. J., Santos, G. W., and Colvin, M. (1980). Elimination of acute myelogenous leukemic cells from marrow and tumor suspensions in the rat with 4-hydroperoxycyclophosphamide. *Blood* **55**, 521–523.

127. Sieff, C., Bicknell, D., Caine, G., Robinson, J., Lam, G., and Greaves, M. F. (1982). Changes in cell surface antigen expression during hemopoietic differentiation. *Blood* **60**, 702–713.

128. Siena, S., Castro-Malaspina, H., Gulati, Sc, Lu, L., Colvin, M. O., Clarkson, B. D., O'Reilly, R. J., and Moore, M. A. S. (1985). Effects of *in vitro* purging with 4-hydroperoxycyclophosphamide on the hematopoietic and microenvironmental elements of human bone marrow. *Blood* **65**, 655–662.

129. Singh, L., Matsukuma, S., and Jones, K. W. (1987). The use of Y-chromosome-specific repeated DNA sequences in the analysis of testis development in an XX/XY mouse. *Development* **101** (suppl.), 143–149.

130. Spangrude, G. J., Heimfeld, S., and Weissman, I. L. (1988). Purification and characterization of mouse hematopoietic stem cells. *Science* **241**, 58–62.

131. Sparrow, R. L., and Williams, N. (1986). The pattern of HLA-DR and HLA-DQ antigen expression on clonable subpopulations of human myeloid progenitor cells. *Blood* **67**, 379–384.

132. Spooncer, E., Lord, B. I., and Dexter, T. M. (1985). Defective ability to self-renew *in vitro* of highly purified primitive haematopoietic stem cells. *Nature* **316**, 62–64.

133. Strauss, L. C., Rowley, S. D., LaRuss, V. F., Sharkis, S. J., Stuart, R. K., and Civin, C. I. (1986). Antigenic analysis of hematopoiesis. V. Characterization of MY-10 antigen expression by normal lymphohematopoietic progenitor cells. *Exp. Hematol.* **14**, 878–886.

134. Suda, T., Suda, J., and Ogawa, M. (1983). Single-cell origin of mouse hemopoietic colonies expressing multiple lineages in variable combinations. *Proc. Natl. Acad. Sci. USA* **80**, 6689–6693.

135. Suda, T., Suda, J., Ogawa, M., and Ihle, J. N. (1985). Permissive role of interleukin 3 (IL-3) in proliferation and differentiation of multipotential hemopoietic progenitors in culture. *J. Cell. Physiol.* **124**, 182–190.

136. Sugiura, K., Inaba, M., Ogata, H., Yasumizu, R., Inaba, K., Good, R. A., and Ikehara, S. (1988). Wheat germ agglutinin-positive cells in a stem cell-enriched fraction of mouse bone marrow have potent natural suppressor activity. *Proc. Natl. Acad. Sci. USA* **85**, 4824–4826.

137. Sutherland, H. J., Eaves, C. J., Eaves, A. C., Dragowska, W., and Lansdorp, P. M. (1989). Characterization and partial purification of human marrow cells capable of initiating long-term hematopoiesis in vitro. *Blood* **74**, 1563–1570.

138. Szer, J., Deeg, H. J., Appelbaum, F. R., and Storb, R. (1985). Failure of autologous marrow reconstitution after cytolytic treatment of marrow with anti-Ia monoclonal antibody. *Blood* **65**, 819–822.

139. Szilvassy, S. J., Lansdorp, P. M., Humphries, R. K., Eaves, A. C. and Eaves, C. J. (1989). Isolation in a single step of a highly enriched murine hematopoietic stem cell population with competitive long-term repopulating ability. *Blood* **74**, 930–939.

140. Thierfelder, S. (1977). Hemopoietic stem cells of rat but not mice express Thy 1.1 alloantigen. *Nature* **269**, 691–693.

141. Till, J. E., and McCulloch, E. A. (1961) A direct measurement of the radiation sensitivity of normal mouse bone marrow cells. *Radiat. Res.* **14**, 213–222.

142. Tindle, R. W., Nichols, R. A. B., Chan, L., Campana, D., Catowski, D., and Birnie, G. D. (1985). A novel monoclonal antibody BI-3C5 recognizes myeloblasts and non-B non-T lymphoblasts in acute leukemias and CGL blast crises, and reacts with immature cells in normal bone marrow. *Leuk. Res.* **9**, 1–9.

143. Töksoz, D., Dexter, T. M., Lord, B. I., Wright, E. G., and Lajtha, L. G. (1980). The regulation of hemopoiesis in long-term bone marrow cultures. II. Stimulation and inhibition of stem cell proliferation. *Blood* **55**, 931–936.

144. Torok-Storb, B. (1988). Hematopoietic cellular interactions: role of the major histocompatibility complex (MHC). *Blood Cells* **14**, 497–504.

145. Torok-Storb, B., and Hansen, J. A. (1982). Modulation of *in vitro* BFU-E growth by normal Ia-positive T cells is restricted by HLA-DR. *Nature* **298**, 473–474.

146. Torok-Storb, B., Nepom, G. T., Nepom, B. S., and Hansen, J. A. (1983). HLA-DR antigens on lymphoid cells differ from those on myeloid cells. *Nature* **305**, 541–543.

147. Trask, B., and Engh, G. J. van den (1980). Antigen expression of CFU-s determined by light activated cell sorting. *In* "Experimental Hematology Today 1980" (S. J. Baum, G. D. Ledney, D. W. van Bekkum eds), pp. 299–307. Karger, Basel.

148. Turner, R W. A., Siminovitch, L., McCulloch, E. A., and Till, J. E. (1967). Density gradient centrifugation of hemopoietic colony forming cells. *J. Cell Physiol.* **69**, 873–884.

149. Visser, J. W. M., and Bol, S. J. L. (1981). A two-step procedure for obtaining 80-fold enriched suspensions of murine pluripotent hemopoietic stem cells. *Stem Cells* **1**, 240–249.

150. Visser, J. W. M., and Vries, P. de (1988) Isolation of spleen-colony forming cells (CFU-s) using wheat germ agglutinin and rhodamine 123 labeling. *Blood Cells* **14**, 369–384.

151. Visser, J. W. M., and Vries, P. de (1990). Identification and purification of murine hemopoietic stem cells by flow cytometry. *In* "Methods in Cell Biology Vol. 33. Flow Cytometry" (Z. Darzynkiewicz, and H. A. Crissman, eds), pp. 451–468, Academic Press, Orlando.

152. Visser, J. W. M., and Bekkum, D. W. van (1990). Purification of pluripotent hemopoietic stem cells; past and present. *Exp. Hematol.* **18**, 248–256.

153. Visser, J. W. M., Engh, G. J. van den, Williams, N., and Mulder, A. H. (1977). Physical separation of the cycling and noncycling compartments of murine hemopoietic stem cells. *In* "Experimental Hematology Today" (S. J. Baum, and G. D. Ledney, eds), pp. 21–27. Springer-Verlag, New York.

154. Visser, J. W. M., Cram, L. S., Martin, J. C., Salzman, G. C., and Price, B. J. (1978). Sorting of a murine granulocytic progenitor cell by use of laser light scattering measurements. *In* "Pulse Cytophotometry" (D. Lutz, ed.), pp. 187–192. European Press, Ghent.

155. Visser, J. W. M., Engh, G. J. van den, and Bekkum, D. W. van (1980). Light scattering properties of murine hemopoietic cells. *Blood Cells* **6**, 391–407.

156. Visser, J. W. M., Bol, S. J. L., and Engh, G. J. van den (1981). Characterization and enrichment of murine hemopoietic stem cells by fluorescence activated cell sorting. *Exp. Hematol.* **9**, 644–655.

157. Visser, J. W. M., Bauman, J. G. J., Mulder, A. H., Eliason, J. F., and Leeuw, A. M. de (1984). Isolation of murine pluripotent hemopoietic stem cells. *J. Exp. Med.* **59**, 1576–1590.

158. Visser, J. W. M., Hogeweg-Platenburg, M. G. C., Vries, P. de, Bayer, J., and Ploemacher, R. E. (1990). Culture of purified pluripotent haemopoietic stem cells. *In.* "The Biology of Hematopoiesis" (N. Dainiak, *et al.* eds), pp. 1–8. Wiley-Liss Inc., New York.

159. Vries, P. de (1988). Identification of murine haemopoietic stem cells with monoclonal antibodies. Ph.D. Thesis, Erasmus University, Rotterdam.

160. Vries, P. de, Pronk, G. J., and Visser, J. W. M. (1988). A multiple staining procedure for the flow cytometric screening of monoclonal antibodies directed against murine pluripotent stem cells (CFU-S). *Blood Cells* **14**, 561.

161. Wagemaker, G., Visser, T. P., and Bekkum, D. W. van (1986). Cure of murine thalassemia by bone marrow transplantation without eradication of endogenous stem cells. *Transplantation* **42**, 248–251.

162. Watt, S. M., Gilmore, D. J., Davis, J. M., Clark, M. R., and Waldmann, H. (1987a). Cell-surface markers on haemopoietic precursors. Reagents for the isolation and analysis of progenitor cell subpopulations. *Molecular and Cellular Probes* **1**, 297–326.

163. Watt, S. M., Karhi, K., Gatter, K., Furley, A. J. W., Katz, F. E., Healy, L. E., Atass, L. J., Bradley, N. J., Sutherland, D. R., Levinsky, R. J., and Greaves, M. F. (1987b). Distribution and epitope analysis of the cell membrane glycoprotein (HPCA-1) associated with human hemopoietic progenitor cells. *Leukemia* **1**, 417–426.

164. Wielenga, J. J. (1990). Hemopoietic stem cells in rhesus monkeys. Surface antigens, radiosensitivity, and responses to GM-CSF. Ph.D. Thesis, Erasmus University, Rotterdam.

165. Williams, D. E., Straneva, J. E., Shen, R-N, and Broxmeyer, H. E. (1987). Purification of murine bone-marrow-derived granulocyte-macrophage colony-forming cells. *Exp. Hematol.* **15**, 243–250.

166. Winchester, R. J., Ross, G. D., Jarowsky, C. I., Wang, C. Y., Halper, J., and Broxmeyer, H. E. (1977). Expression of Ia-like antigen molecules on human granulocytes during early phases of differentiation. *Proc. Natl. Acad. Sci. USA* **74**, 4012–4016.

167. Winchester, R. J., Meyers, P. A., Broxmeyer, H. E., Wang, C. Y., Moore, M. A. S., and Kunkel, H. G. (1978). Inhibition of human erythropoietic colony formation in culture by treatment with Ia antisera. *J. Exp. Med.* **148**, 613–618.

168. Winton, E. F., and Colenda, K. W. (1987). Use of long-term human marrow cultures to demonstrate progenitor cell precursors in marrow treated with 4-hydroperoxycyclophosphamide. *Exp. Hematol.* **15**, 710–714.

169. Winton, E. F., Colenda, K. W., McClure, H. M., Chan, W. C., and Gordon, D. S. (1985). Studies on the use of nonhuman primates to determine the DR status of the human hematopoietic stem cell. *Exp. Hematol.* **13**, 869–873.

170. Witte, T. de, Plas, E., Koekman, E., Blankenborg, G., Salden, M., Wessels, J., and Haanen, C. (1983). Cell size monitored counterflow centrifugation of human bone marrow resulting in clonogenic cell fractions substantially depleted of small bone marrow lymphocytes. *J. Immunol. Meth.* **65**, 171–182.

171. Worton, R. G., McCulloch, E. A., and Till, J. E. (1969). Physical separation of hemopoietic stem cells differing in their capacity for self-renewal. *J. Exp. Med.* **130**, 91–103.

172. Yeager, A. M., Kaizer, H., Santos, G. W., *et al.* (1986). Autologous bone marrow transplantation in patients with acute nonlymphocytic leukemia, using *ex vivo* marrow treatment with 4-hydroperoxycyclophosphamide. *New Engl. J. Med.* **315**, 141–147.

1.3

Characterization of Erythroid, Lymphoid and Monomyeloid Lineages in Normal Human Bone Marrow

MICHAEL R. LOKEN,[1] CURT I. CIVIN,[2] VIREN O. SHAH,[1] MARY JO FACKLER,[2] INE SEGERS-NOLTEN[3] and LEON W. M. M. TERSTAPPEN[1]

[1] *Becton Dickinson Immunocytometry Systems*
San Jose, CA, USA
[2] *Johns Hopkins Oncology Center*
Baltimore, MD, USA
[3] *Twente University*
Enschede, The Netherlands

I. INTRODUCTION

In hematology, the goal of examining bone marrow aspirates or biopsies is to identify abnormalities which are related to dysfunctional hematopoiesis. This analysis requires a basic understanding of the composition of normal bone marrow against which to compare potential disease states. Currently the analysis of bone marrow is performed by microscopic review of cells on a slide stained for morphologic or cytochemical characteristics. The person performing the analysis must be highly skilled to be able to distinguish normal from abnormal cells. Abnormalities are usually identified by an increase in the relative number of cells which are infrequent or not detectable in normal bone marrow aspirates.

Visual examination of slides, by its very nature, is subjective with the results reported in a qualitative or semiquantitative manner. Since the preparation of slides for Romonowski stains (used for morphology) is different from

FLOW CYTOMETRY IN HEMATOLOGY
ISBN 0-12-432940-3

the preparation of enzymatic-based cytochemical stains, simultaneous correlation between morphology and cytochemistry is not usually performed on individual cells. As a result, cells which appear infrequently in the bone marrow are skipped and are not identified. Any cell which appears at a frequency of less than 5% may be overlooked unless it has very distinctive characteristics which discriminate it from the remainder of the cells.

Flow cytometric analysis of bone marrow offers significant advantages over the microscope-based technology. Flow cytometric analysis is quantitative both in its ability to count large numbers of cells precisely as well as in its discrimination based on quantitative differences in cellular characteristics. The discrimination between cells is not based simply on the presence or absence of a particular antigen but can be based on the amount of antigen relative to the other cells. Current instrumentation has a dynamic range of four decades over which to distinguish differences in characteristics.

Multiple characteristics can be identified simultaneously on an individual cell. This allows correlation of size (forward light scattering), granularity (orthogonal light scattering), and three (or more) cell surface or cytoplasmic antigens. The simultaneous measurement of four or five independent parameters on the cells passing through the laser beam creates a multidimensional data space in which cell populations with dissimilar properties emerge in different locations as clusters of cells. One of the difficulties in multi-parameter flow cytometry is to display and to identify cell subpopulations in these multi-dimensional data files. An attempt has been made to facilitate the identification of cell populations in multidimensional space by the use of color (*35–38*). Populations of cells identified in a single two-dimensional projection of the four- or five-dimensional data can be depicted with a primary color and appear as the same color in other projections of that data. In this manner, other primary colors can be applied in the data projections to improve the definition, i.e. purity of a cell population. The cell population might then be characterized by the overlap of the two primary colors yielding a secondary color. This cell population once again can be further defined with other primary colors. Once a cell cluster is identified with respect to all measured parameters, the definition of the cell cluster is verified by checking all possible two-parameter projections. Now a specific color can be assigned to this cell population which is based on the intersection of multiple projections of the multi-dimensional data. In this way populations can be distinguished in multi-dimensional space which may not be observed in any two-dimensional projection.

In normal peripheral blood only the most mature stages of cell development are present. These form well-defined cell clusters of each lineage by correlating cell surface antigens, forward and orthogonal light scattering (*39*). Although the analysis of peripheral blood cell clusters shows them to be discrete, they elongate in the analysis of bone marrow since precursors of each of the cell lineages are present. Gradual changes in light scattering properties and cell surface antigens on the cells permit the identification of the individual cell lineages and the various maturational stages within each lineage (*22,23,40,41*). By varying the cell surface antigens in the combination of parameters the identification of discrete maturational stages within each cell lineage can be optimized.

In this article we will relate the multidimensional flow cytometric data with standard morphological examination. Cells identified by flow cytometry were sorted for further morphological or cytochemical analysis. In this manner it is possible to illustrate the relationship between cell surface antigen expression with the more conventional identification of morphological changes which occur both in the cytoplasm and in the nucleus of cells in normal bone marrow. This provides clinicians with a means of establishing the relationship between flow cytometric data and the more familiar morphologic distinction.

The data presented here are based entirely on

normal samples in order to provide the groundwork against which to evaluate abnormal samples. Many previous morphological studies have made use of leukemia or other abnormal samples where infrequent immature stages are amplified to become predominant in the sample. We have taken an alternative approach of enriching cell populations based on their cellular characteristics by flow cytometry for subsequent microscopic analysis. By purifying a population to homogeneity, it is possible to discern subtle differences between cells which may have been overlooked in the examination of the total heterogeneous bone marrow sample. In this way the difficulty of characterizing infrequent cell types is overcome.

In multi-dimensional flow cytometry the mature cells of each lineage that appear in blood can be clearly distinguished based on cell surface and light scattering characteristics (*39*). As a result, it is possible to work backwards from the mature stages to the less mature stages within an individual lineage. There are two basic assumptions which are made in this approach. First, it is assumed that the normal bone marrow is in a steady state with constant turnover of the cells. As a consequence, all stages of development are present in the bone marrow aspirate. Second, it is assumed that the changes in cellular characteristics are gradual as the cells mature to become fully functional end-stage cells. This precludes discontinuous changes in which antigens appear and disappear rapidly.

The three major lineages of cells which comprise the bone marrow, erythroid, lymphoid, and monomyeloid, will be discussed separately. A limited number of monoclonal antibodies will be used as illustration in the comparison of the flow data with the morphologic analysis. Once the relationship between morphology and flow cytometric characteristics has been established using one set of monoclonal antibodies, it is possible to relate other cell surface antigens in this scheme.

Several review articles and primary references illustrate the use of multiple cellular markers to define maturational differences within bone marrow (*1,6,22–24,30,31,33,38, 40–43*). The striking feature of these studies is the amazing consistency of the maturational patterns demonstrated in the analysis of bone marrow samples of different individuals. The consistent pattern observed in normal tissue illustrates the highly regulated nature of the hematopoetic system since there is very little variation from individual to individual.

II. THE ERYTHROID LINEAGE

The identification of the erythroid lineage in normal bone marrow was facilitated by the use of a lineage restricted antigen, glycophorin A (*9,17,22,29*). This antigen appears only on the erythroid lineage and can be paired with other markers which may have a wider distribution on other lineages (lineage associated antigens) (*21,22*). By combining cell surface antigen expression with light scattering characteristics, it is possible to demonstrate the progressive changes during cell maturation. This is illustrated for the erythroid lineage in Plate 1. Low density cells from a normal bone marrow aspirate were stained with anti-glycophorin A FITC and CD71-PE (anti-transferrin receptor). Four two-parameter projections of anti-glycophorin A FITC, CD71-PE, forward light scattering and orthogonal light scattering are shown. (Note that the orthogonal light scattering parameter was transformed to increase the resolution between cell populations (*37*).) The erythroid cells identified by multi-dimensional flow cytometric analysis were assigned colors, whereas cells not belonging to the erythroid lineage were depicted gray (Plate 1). The populations indicated with colors in Plate 1 were sorted for morphological characterization and are illustrated in Plate 2.

Red. (Anti-glycophorin A bright, $CD71^-$, low forward light scattering and intermediate orthogonal light scattering (Plate 1).) The red colored cells in Plate 1 were clearly the most mature erythrocytes by morphology (Plate 2(D)). The mature erythrocytes are relatively small by forward light scattering, but have significant orthogonal light scattering signals. The relatively large orthogonal light scattering is most likely a result of the unique shape of the erythrocyte rather than granularity. These mature erythrocytes express high levels of glycophorin A and do not express CD71 (Plate 1(B), (C) and (D)).

To distinguish the reticulocytes from the mature erythrocytes or the nucleated erythrocytes it is necessary to use additional parameters, i.e. a nucleic acid stain, reactive with RNA to distinguish the reticulocytes from the mature erythrocytes and a nucleic acid stain reactive with DNA to distinguish the reticulocytes from the nucleated erythroid cells (*36*,*44*).

Gold. (Anti-clycophorin A bright, CD71 dim to bright, low to intermediate forward light scattering and low orthogonal light scattering (Plate 1).) The gold colored cells in Plate 1 contained predominantly late normoblasts (Plate 2(C)). These erythroid cells express high levels of glycophorin and increasing levels of transferrin receptor (Plate 1(B), (C) and (D)). By light scattering these cells are slightly larger by forward light scattering and have lower orthogonal light scattering signals than the red colored mature enucleated cells. The significant differences in orthogonal light scattering of both populations suggest that during the enucleation process, the forward light scattering decreases while the orthogonal light scattering increases.

Blue. (Anti-glycophorin A intermediate to dim, CD71 bright to intermediate, intermediate to large forward light scattering and low to intermediate orthogonal light scattering (Plate 1).) The blue colored cells in Plate 1 were mainly early normoblast by morphology (Plate 2(B)). The early normoblast displayed significantly larger forward light scattering signals as compared with the later stages (Plate 1(A), (B) and (D)). The blue colored early normoblast cells express intermediate to high levels of glycophorin A and a high amount of transferrin receptor.

Black. (Anti-glycophorin A^-, CD71 intermediate to dim, large forward light scattering and low orthogonal light scattering (Plate 1).) The black colored cells in Plate 1 contained predominantly erythroblast (Plate 2(A)). The cells were large by forward light scattering and express transferrin receptor, but do not express glycophorin A on their surface (Plate 1(B), (C) and (D)). The cells were defined as erythroid not only by cell morphology but also by the absence of expression of specific lymphoid markers, CD19 (B cell), CD3 (T cell), natural killer cell markers, CD16 and CD56 (NK cells) and myeloid markers, CD33 as determined in three color immunofluorescence experiments (*22*,*31*).

The multi-dimensional assessment of erythrocyte maturation as illustrated in Plate 1 can serve as a model to further characterize the erythroid lineage. The additional measurements of CD45 illustrated that this antigen is progressively lost as the erythroid cells develop from progenitors to mature erythrocytes (*31*). Other antigens as CD35 or blood group antigens can be used in a similar manner to further characterize erythroid maturation. The identification of the earliest committed erythroid progenitor cells (CFU-E, BFU-E) at this time requires the measurement the CD34 antigen in any combinations of cell surface markers. The cells expressing the CD34 antigen constitute 1% of the total bone marrow and contain the progenitors of all cell lineages (*1*,*7*,*8*,*18*,*19*,*43*). The earliest committed erythroid progenitor cells can in fact be identified by pairing the CD34, CD38 and CD71 antigens in concordance with forward and orthogonal light scattering (*43*).

III. THE LYMPHOID LINEAGE

A. T and NK Cells

Both immature and mature lymphoid cells can easily be identified within the bone marrow (*23*,*31*). Mature T cells ($CD3^+$) and NK cells ($CD16^+$ and/or $CD56^+$) comprise a resident population and are not simply contaminants of blood into the marrow aspirate. Since T cell maturation occurs within the thymus, it's expected that immature forms are not readily observed within the bone marrow. Recent studies, however, have identified candidates for the precursor T cell which migrates to the thymus (*3,11*). The disclosure of CD34 expression on early human thymocytes strongly supports the finding of the T lymphoid precursor in human bone marrow (*28*).

A potential precursor of the NK cells within the bone marrow has not yet been identified. Recent studies however identified distinct maturational stages of NK cells within the peripheral blood (*27*). The latter study provides the groundwork for further investigation of potential bone marrow NK cell precursors.

B. B CELLS

In contrast, it is well known that B cells undergo maturation within the bone marrow (*10*). Studies of B cells using flow cytometry have been facilitated by a lineage restricted marker, CD19, which is present both on immature and mature B cells and has not been found on any other hematopoietic cells (*26*). Commitment of the progenitor cells into the B lymphoid lineage at the cell surface antigen level is indicated by the presence of the immature CD34 antigen and the simultaneous expression of both CD10 and CD19. In Plate 3 this is illustrated in four two-parameter projections of a bone marrow aspirate stained with CD34 FITC and CD10 PE. Immature B lymphoid and $CD34^+$ progenitor cell populations are depicted in the correlative analysis of the four parameters and each assigned a color.

Gold. ($CD34^+$, $CD10^-$, large forward, low to intermediate orthogonal light scattering.) The majority of the $CD34^+$ progenitor cells did not express the CD10 antigen and were blasts of heterogeneous morphology. Approximately half of these $CD34^+$, $CD10^-$ cells were reactive with either α-naphthol AS-D chloroacetate esterase of α-naphthyl acetate esterase and therefore could be classified as being differentiated into the monomyeloid cell lineage. No TdT positive cells were found in this cell fraction indicating that the $CD34^+$, $CD10^-$ cells were not committed to the lymphoid lineage.

Red. ($CD34^+$, $CD10^+$, low forward, low orthogonal light scattering (stage BI (*23*)).) Approximately 25% of the $CD34^+$ cells coexpressed the CD10 antigen. Three color immunofluorescent experiments showed that all $CD10^+$, $CD34^+$ cells coexpressed the CD19 antigen indicating the commitment into the B lymphoid lineage. Cell sorting of these cells revealed a rather uniform morphologic appearance (Plate 4(A), and the presence of the nuclear stain terminal deoxynucleotidyl transferase (TdT) in all $CD10^{++}$, $CD34^+$ cells (Plate 4(D)). This population of cells did not contain detectable quantities of naphthol AS-D chloroacetate esterase or α-naphthyl acetate esterase.

Blue. ($CD34^-$, $CD10^+$, low to large forward, low orthogonal light scattering (stage BII and III (*23*).) This cell population does not express CD34 and expresses less CD10 as compared with the red colored cells. All blue colored $CD10^+$, $CD34^-$ cells were B lymphocytes as defined by the coexpression of CD19. (Note that the CD10 antigen also stains neutrophils but they are not depicted blue since they are distinguishable from the $CD19^+$, $CD10^+$ B lymphocytes by orthogonal light scattering.) The morphology of the cells was typical for lymphoblasts with round or slightly indented nuclei, fine homogeneously distributed chromatin, one to four nucleoli and scanty blue agranular cytoplasm (Plate 4(B)). All of the red and blue colored immature B lineage cells (stage BI, II, III) are rather heterogeneous in size as identified by forward light scattering in Plate 3(A), (B) and (D). The majority of the $CD10^+$ B lymphocytes are small by forward light scattering, yet a significant fraction of these cells extend into the region with large forward and orthogonal light scattering (Plate 3(A), (B) and (D), red and blue). Since it has been shown that approximately 16% of the immature, $CD10^+$, B lineage cells are proliferating, this size heterogeneity may reflect the expansion of these populations (*14*).

The relationship between the CD19, CD34 and CD10 antigens was used to develop a scheme for antigen expression during B lymphocyte development (*23*). Four stages of maturation, stages BI–IV, were identified. Multi-dimensional flow cytometric analysis was used to further address the expression of antigens during B cell development. Expression of a variety of antigens during B cell differentiation was assessed in three color immunofluorescence experiments and achieved by exchanging one of the staging defining antibodies for the antigen studied.

(1) Stage BI. The earliest B lineage cell (red in Plate 3) was identified by the simultaneous expression of CD34, CD19 and high levels of CD10. Cell sorting revealed a homogeneous morphology and the presence of TdT (Plate 4(A) and (D)). These cells are also observed in fetal bone marrow and include essentially all of the TdT positive cells resident in the bone marrow (*20*). The stage BI cells constituted approximately 10% of bone marrow B cells, and expressed the CD45 and HLA-DR antigens dimly and the CD38 antigen brightly (*23*,*24*,*42*). The CD20, CD21, CD22, HLA-DP and HLA-DQ antigens are not expressed at this stage (*23*,*24*), neither is cell surface or cytoplasmic lgM and lgD (V. O. Shah, unpublished observations).

(2) Stage BII. This stage is characterized by the absence of CD34 and the presence of CD19 and CD10 (*23*). This, in fact, represents the classical pre-B cell as it has cytoplasmic IgM, but lacks surface IgM (V. O. Shah, unpublished observations). The cells in this stage constituted 40% of the bone marrow B cells and are further characterized by the absence of nuclear TdT and both cytoplasmic and cell surface IgD (V. O. Shah, unpublished observations). The antigens CD22, CD21, and HLA-DQ were not yet expressed at this stage. HLA-DP however appears on the cell surface of these cells and the density of HLA-DR increases. No difference in the density of the CD38 antigen was observed.

(3) Stage BIII. This stage is characterized by the absence of CD34, the presence of CD19 and a decrease of the CD10 antigen (*23*). Stage BIII would be classically defined as an immature B cell population since it expressed cIgM, sIgM, cIgD but not sIgD (V. O. Shah, unpublished observations). This population of cells is relatively small within the adult (approximately 10% of the B lineage cells) but is a significant population in fetal tissue (approximately 40% of B lineage cells) (*20*). By morphology the cells in this stage had moderately condensed smooth chromatin, few agranular cytoplasm and slightly indented nuclei (Plate 3(C)) and were not distinguishable from the more mature stage BIV cells by morphology. This population, like the mature B cells, expresses high levels of the CD45 antigen. The CD38 antigen decreased concurrently with an increase of the CD20 and HLA-DQ antigen (*23*,*24*,*42*).

(4) Stage BIV. This stage is characterized by the absence of both CD34 and CD10 and the presence of CD19 (*23*). This population of mature bone marrow B cells is indistinguishable from the B cells found in peripheral blood. These resting, mature B cells comprise approximately 40% of the B lineage cells ($CD19^+$) in the adult marrow (*23*). Interestingly, this mature population is conspicuously absent from fetal bone marrow (*20*). This population can be characterized by the appearance of CD21, CD22 on the cell surface with most cells expressing sIgD. Further the cells in this stage express CD20, HLA-DQ, HLA-DP, HLA-DR and sIgM. These resting B lymphocytes are small by forward light scattering and have low orthogonal light scattering. These cells appear morphologically as fully mature lymphocytes and are not distinguishable from the earlier stage BIII cells by morphology.

C. Plasma Cells

B cell differentiation does not end with cells in stage BIV, but evolves into an immunoglobulin producing plasma cell following antigen stimulation (*2*,*4*,*34*). The plasma cells appeared to be relatively infrequent in normal bone marrow, but could be identified using the combination of CD38 and forward and orthogonal light scattering (Plate 5) (*42*). In Plate 5(A) and (B) forward and orthogonal light scattering and the expression of the CD38 and CD22 antigens of a bone marrow aspirate are shown. The immature stages of B lymphocyte development, stages BI and BII ($CD22^-$, $CD38^+$) B cells are depicted red, while the mature B cells, stage BIV ($CD22^+$, $CD38^-$) are depicted green. The NK cells, monocytes and immature monomyeloid cells, all expressing the CD38 antigen, are depicted blue. Cells not expressing CD38 and CD22 containing predominantly T lymphocytes are depicted gold.

The plasma cells all express the CD38 antigen extremely brightly and have a typical location in the correlation of both light scattering parameters as visualized by gating on both light scattering and CD38 fluorescence intensity, in Plate 5(C) and (D). The confirmation that the cells colored blue and red in Plate 5(C) and (D) are plasma cells was made by cell sorting for morphological characterization (Plate 6). The purity of the plasma cells in these sorted cell fractions was greater than 95%. The sorted plasma cells have high levels of intracytoplasmic Ig of either kappa or lambda types. The frequency of these cells is low, 0.25%, but is relatively constant from individual to individual (SD = 0.05) (*42*). This number correlates quite well with the frequency identified by cytoplasmic Ig staining (*34*). Interestingly, these bright $CD38^+$ cells no longer express the B lineage restricted antigen, CD19. In contrast some of the plasma cells expressed the early B cell antigen CD10 and/or the myeloid antigen CD33 (*42*).

Two distinct populations of plasma cells were identified, one expressing CD22 which is relatively small by forward light scattering, depicted blue in Plate 5(C) and (D), and the other which did not express the CD22 antigen, was larger by forward light scattering, depicted red in Plate 5(C) and (D). Both populations represented distinct maturational stages. The $CD22^+$ plasma cells had the typical morphology of lymphoplasmacytoid cells, whereas the $CD22^-$ plasma cells were the fully matured plasma cells. Microscopic examination of cytoplasmic immunoglobulin light chain immunofluorescence confirmed the difference in maturation by the lower amounts of cytoplasmic Ig in the $CD22^-$ as compared with the $CD22^+$ plasma cells.

These data suggest that the development of plasma cells in human bone marrow can be divided into two stages, the "early" lymphoplasmacytoid plasma cell and the classic "late" plasma cell which secretes large amounts of immunoglobulin. The end stage of B cell differentiation, i.e. plasma cells, is not directly linked to the mature B lineage cells in the bone marrow. No B lymphocytes could be detected in the human bone marrow transient from the mature small stage BIV $CD22^+$, $CD38^-$ B cells to the "early" large $CD22^+$, CD38 bright "early" lymphoplasmacytoid plasma cells. We presume that once the bone marrow B lymphocytes mature, they migrate out of the marrow to peripheral tissues, where they can undergo activation in response to antigen and return to the bone marrow as "early" lymphoplasmacytoid plasma cells (*42*).

IV. THE MONOMYELOID LINEAGE

The monomyeloid lineages include monocytes, eosinophils, basophils and the predominant neutrophils. These cells can be uniquely distinguished in their mature forms in blood based on their light scattering properties in combination with differential antigen expression (*39*). Of the monomyeloid lineages the neutrophils have been extensively studied in correlating their cell surface antigen expression with morphology (*40*).

Whereas identification of the erythroid and B lymphoid lineages was facilitated by the availability of lineage-specific monoclonal antibodies, few monoclonal antibodies exist which are specific for the monomyeloid compartment (*25,32,39*). Cell differentiation within the monomyeloid lineage is even more complicated since for the neutrophils only lineage-associated monoclonal antibodies are available with no neutrophil-specific markers yet described (*25,32,39*). This impediment was overcome by the use of a combination of lineage-associated antibodies which place the neutrophil lineage in a unique position in multi-dimensional space separate from the monocytic and eosinophilic lineages. The combination which gave the best discrimination of both lineage and maturation was CD11b, CD15, and CD16. Changes in light scattering were as important as antigen expression in defining neutrophil maturation.

A typical example of five-dimensional flow cytometric analysis of the neutrophil maturation within normal human bone marrow is illustrated in Plate 7. The plate shows four two-parameter projections of the data. Only the neutrophilic lineage was colored, all other cells remained gray. Gradual changes in both light scattering properties and cell surface antigen expression denote maturational differences along the neutrophilic cell lineage. The black, blue, gold, red and green colors represent neutrophil populations in increasingly mature developmental stages. In none of the two parameter projections can all the five stages be distinguished; only simultaneous analysis of the five parameters in five-dimensional space permitted the discrimination.

It is clear from Plate 7 that the distinction of maturational stages among the neutrophils is gradual, however boundaries between stages can be set reproducibly since objective criteria can be defined. The cells present in five locations of the neutrophil differentiation track were sorted for comparison with the classical light microscopic methodology to identify maturational stages within the neutrophilic lineage. Microscope slide preparations were made using a Shandon Cytocentrifuge and stained with either Wright stain, myeloperidase, α-naphthol As-D chloroacetate esterase or α-naphthyl acetate esterase. Each of the five stages was $>95\%$ pure neutrophils as confirmed by the presence of myeloperoxidase and α-naphthol AS-D chloroacetate esterase and the absence of the monocyte-specific α-naphthyl acetate esterase.

The five identified maturational stages have unique characteristics:

(1) Black, large forward light scattering, low orthogonal light scattering $CD16^-$ and $CD11b^-$, dimly $CD15^+$ (Plate 7). This stage contained predominantly myeloblast, with an abundance of myeloperoxidase staining, α-naphthol AS-D chloroacetate esterase staining and the absence of the monocyte specific α-naphthyl acetate esterase (Plate 8(E)).

(2) Blue, intermediate to large forward light scattering, intermediate to large orthogonal light scattering, $CD16^-$ and $CD11b^-$ to dim, dim to bright CD15 (Plate 7). This stage contained predominantly promyelocytes, with an abundance of myeloperoxidase staining, α-naphthol AS-D chloroacetate esterase staining and the absence of the monocyte-specific α-naphthyl acetate esterase (Plate 8(D)).

(3) Gold, intermediate forward light scattering, large orthogonal light scattering, $CD16^-$, $CD11b^+$, bright CD15 (Plate 7). This stage contained predominantly myelocytes and metamyelocytes with an abundance of both myeloperoxidase staining and α-naphthol AS-D chloroacetate esterase staining and the absence of the monocyte-specific α-naphthyl acetate esterase (Plate 8(C)).

(4) Red, intermediate forward light scattering, large to intermediate orthogonal light scattering, dim CD16, bright CD11b and bright CD15 (Plate 7). By morphology the cells mainly consisted of metamyelocytes and bands. The monocyte-specific α-naphthyl acetate esterase could not be found in this stage the cells stained with myeloperoxidase and weakly with α-naphthol AS-D chloroacetate esterase (Plate 8(B)).

(5) Green, low forward light scattering, large to intermediate orthogonal light scattering, bright $CD16^+$, CD11b bright and intermediate CD15 (Plate 7). This sorted cell fraction contained only bands and segmented neutrophils. As expected the monocyte-specific α-naphthyl acetate esterase could not be found on the mature neutrophils, whereas all cells stained with myeloperoxidase and weakly with

α-naphthol AS-D chloroacetate esterase (Plate 8(A)).

It must be noted that not all antibodies assigned to a CD number are equivalent (*32*). In these studies, clone L16 was preferable to anti-Leu-M1 as the CD15 antibody since it had minimal reactivity with monocytes and eosinophils as compared with neutrophils (*35,39,40*). In a similar manner, My23 was used in preference to anti-Leu-11a as the antibody recognizing the CD16 antigen because of its relatively bright staining of neutrophils.

Initial studies have been made to characterize the monocytes separately from the neutrophils. By combining forward light scattering, orthogonal light scattering and the expression of the antigens CD14, CD11b and CD33 the monocytic lineage could be uniquely depicted and different maturational stages identified (*41*). These studies however have not been completed to the same extent for the neutrophils. The difficulties in defining maturation within the monocyte lineage include the very low frequency of these cells and the lack of clear morphological differences between maturational stages within the bone marrow monocytes. Thus additional confirmation is required to prove that the monocyte stages defined by cell surface antigen expression are indeed different maturational stages.

The assessment of eosinophil and basophil maturation is even more difficult as a result of their low frequency.

V. SUMMARY

Morphological analysis of bone marrow is a cornerstone of diagnostic procedures used by the hematologist. There is a large knowledge base developed by the morphologist in order to discriminate between normal and abnormal cells and to place the cells in a larger context in relation to other formed elements within the bone marrow. This technique, however, is semi-quantitative at best and subtle differences between cells may be scored differently based on subjective criteria.

In this article we have demonstrated the relationship between flow cytometric identification of cell populations and that of the classical cell morphology. Different stages of cellular development could be defined on a flow cytometer by combining four or five characteristics simultaneously: first to identify each of the cell lineages; and second, to identify stages of development within each cell lineage in normal bone marrow. Confirmation of these studies has utilized morphological differences which can be studied after sorting the cell populations identified by flow cytometry.

Multi-dimensional flow cytometric analysis of bone marrow is particularly powerful for studying infrequent populations of cells in a heterogeneous mixture and has some distinctive advantages over standard morphological techniques. This technique is quantitative in terms of characterizing populations of cells based on intensity differences in light scattering and/or cell surface antigen expression. Flow cytometry is also quantitative in terms of accurately counting the number of cells identifiable in each population. The identification of populations using flow cytometry can be objective with specific criteria used to define differences between populations of cells. In addition, multiple characteristics can be assessed simultaneously on individual cells, thereby providing maximal discrimination between cells based on several parameters.

Indeed, the results of multiple flow cytometric studies of normal bone marrow indicate that the expression of cell surface antigens and light scattering characteristics are highly reproducible and distinct for each lineage of cells. The typical position of the normal cells in the simultaneous analysis of multiple parameters provides the basis on which to discriminate the abnormal cells. The cells with abnormal antigen expression and/or light scattering characteristics can easily be distinguished from normal. Such cells can be quantified to monitor residual disease upon treatment (*5,12,13,15,16,38,41,45*).

Currently flow cytometry and light microscopy are complementary techniques when used in a clinical setting. The current method of training hematologists and morphologists focuses on visual discrimination between cells. The application of flow cytometry in the clinical setting, then, will require a significant amount of education correlating the flow analysis with the visual analysis by microscope. The flow cytometric data could be used to screen normal from abnormal, thereby limiting the number of slides morphologists must review. Once an abnormal specimen has been identified, the flow cytometric data can be used to focus the attention of the morphologists on a specific type of cell with specific characteristics. If needed, these cells can be sorted for more careful and precise morphological examination.

Presently the most significant disadvantage of flow cytometry, as compared with traditional morphology by light microscopy, is the complexity of the data and the difficulty in data interpretation. The human mind is not easily adapted to thinking in more than two-dimensional space. Since the power of flow cytometry is in defining populations based on multiple dimensions, it is essential to distil the information into a form that is readily understandable by a clinician. For future acceptance of flow cytometry in assessment of hematological specimen, significant emphasis must be placed on computer analysis of the multi-dimensional data with extensive interpretation included in the software package.

REFERENCES

1. Andrews, R. G., Singer, J. W., and Bernstein, I. D. (1989). Precursors of colony-forming cells in humans can be distinguished from colony-forming cells by expression of the CD33 and CD34 antigens and light scattering properties. *J. Exp. Med.* **169**, 1721–1731.
2. Benner, R., Hijmans, J., and Haaijmans, J. J. (1981). The bone marrow: The major source of serum immunoglobulins, but still a neglected site of antibody formation. *Clin. Exp. Immunol.* **46**, 1–8.
3. Bertho, J. M., Mossalayi, M. D., Dalloul, A. H., Mouterde, G., and Debre, P. (1990). Isolation of an early T-cell precursor (CFU-TL) from human bone marrow. *Blood* **75**, 1064–1068.
4. Boyd, A. W., Anderson, K. C., Freedman, A. S., Fisher, D. C., Slaughenhoupt, B., Schlossman, S. F., and Nadler, L. M. (1985). Studies of *in vitro* activation and differentiation of human B lymphocytes: I. Phenotypic and functional characterization of the B cell population responding to anti-Ig antibody. *J. Immunol.* **134**, 1516–1523.
5. Campana, D., Coustan-Smith, E., and Janossy, G. (1990). The immunologic detection of minimal residual disease in acute leukemia. *Blood* **76**, 163–171.
6. Civin, C. I., and Loken, M. R. (1987). Cell surface antigens on human marrow: dissection of hematopoietic development using monoclonal antibodies and multiparameter flow cytometry. *Int. J. Cell Cloning* **5**, 267–288.
7. Civin, I., Banquerigo, M. L., Strauss, L. C., and Loken, M. R. (1987). Antigenic analysis of hematopoiesis. IV. Characterization of MY10-positive progenitor cells in normal human bone marrow. *Exp. Hematol.* **15**, 10–17.
8. Civin, C. I., Trischmann, T. M., Fackler, M. J., Bernstein, I. D., Bühring, H., Campos, L., Greaves, M. F., Kamoun, M., Katz, D. R., Lansdorp, P. M., Look, A. T., Seed, B., Sutherland, D. R., Tindle, R. W., and Uchanska-Ziegler, B. (1990). Report on the CD34 cluster workshop. *In* "Leucocyte Typing IV White Cell Differentiation Antigens" (W. Knapp, B. Dorken, W. R. Gilks, E. P. Rieber, R. E. Schmidt, H. Stein, and A. E. G. Kr. von dem Borne, eds), pp. 818–825. Oxford University Press, Oxford.
9. Gahmberg, C. G., Jokinen, M., and Andersson, L. C. (1978). Expression of the major sialoglycoprotein (glycophorin) on erythroid cells in human bone marrow. *Blood* **52**, 379–387.
10. Gobbi, M., Caligaris-Cappio, F., and Janossy, G. (1983). Normal equivalent of B cell malignancies: analysis with monoclonal antibodies. *Br. J. Haematol.* **54**, 393–403.
11. Gore, S. D., Kasten, M. B., and Civin, C. I. (1989). Normal human bone marrow cells which express CD34 and terminal deoxynucleotidyl transferase (TdT) include T cells and putative lymphoid stem cells. *Bood* **74**, 9a.
12. Greaves, M. F., Delia, D., Robinson, J., Sutherland, R., and Newman, T. R. (1981). Exploitation of monoclonal antibodies: A "who's who" of haematopoetic malignancy. *Blood Cells* **7**, 257.
13. Greaves, M. F., Chan, L. C., Furley, A. J. W., Watt, S. M., and Molgaard, H. V. (1986). Lineage promiscuity in hemopoietic differentiation and leukemia. *Blood* **67**, 1–11.
14. Hollander, Z., Shah, V. O., Civin, C. I., and Loken, M. R. (1988). Assessment of proliferation during maturation of the B lymphoid lineage in normal human bone marrow. *Blood* **71**, 528–531.
15. Hurwitz, C. A., Loken, M. R., Graham, M. L.,

Karp, J. E., Borowitz, M. J., Pullen, D. J., and Civin, C. I. (1988). Asynchronous antigen expression in B lineage acute lymphoblastic leukemia. *Blood* **72**, 299–307.
16. Janossy, G., Coustan-Smith, E., and Campana, D. (1989). The reliability of cytoplasmic CD3 and CD22 antigen expression in the immunodiagnosis of acute leukaemia: a study of 500 cases. *Leukemia* **3**, 170–181.
17. Langlois, R. G., Bigbee, W. L., and Hensen, R. H. (1985). Flow cytometric characterization of normal and variant cells with monoclonal antibodies specific for glycophorin A. *J. Immunol.* **134**, 4009–4017.
18. Lansdorp, P. M., Dougherty, Humphries, R. K. (1990). CD34 epitopes. *In* "Leucocyte Typing IV White Cell Differentiation Antigens" (W. Knapp, B. Dorken, W. R. Gilks, E. P. Rieber, R. E. Schmidt, H. Stein, A. E. G. Kr. von dem Borne, eds), pp. 826–827. Oxford University Press, Oxford.
19. Lansdorp, P. M., Sutherland, H. J., and Eaves, C. J. (1990). Selective expression of CD45 isoforms on functional subpopulations of CD34 positive hemopoietic cells from human bone marrow. *J. Exp. Med.* **172**, 363–366.
20. Le Bien, T. W., Wörmann, B., Villablanca, J. D., Law, L. M., Steinber, L. M., Shah, V. O., and Loken, M. R. (1990). Multiparameter flow cytometric analysis of human fetal bone marrow B cells. *Leukemia* **4**, 354.
21. Lebman, D., Trucco, M., Bottero, L., Lange, B., Pessano, S., and Rovera, G. (1982). A monoclonal antibody that detects expression of transferin receptor in human erythroid precursor cells. *Blood* **59**, 671.
22. Loken, M. R., Shah, V. O., Dattilio, K. L., and Civin, C. I. (1987). Flow cytometric analysis of human bone marrow: Normal erythroid development. *Blood* **69**, 255–263.
23. Loken, M. R., Shah, V. O., Dattilio, K. L., and Civin, C. I. (1987). Flow cytometric analysis of human bone marrow: II. Normal B lymphocyte development. *Blood* **70**, 1316–1324.
24. Loken, M. R., Shah, V. O., Hollander, Z., and Civin, C. I. (1988). Flow cytometric analysis of normal B lymphoid development. *Pathol. Immunopathol. Res.* **7**, 357–370.
25. Loken, M. R., Shah, V. O., and Terstappen, L. W. M. M. (1990). Multicolor analysis of myeloid antigens on normal human bone marrow. *In* "Leucocyte Typing IV White Cell Differentiation Antigens" (W. Knapp, B. Dorken, W. R. Gilks, E. P. Rieber, R. E. Schmidt, H. Stein, A. E. G. Kr. von dem Borne, eds), pp. 859–862. Oxford University Press, Oxford.
26. Nadler, L. M., Korsmyer, S. J., Anderson, K. C., Boyd, A. W., Slaughenhoupt, B., Park, E., Jensen, J., Coral, F., Mayer, R. J., Sallen, S. E., Ritz, J., and Schlossman, S. F. (1984). B cell origin of non-T cell acute lymphoblastic leukemia. *J. Clin. Invest.* **74**, 332–340.
27. Nagler, A., Lanier, L. L., Cwirla, S., and Phillips, J. H. (1990). Comparative studies of human FcRIII-positive and negative Natural Killer Cells. *J. Immunol.* **143**, 3183–3191.
28. Picker, L., Terstappen, L. W. M. M., Roth, L., Streeter, P., and Butcher, E. (1990). Differential expression of homing-associated adhesion molecules by T-cell subsets in man. *J. Immunol.*, **145**, 3247.
29. Robinson, F. J., Sieff, C., Deilia, D., Edwards, P. A. W., and Greaves, M. (1981). Expression of cell surface HLA-DR, HLA-ABC and glycophorin during erythroid differentiation. *Nature* **289**, 68–71.
30. Ryan, D., Kossover, S., Mitchel, S., Frantz, C., Hennessy, L., and Cohen, H. (1986). Subpopulations of common acute lymphoblastic leukemia antigen-positive lymphoid cells in normal bone marrow identified by hematopoietic differentiation antigens. *Blood* **68**, 417–425.
31. Shah, V. O., Civin, C. I., and Loken, M. R. (1988). Flow cytometric analysis of human bone marrow: IV. Differential quantitative expression of T-200 common leukocyte antigen during normal hematopoiesis. *J. Immunol.* **140**, 1861–1867.
32. Shah, V. O., Safford, M. G., Terstappen, L. W. M. M., and Loken, M. R. (1990). Quantitative comparison of myeloid antigens on peripheral blood lymphocytes, monocytes, neutrophils, eosinophils, and basophils. *In* "Leucocyte Typing IV White Cell Differentiation Antigens" (W. Knapp, B. Dorken, W. R. Gilks, E. P. Rieber, R. E. Schmidt, H. Stein, A. E. G. Kr. von dem Borne, eds), pp. 855–858. Oxford University Press, Oxford.
33. Sieff, C., Bicknell, D., Caine, G., Robinson, J., Lam, G., and Greaves, M. F. (1982). Changes in cell surface antigen expression during hemopoietic differentiation. *Blood* **60**, 703–713.
34. Tedder, T. F., Clement, L. T., and Cooper, M. D. (1984). Discontinuous expression of a membrane antigen (HB-7) during B lymphocyte differentiation. *Tissue Antigens* **24**, 140–149.
35. Terstappen, L. W. M. M., Shah, V. O., Conrad, M. P., Recktenwald, D., and Loken, M. R. (1988). Discriminating between damaged and intact cells in fixed flow cytometric samples. *Cytometry* **9**, 477–484.
36. Terstappen, L. W. M. M., and Loken, M. R. (1988). 5-Dimensional flow cytometry as a new approach for blood and bone marrow differentials. *Cytometry* **9**, 548–556.
37. Terstappen, L. W. M. M., Mickaels, R., Dost, R., and Loken, M. R. (1990). Increased light scattering resolution facilitates multidimensional flow cytometric analysis. *Cytometry* **11**, 506–512.
38. Terstappen, L. W. M. M., Shah, V. O., Civin, C. I., Hurwitz, C. A., and Loken, M. R. (1990). Multidimensional flow cytometry as a new approach for

discrimination between normal and leukemic cells in peripheral blood and bone marrow. *In* "Progress in Cytometry—1989", pp. 4–29. BDIS, Erebodegem.

39. Terstappen, L. W. M. M., Hollander, Z., Meiners, H., and Loken, M. R. (1990). Quantitative comparison of myeloid antigens on five lineages of mature peripheral blood cells. *J. Leuk. Biol.* **48**, 138–148.

40. Terstappen, L. W. M. M., Safford, M., and Loken, M. R. (1990). Flow cytometric analysis of human bone marrow: III. Neutrophil development. *Leukemia* **4**, 657–663.

41. Terstappen, L. W. M. M., and Loken, M. R. (1990). Myeloid cell differentiation in normal bone marrow and acute myeloid leukemia assessed by multidimensional flow cytometry. *Anal. Cell. Pathol.* **2**, 229–240.

42. Terstappen, L. W. M. M., Johnsen, S., Segers-Nolten, I., and Loken, M. R. (1990). Identification and characterization of normal human plasma cells by high resolution flow cytometry. *Blood*, **76**, 1739–1747.

43. Terstappen, L. W. M. M., Huang, S., Safford, M., and Loken, M. R. (1991). Sequential generations of hematopoietic colonies derived from single non lineage committed $CD34^+$, $CD38^-$ progenitor cells. *Blood* **77**, 1218–1227.

44. Terstappen, L. W. M. M., and Loken, M. R. (1992). Leukocyte differential counting. *In* "Flow Cytometry in Hematology" (O. D. Laerum, and R. Bjerknes, eds), *Anal. Cytol. Ser.*, pp. 95–110. Academic Press, London.

45. Wörmann, B., Safford, M., Könemann, S., Büchner, Th., Hiddeman, W., Loken, M. R., and Terstappen, L. W. M. M. (1989). Identification of clonal subpopulations in ANLL by multidimensional flow cytometry. *Blood* **74**, 22a.

1.4

Megakaryocytopoiesis: Analysis and Cell Sorting

ALEXANDER NAKEFF

Division of Hematology and Oncology
Wayne State University School of Medicine
Detroit, MI 48201, USA

I. INTRODUCTION

Megakaryocytopoiesis is a cytokinetic process that is ideally suited to analysis by flow cytometry. It lends itself to correlated, multivariate analysis since numerous, quantitative measurements can be made simultaneously on the same cell. These parameters include DNA content (indicative of polyploidization, a property unique to this cell system), low-angle forward light scatter (FLS) as an approximate measure of cell size, 90° (side) light scatter (SLS) as reflecting the complexity of cytoplasmic fine structure, and platelet-associated proteins (PAP) as measured by (a) surface membrane immunofluorescence (SMI) using polyclonal or specific monoclonal antibodies (MoAb), (b) cytoplasmic staining using some of these same antibodies combined with

FLOW CYTOMETRY IN HEMATOLOGY
ISBN 0–12–432940–3

membrane solubilization to facilitate intracellular antibody entry and (c) ultrastructural PAP localization using immunogold probing in megakaryocyte subclasses sorted on the basis of other quantitative parameters.

Megakaryocytes (morphologically identified) and their progenitor cells (small, acetylcholinesterase positive, $SAChE^+$ (particularly in murine tissue) and colony forming unit, megakaryocyte, CFU-Meg) present special problems to flow cytometric analysis. The former are problematic due to (a) their infrequency (approx. $1:2 \times 10^3$ nucleated cells) in easily quantified organs, such as bone marrow or spleen, (b) their large cell volume (range 20–50 μm (*35*)) as compared to the majority of other cell types present in these organs, (c) their fragility and lack of viability in adequately monodispersed cell suspensions, and (d) their tendency to clump upon initial monodispersion. Progenitor cells present problems due to their relative infrequency ($SAChE^+$ at 1 per 10^3 and CFU-Meg at 1 per 10^4 nucleated cells in these organs) and relatively few unique fluorescent cell or membrane markers expressed by these subsets, as documented in several recent reviews (*6,15,20,41*).

Clearly, purified megakaryocytes would be the most ideal cell suspension to begin their flow analysis, especially when cell sorting is used for the purpose of obtaining relatively large numbers of various subpopulations for subsequent biochemical analysis. Numerous batch isolation techniques have been applied successfully to obtain such starting material using: (a) unit gravity sedimentation (*8,43,53,57,60*) and centrifugal elutriation (CE) (*31,44,45,71–73*) for separating megakaryocytes by cell size; (b) various types of continuous or discontinuous density gradient centrifugation using either albumin (*33,44,54,55*), Ficoll (*53*) or Percoll (*22,23,25,55,58,61*) for separation by buoyant density differences; (c) magnetic beads together with ristocetin-induced aggregation–disaggregation of clumped megakaryocytes (*26*); and (d) combinations of the above isolation techniques (*2,17,29,33,34,54,56*), some of which can yield relatively large numbers of essentially homogeneous megakaryocytes from spleen and bone marrow. The major problem, in addition to reduced cell yields, is any substantial selective loss of cell subpopulations. Thus, independent means of verifying cell loss (e.g. by Feulgen staining on tissue sections of smears and scanning microspectrophotometry) are important controls. Furthermore, most isolation techniques are laborious and time consuming and require sophisticated equipment, such as a centrifugal elutriator. The pitfalls and caveats implicit in the batch isolation of megakaryocytes and their progenitor cells prior to flow cytometric analysis have been reviewed extensively (*47,48*) and will not be discussed further here.

If analysis only is required of relatively small numbers of megakaryocytes (< 1000), unfractionated cell suspensions of marrow and spleen have been used effectively with appropriate megakaryocyte-specific markers, DNA content, FLS and SLS (*24*). This type of correlated, multivariate (e.g. SMI vs DNA content) approach has been applied to megakaryocytes in unfractionated marrow and marrow cell liquid cultures (*3,12,13,25,29,30,56,63–65*). These data indicate 16C megakaryocytes to be the majority ploidy class rather than 8C as determined from univariate analysis (*4,34*).

The focus of this chapter will be to present an updated review of the promise and limitations of flow cytometry and cell sorting as applied to the cytokinetics of megakaryocyte (and ultimately platelet) production.

II. CYTOKINETICS OF MEGAKARYOCYTE AND PLATELET PRODUCTION

Advances in our knowledge of the cytokinetics in this cell lineage reviewed recently (*20,41,42,59*) highlight the unique role that flow cytometry and cell sorting play in unravelling correlative events involved in megakaryocyte production.

Platelets are anucleate pieces of megakaryocyte cytoplasm whose function in the peripheral blood is to maintain vascular integrity and participate in hemostasis. This involves the initiation of blood clot formation and retraction, in addition to a role in tumor cell metastasis. Although no exact measurements have ever been made directly, it is generally accepted that between 2000 and 4000 platelets are produced from a single megakaryocyte such that the "death" of the megakaryocyte heralds the "birth" of platelets which are released subsequently into the peripheral blood. Although there is substantial electron microscopic evidence for platelet formation within megakaryocytes, the exact mechanism and anatomical site of platelet release remain controversial areas under active investigation. Newer techniques such as micropipette aspiration of single, isolated megakaryocytes (*61*) indicate that this is a late maturational event not yet initiated in otherwise fully mature megakaryocytes. A central, and as yet unresolved, question relates to what degree the heterogeneity of platelet function (e.g. stickiness, hemostatic efficiency) reflects megakaryocytic heterogeneity, in terms of potentially correlative properties including degree of nuclear endoreduplication (DNA content) and cytoplasmic maturation (*49*).

The distribution of DNA content in megakaryocytes reflects a degree of polyploidization that involves successive cycles of endoreduplicative DNA synthesis in the absence of cytokinesis. Polyploidization responds to platelet demand, being upregulated by increased platelet demand following platelet depletion and down-regulated by decreased platelet demand following platelet hypertransfusion or rebound thrombocytosis (*12,24,30,48,65*). The importance of polyploidization to the final number and quality of platelets produced from each ploidy class remains to be demonstrated definitively (*48*). Recent information on flow cytometrically determined DNA histograms of megakaryocytes induced in response to various degrees and lengths of immune-induced thrombocytopenia indicate either a loose (*12,62*) or tight (*30*) correlation between platelets and megakaryocytic nuclear endoreduplication. The latter study provides evidence of a soluble factor inducing DNA shifts whose level is inversely related to the platelet count. Stages of megakaryocytic maturation have been difficult to characterize quantitatively by surface membrane-directed antibodies and immunofluorescence although recent attempts to establish these criteria are encouraging, using the glycoprotein (GP) IIb/IIIa complex (*6,56,63,64*) in human megakaryocytes and polyclonal anti-platelet antisera (*13,24,73*) in murine megakaryocytes. An approach entitled multivectorial analysis (*49*) has been proposed to quantitate differentiation in terms of vectors relating cytoplasmic maturation to nuclear endoreduplication.

Morphologically recognized megakaryocytes are easily among the largest cells in marrow. Cell size is a complex parameter that reflects both degree of cytoplasmic maturation (the largest cells being among the most "mature") and DNA content (the largest cells attaining levels $>64C$ in response to platelet demand) (*14,35,48*). This unique property has been utilized to isolate large numbers of megakaryocytes from marrow and spleen for fluorescence-activated cell sorting using unit velocity sedimentation and CE (*48*).

Megakaryocytes also are much lighter in cell density in Percoll (the majority <1.05 g/ml) than most nucleated cells and this parameter has been used effectively to batch separate (*2,31,48,61,64,65*) or further subdivide subpopulations of these cells as possibly reflective of their state of differentiation (*22,23,25*), including light density clonogenic progenitor cells (*10*).

Megakaryocytes are extremely rare in mammalian hematopoietic tissue, constituting from 0.2 to 0.02% of total nucleated cells as surveyed recently from flow cytometric data obtained using megakaryocytes enriched by various physical separation techniques, as well as from unfractionated marrow (*13*). The rarity of these cells and the limited number of specific second markers (in addition to DNA) renders the analysis of sufficient numbers of cells for good statistical confirmation of parameters (such as distribution of ploidy) somewhat difficult and time consuming. For example, at a starting concentration of 1 : 2000 nucleated cells in unfractionated rat bone marrow using dual DNA (propidium iodide) and SMI (polyclonal anti-rat platelet antiserum) staining, it has been estimated to take approximately 30 min of run time to obtain data on 5×10^3 megakaryocytes (*29,30*).

Lastly, there is a substantial body of data available on the ability of megakaryocytes to synthesize specific PAP for inclusion into membranes and specific granules of platelets being formed within the cytoplasm of differentiating megakaryocytes (*6*). Specific monoclonal antibodies directed to target antigens such as GPIIb/IIIa that are expressed on the membranes of developing megakaryocytes, have been used as second markers for megakaryocytic identification in correlated, multivariate analyses of unfractionated marrow (*56,63–65*).

III. PROGENITOR CELLS

Murine megakaryocytes and platelets are derived ultimately from progenitor cells that, though morphologically unrecognized, can be quantitated by *in vitro* clonogenic cell assays (CFU-Meg) and staining for acetylcholinesterase activity ($SAChE^+$). Properties of these progenitors and their potential role in the humoral and cellular regulation of megakaryocyte production have been reviewed (*47*) and updated most recently (*15,20,41, 42,59*).

A. CFU-S

This class of progenitor represents the most primitive stem cell in murine marrow whose extensive proliferative capacity and ability to self-renew and differentiate supports all of hematopoiesis (see detail elsewhere in this volume). CFU-S are generally considered to constitute the population from which the more differentiation restricted CFU-Meg are derived, an influx pathway that may be regulated by the size of the CFU-Meg cell pool (*42,50*). CFU-S give rise to megakaryocytic spleen cell colonies whose number and size may be under humoral (thrombopoietin) regulation (*42*).

B. CFU-Meg

This class of mitotic progenitor is limited to megakaryocytic differentiation and is assayed *in vitro* by the ability to form colonies of megakaryocytes in semi-solid culture media, including plasma clot, agar and methylcellulose (*47*). Cells within individual colonies undergo nuclear endoreduplication to give rise to morphologically recognized megakaryocytes characterized by high polyploidization. CFU-Meg respond to cytokines including IL-3 and GM-CSF (*9,20,39,69*) and their proliferative activity can be modulated by NK cells (*16,52*) and T lymphocytes (*11,36*). Whether megakaryocytes in *in vitro* colonies can make platelets remains to be shown definitely despite the electron micrographic evidence to the contrary (*47*). In addition, whether CFU-Meg respond directly to humoral regulators such as thrombopoietin remains controversial (*20,42*). The physiologic role that CFU-Meg play in the regulation of platelet production is also controversial, particularly whether these clonogenic progenitors respond directly to an induced platelet demand or to changes induced in the megakaryocyte cell compartment secondary to changes in blood platelets. Lastly, there is information (*7,37*), similar to that demonstrated in erythropoiesis, that under defined culture conditions, an earlier, more immature progenitor (burst forming unit, megakaryocyte, BFU-Meg) can be identified with a high proliferative potential that developmentally precedes and eventually gives rise to more proliferation restricted CFU-Meg.

C. SAChE$^+$ Cells

Demonstrated in murine marrow, SAChE$^+$ cells have essentially lost the proliferative mitotic potential of CFU-Meg which developmentally precede them. They have retained the ability to undergo limited nuclear endoreduplication (*47,74*). SAChE$^+$ cells are generally considered to be the progenitors that differentiate to early, recognized megakaryocytes by undergoing one or more rounds of nuclear endoreduplication (*74*) and may themselves be polyploid (possibly 4C and 8C) (*46*) and play a physiologic role in the regulation of megakaryocyte production (*42*).

IV. ROLE OF FLOW CYTOMETRY IN ELUCIDATING MEGAKARYOCYTE CYTOKINETICS

Flow cytometry can make potentially important contributions to our understanding of this developing cell system although caveats must be considered in extrapolating flow data to the development of new cytokinetic concepts.

A. Advantages

(1) The rapidity of analysis at rates of $2\text{–}5 \times 10^3$ cells/s permits a sufficiently large data base to be accumulated on each sample that good statistical verification can be applied to flow histograms (*24,73*).

(2) The ability to analyze multiple, independent parameters on single cells in a correlated fashion is a powerful new tool with which to discover hitherto unknown cytokinetic relationships (*51*).

(3) Quantitative measurements of DNA content or SMI, for example, are much more "objective" than the subjectivity inherent in microscopic-based morphometry (*13,47*).

(4) Data collected in list mode is "immortalized" such that reanalysis can be attempted on the same data base at a later time using different analytic criteria.

(5) Cells identified by unique flow parameters can be physically sorted at relatively high rates for morphometry by scanning (*70*) and transmission electron microscopy with immunogold probing for specific intracellular proteins or substructures (*18,19*).

(6) The availability of two or more unique parameters permits the analysis (and possible sorting) of infrequent target cell populations without their prior enrichment, e.g. megakaryocyte DNA profiles from unfractionated marrow cell suspensions (*13,24,29–31,56, 63–65*), thus obviating problems inherent in some of the enrichment techniques that may lead to the selective loss of target subpopulations (*47,48*).

(7) The analysis of either fixed or viable target cells permits, in the former case, the accumulation of a large data file for rather rare cells (such as megakaryocytes or their progenitors) while, in the latter case, specific viable cells can be analyzed and sorted sterilely for subsequent clonogenic cell assay (*7,21,28,38, 39,51,69*) or analysis of nuclear endoreduplication in cell culture (*29,30*).

(8) Flow technology applies potentially new parameters such as SLS for analysis of internal fine structure (e.g. granules) and fluorescein diacetate as a non-fluorogenic substrate activated by viable cells to release free fluorescein intracellularly as a measure of cell viability (*48*). All measurements are correlated on a cell by cell basis that measures target cell heterogeneity.

B. Limitations

(1) The organ to be analyzed must be completely monodispersed, a prerequisite that poses a problem for megakaryocytes that are rare and fragile (*13*).

(2) Following monodispersion, cell aggregation must be minimized, a problem that becomes a more serious concern following alcohol fixation or during long (> 30 min) analysis or sorting times. Numerous complex

cell suspension media have been formulated to reduce cell aggregation (*12,24,29,31,33,56, 63,73*).

(3) The binding of fluorescently labeled markers to target cells measured as total fluorescence/cell is often difficult to relate to an absolute number of target molecules per cell. For SMI, lack of knowledge of the target antigenic structure (especially using polyclonal as compared to monoclonal reagents) can be problematic. In addition, the designation of ploidy classes is best expressed by C rather than N (e.g. 8C vs 8N) since the latter relates to chromosome content while the former is a measure of the amount of DNA that stoichiometrically binds a DNA-specific fluorescent dye (*48*).

(4) Cells must permit the nuclear entry of some dyes such as propidium iodide (PI) either following hypotonicity or prior fixation, procedures that themselves can affect staining from cell type to cell type and experiment to experiment. Thus, in the absence of a "spiked" absolute DNA "standard" run at the same time, valid comparisons can only be made of shifts in certain ploidy classes relative to others within the same cell population. Furthermore, alcohol fixation results in the degradation of FLS and SLS as useful independent parameters. It also induces severe cell clumping unless specific precautions are observed, such as resuspending cell pellets before the stepwise addition of increasing ethanol concentrations at 4°C. Problems with fixation artifacts are compounded in studies in which ultrastructural analysis (*70*) and immunogold probing (*18,19*) are applied to sorted cells. Ethanol fixation severely disrupts internal structure such as cytoplasmic granules. An empirically determined compromise involves *in situ* perfusion fixation of marrow with a paraformaldehyde–glutaraldehyde mixture followed by a short fixation in 1% glutaraldehyde and DNA staining with Hoechst 33342 (*18,19*).

(5) Some practical problems impose substantial deterrents to the sorting of viable megakaryocytic cells, including the heterogeneity of SMI for specific antigens for CFU-Meg, the survivability of CFU-Meg following sorting, changes induced in their subsequent kinetic properties and their potential loss following hypotonic lysis of erythrocytes and Ficoll density gradient centrifugation. Viable megakaryocytes, on the other hand, are large enough to block the sample stream at the nozzle tip and interfere both with signal transmission (since they occupy a substantial portion of the sample stream) and droplet formation without machine modifications that permits the use of low resonance frequencies. They settle out rapidly in the sample tube without continuous, gentle agitation and lose viability readily.

(6) Numerous problems must be addressed for sorting fixed cells, including: perfusion fixation *in vivo* vs fixation *in vitro*; alcohol vs paraformaldehyde (PFA) ± Triton X-100 for DNA plus SMI; isotonic vs hypotonic PI staining ± RNase digestion for DNA measurements (*13*); the sticking of fixed cells to sample tubes and tubing; proper adhesion of sorted cells on to glass and plastic surfaces and the effect of poly-L-lysine pretreatment (*27*); the lack of a fluorescent assay for AChE activity with which to determine the distribution of DNA within $SAChE^{+}$ cells; clogging of the nozzle tip and potential selective losses of megakaryocytic cells following (a) fixation protocols for unfractionated marrow (*13*), (b) physical enrichment techniques prior to sorting (*48*) or (c) protocols that necessitate multiple centrifugal cell washings (*13*).

(7) There is an inherent difficulty in relating the heterogeneity of a fluorescent signal to true biological variability reflecting changes in proliferation, endoreduplication or differentiation (*47,48*).

(8) Flow analysis and especially cell sorting or gating of specific subpopulations necessitates verification by phase contrast or fluorescence microscopy or other independent single-cell measurements to have confidence that the signal derived from the flow cytometer as a "black box" has biologic importance.

V. MULTIVARIATE ANALYSIS AND SORTING OF PROGENITOR CELLS

The main applicability of flow cytometry to progenitor cell cytokinetics has been twofold, namely (a) the use of SMI, preferably with monoclonal antibodies directed to known antigenic sites (although polyclonals have also been used) to define the specific membrane characteristics of these progenitors in both unfractionated and enriched marrow cell suspensions and (b) the use of electronic gating with these and other properties (FLS, SLS and DNA) to physically sort progenitor cells sterilely for subsequent clonogenic cell assay in semi-solid or liquid cultures and analysis of their ability to differentiate *in vitro*.

In some ways, this application resembles that directed to the elucidation of the pluripotent stem cell, in the sense that the further back one wishes to examine differentiation, the fewer the uniquely expressed properties that define the cell of interest. An analogy can be made with the picture of Stalin, Churchill and Roosevelt at the Yalta conference: although there is little difficulty in recognizing them in their "fully differentiated state", it would be quite another matter to pick out these three famous personalities by their "phenotypic expression" as reflected in photographs taken in the few days after their respective births. This "catch-22" situation is at the heart of using flow cytometry to define cells in states of growth and development by fluorescence of markers that are weakly expressed early in development.

A. Human CFU-Meg

Bodger *et al.* (*5*) were the first to characterize human CFU-Meg by combining SMI with FACS-sorting using a FACS-1 and complement-mediated cytotoxicity. Light density (< 1.077 g/ml Ficoll–Hypaque) bone marrow cells from normal volunteers were labeled indirectly with either RFB-1 (IgG1), a monoclonal in ascites form reacting with CFU-GM and immature marrow lymphoid cells, or RFB-HLA-DR (IgM), another monoclonal that reacted with a 28–33 kD antigen identical with an Ia-like antigen, followed by an FITC-labeled goat anti-mouse IgG. These target cells were then sorted sterilely and assayed for CFU-Meg in methylcellulose cultures. Although the percentages of sorted cells that were RFB-1^+ and RFB-HLA-DR^+ were low (15–18% and 19–23%, respectively), CFU-Meg were derived solely from the positively stained populations that express both the RFB-1 and Ia-like antigens, with enrichments of 3- to 7-fold and 1- to 2-fold, respectively. These monoclonals clearly were not specific for CFU-Meg since CFU-GEMM, CFU-GM, BFU-E and CFU-E also expressed both antigens. Control cells were incubated with an ascites from a non-antibody-producing hybrid clone but not passed through the FACS-1. Unfortunately, no information was provided on the recovery of CFU-Meg following sorting although, from studies to be described, it probably was in the range of only 10–20%.

Watt *et al.* (*69*) used dual color (mouse BI.3C5 (monoclonal My10, IgG1) and FITC-goat anti-mouse IgG for the human HPLA-1 determinant and biotin-labeled mouse monoclonal (clone L243 IgG2a) and phycoerythrin (PE)-avidin for the HLA-DR determinant) sorting of light density, plastic non-adherent human marrow cells to obtain enrichments of 2- to 15-fold in CFU-Meg from cells that expressed a high level of BI.3C5 and a moderate to high level of HLA-DR ($BI.3C5^{++} \times HLA\text{-}DR^{+/++}$). Although functionally not well characterized, the importance of these determinants may potentially involve some role in growth regulation. Again, the high level of coexpression of these markers on other clonogenic progenitors prevented their use as unique probes for isolating CFU-Meg by FACS-sorting.

Lu *et al.* (*38*) used the same two antigenic determinants (HPLA-1 (My10) vs HLA-DR) in a correlated, dual color sterile sort of

CFU-Meg from normal, light density (< 1.077 g/ml), plastic non-adherent and E-rosette negative ($NALT^-$) human bone marrow cell suspensions. These were dually stained with HPLA-1 (My10) MoAb followed by biotinylated goat anti-mouse IgG plus Texas Red Avidin D and anti-HLA-DR MoAB conjugated with PE. An Epics 753 (Coulter) was used to excite these dual stained cells using the 590 nm line from a rhodamine 6G primed dye laser pumped by a 6W argon ion laser for Texas Red and the 488 nm excitation line for PE from a second 6W argon ion laser. In support of the previous findings (*69*), the majority of CFU-Meg (18-fold enrichment over unsorted marrow) were found in the high My10 $\times$ low DR expression fraction, a fraction that was enriched 2- to 3-fold over any other fraction sorted as a function of My10 $\times$ HLA-DR expression. Here again, there was no selective staining for CFU-Meg since this same fraction also supported the growth of CFU-GM, BFU-E and CFU-GEMM. Even though a serum-free fibrin clot assay was used in conjunction with recombinant growth factors, the coincident growth of other cell types may modulate CFU-Meg growth by elaboration of other cytokines. Interestingly, the addition of either IL-3 or GM-CSF increased the plating efficiency of CFU-Meg, as well as the average size of individual colonies. Their addition together resulted in no more than an additive effect whereas the addition of erythropoietin, granulocyte colony stimulating factor (G-CSF), colony stimulating factor-1 (CSF-1) and thrombocyte stimulating factor had no effect on either parameter. Although the mechanism for the IL-3 and GM-CSF additive effect remains unknown, the authors speculate that different subpopulations of CFU-Meg may be stimulated by one or the other recombinant growth factor or that one factor may affect the number or affinity of specific receptors for the second factor. A caveat to this speculation concerns the generally low sorting recovery ($< 20\%$) of positive cells which limits generalizations since sorted cells constitute a minority functional cell population that may or may not be representative of all the labeled progenitors.

Extending the above approach, the same group (*7*) applied electronic gating to CE-enriched marrow (12–14 ml/min $\times$ 1950 rpm) to subdivide $CD34^+$ cells (double labeled for CD34 (FITC) and HLA-DR (PE)) and sort CFU-Meg on the basis of low HLA-DR expression in which $> 95\%$ of cells were $HLA\text{-}DR^+$ or no HLA-DR expression in which $< 5\%$ of cells were $HLA\text{-}DR^+$. CFU-Meg were immunophenotyped as expressing low HLA-DR activity and gave rise to pure colonies with multi-foci averaging approx. 12 megakaryocytes after 12 days in culture. In addition, a more immature progenitor (burst forming unit, megakaryocyte or BFU-Meg) was identified that expressed no HLA-DR activity and that gave rise to larger (> 100 megakaryocytes), single foci colonies after 21 days in culture. This latter cell's clonogenic potential was entirely dependent upon the addition of both hrIL-3 and hrGM-CSF. BFU-Meg may play a role in megakaryocytopoiesis analogous to that described for BFU-E in erythroid cell development in that they are antecedent to CFU-Meg (possibly its precursor?) and exhibits a high proliferative potential *in vitro*. Its refractoriness to 5-fluorouracil exposure *in vitro* (in contradistinction to CFU-Meg) may have implications for explaining the cytokinetics of megakaryocyte recovery from drug-induced toxicity *in vivo*.

Kanz *et al.* (*28*) used MoAbs for GPIIIa (C17 (IgG1) and AP3 (IgG1)) and GPIIb/IIIa (PA3 (IgG2a) and M148 (IgG1)) as SMI probes in a systematic analysis of their expression on CFU-Meg. CFU-Meg were sorted on a BD FACS 440 from normal human marrow cell suspensions following their enrichment by sequential three-step discontinuous Percoll density gradient centrifugation (CFU-Meg fraction $= 1.05 > p < 1.077$ g/ml) and either positive ($GPIIIa^+$) or negative ($CD15^-$, $CD11b^-$, $CD3^-$, $VIMD5^-$, anti-glycophorin A^-, 2.5-fold enrichment) selection by immunopanning. CFU-Meg that were sorted initially from an electronic gate comprising a high FLS × low SLS fraction that was progenitor cell-rich (approx. 2-fold enrichment) and plated in methylcellulose culture were consistently GPIIIa$^-$; GPIIIa$^+$ cells comprised only about 2% of all marrow cells. HLA-DR$^+$ cells (using Clonab LN3 (IgG2a)) constituted about 22% of the starting (CFU-Meg enriched) marrow cell suspension and included 50–100% of CFU-GEMM, CFU-GM, BFU-E, CFU-E and approx. 80% of CFU-Meg. HLA-DR$^+$ CFU-Meg formed typical pure megakaryocyte colonies that were large and loose in structure with different sized megakaryocytes whereas HLA-DR$^-$ CFU-Meg formed small colonies composed predominantly of large megakaryocytes. These differences indicate that immature CFU-Meg (those forming the large colonies and thus exhibiting the highest proliferative potential) express detectable levels of HLA-DR on their surface. Positive immunopanning using a panel of anti-GpIIIa MoAbs directed to a spectrum of different epitopes supported the FACS finding that CFU-Meg were GPIIIa$^-$. Thus, the correlation between immunopanning (where there was no preselection for CFU-Meg) and FACS sorting (following sequential density gradient centrifugation, indirect negative immunopanning and FLS vs SLS) served as an important control to document that no major CFU-Meg subpopulation or potentially important accessory cells had been lost following progenitor cell enrichment prior to FACS sorting. Nevertheless, the finding that CFU-Meg were GPIIIa$^-$ by FACS analysis conflicted with previous data utilizing complement-mediated anti-GPIIIa cytotoxicity (*32*) that demonstrated CFU-Meg to be GPIIIa$^+$. Nevertheless, the former findings were supported by *in vitro* clonogenic data demonstrating that the earliest morphologically identified megakaryocytes are GPIIIa$^-$ HLA-DR$^+$ (*6*). There are several technical limitations to FACS sorting that must be considered before definitive conclusions can be drawn, including (a) variable recoveries of sorted CFU-Meg (16.7–110%) and (b) antigen densities of GPIIIa at the cell surface that may be below SMI detection limits, possibly as a result of insufficient transport to the cell surface from cytoplasmic sites of GPIIIa synthesis. The intriguing implication to megakaryocyte biology reflects the role of GPIIIa in forming the B subunit of the vitronectin receptor and the GPIIb/IIIa heterodimer that participates in the binding to cytoadhesive protein that may modulate the receptor-mediated binding of both circulating and resident CFU-Meg to matrix proteins in marrow interstitium.

B. Murine CFU-Meg

Flow immunophenotyping of human CFU-Meg (see above) was aided by the availability of numerous MoAbs to defined cell surface antigens on immature marrow cell populations, in contrast to the lack of such markers for murine CFU-Meg. The latter has limited the efficacy of positive sorting for CFU-Meg.

Pallavicini *et al.* (51) utilized trivariate analysis and sorting of C3H/He murine marrow CFU-Meg gated on high FLS vs low SLS (immature blast cell window with 75% of starting CFU-Meg) and having low (0.1–0.65 times the model channel) Hoechst 33342 staining (7 μM for 60 min at 37°C, 56% of starting CFU-Meg). This resulted in an overall average enrichment of approx. 60-fold to 0.6% CFU-Meg from 0.009% in starting marrow. The impressive, one-step enrichment depended on short (< 60 min) staining times with concentrations $< 7\ \mu M$ of this DNA supravital dye for clonogenic vs non-clonogenic cell discrimination. However, even these conditions had associated with them a 35% Hoechst 33342-induced toxicity, even though recoveries of sorted CFU-Meg were essentially 100%. Caveats to the usefulness of this enrichment procedure concern the lack of stable Hoechst 33342 staining conditions that would permit the calculation of a reasonable coefficient of variation (CV), in addition to the substantial dye-induced cytotoxicity to CFU-Meg. The transient nature of the "optimal" low Hoechst

33342 staining conditions provides a time window of approx. 30 min during which sorting was possible. As shown by van Zant and Fry (*66*), the cytotoxicity associated with Hoechst 33342 staining of murine marrow that provided stable CVs of approx. 6% (5 μM for 60 min at 37°C) reduced CFU-Meg survival by about 75%. Although the differences between these two studies may reflect differences in cell handling, CFU-Meg culture conditions and mouse strains, the data nevertheless caution against a simplistic interpretation of low Hoechst 33342 staining.

The use of trivariate analysis of CFU-Meg in BC3 (C57BL × C3H) mouse marrow cell suspensions with a high FLS vs low SLS gate ("blast cell window" containing all CFU-Meg) followed by staining for either wheat germ agglutinin (WGA)-FITC or platelet glycoprotein-1 (Pgp-1, using the rat MoAb 142/5.1) expression indicated that the light scatter and cell surface properties of CFU-Meg were similar to CFU-S and other early committed progenitors (*1*). High WGA-FITC binding of CFU-Meg indicated a high sialic acid content of the cell surface and Pgp-1 expression was moderate, both characteristics placing CFU-Meg among the developmentally earliest progenitor cells. Although the recovery of sorted CFU-Meg was low (50%) when compared to unstained and unsorted marrow, there was no difference in the ratio of "pure" vs "mixed" CFU-Meg either before or after sorting, indicating no preferential loss of potential subpopulations. Similar recoveries of CFU-Meg that were sorted and fractionated by light scatter or fluorescence and those that were sorted but unfractionated indicated that the low recovery was probably not the result of technical sorting problems, choice of the light scatter window or non-linearity of the culture assay; reasons for this poor sorting recovery remain unknown.

In summary, human and murine CFU-Meg have been analyzed and sorted by several multivariate approaches that have used supravital DNA staining, FLS, SLS and SMI. Despite these ingenious approaches, there are no methods available for identifying and isolating only CFU-Meg. Not only is there extensive heterogeneity in the expression of these markers among progenitor cells in the major lines of differentiation tested, but the "catch-22" situation of decreasing expression as a function of degree of immaturity limits the direct application of the approach to the progenitor compartment. This is especially relevant with regard to maintaining viability for subsequent *in vivo/in vitro* analysis of clonogenicity (or other cell functions) for sorted cells without the selective loss of minor but possibly important subpopulations; the latter reflects generally low numbers of target progenitors in the starting cell suspension and low sort recovery. A more fruitful approach to progenitor cells may involve negative rather than positive cell sorting. MoAbs for mature marrow cells could be used to exclude these cells and permit the study of individual unlabeled progenitors. This type of approach was first described by Nakeff (*46*). Kuter *et al.* (*29*) applied a similar approach to the isolation of megakaryocyte-free cell suspensions that contained endoreduplicating progenitors for subsequent differentiation analysis *in vitro*. Hitherto unexplored techniques may include enzyme substrate analysis using non-fluorogenic substrates which release fluorogenic intermediates upon enzymatic reaction within individual cells (*68*) and RNA analysis for discriminating clonogenic from non-clonogenic cells (*67*). As more specific labels become available, positive cell sorting for progenitors will take on a new significance which will impact on applications utilizing multiwell culture systems. Here, single progenitor cells may be deposited rapidly and sterilely in individual wells for co-culture with other defined cells and/or soluble mediators in studies to elucidate fundamental cell–cell contact and cell factor-mediated processes.

VI. MULTIVARIATE ANALYSIS AND SORTING OF MEGAKARYOCYTES

This compartment comprises morphologically recognized megakaryocytes. Multivariate analysis and cell sorting have had to be modified for these cells to take advantage of (optimistic view) or circumvent (pessimistic view) properties unique to megakaryocytes, including their large size, infrequency, fragility and tendency to clump in monodispersed cell suspension (*47,48*). From its inception (*34,40,45*), the application of flow cytometry to megakaryocytopoiesis has expanded dramatically over the last few years. This burgeoning literature documents the potential importance of polyploidization to differentiation in this cell lineage (*12,13,18,19,24, 25,27,29–31,56,63–65,70,73*).

A. Univariate vs Multivariate Analysis

Early studies of megakaryocytes by flow cytometry (*4,34,40,45*) used analysis of DNA content as the principal marker of polyploidization; a critical analysis of these studies has been published (*48*). There are limitations to univariate analysis, including a potential for the selective loss of megakaryocytes resulting from their enrichment prior to flow analysis and inherent difficulties in using unfractionated marrow to unequivocally define ploidy peaks (and shifts) following perturbation (*4,34*).

Some of the above problems were circumvented by multivariate analysis. The first successful attempt was reported by Jackson *et al.* (*24*) using DNA content (PI in hypotonic citrate + RNase) vs indirect SMI (polyclonal rabbit anti-rat platelet antiserum (APS) + FITC-goat anti-rabbit IgG) profiles generated for megakaryocytes in unfixed, unfractionated rat marrow. Electronic gating of only high fluorescence intensity APS^+ (green) cells (excluding 99% of cells stained non-specifically with normal rat serum (NRS) and including cells with DNA contents $> 4C$) resulted in reasonable DNA distributions of cells that comprised 0.45% of total marrow; these were recognized morphologically as megakaryocytes following sorting and staining for AChE activity. CVs averaging 1.3% for all ploidy peaks following RNase treatment do not reflect inherent DNA heterogeneity since a linear algorithm was used to calculate % CV on logarithmic data (personal communication). CVs that more reasonably fit linear data for rat megakaryocytes using identical dual labeling were 2–3% (*30*). Surprisingly, megakaryocytes remained preferentially intact in hypotonic citrate with PI for up to 52 h. The release of free DNA from cell disintegration may have caused substantial cell clumping where non-megakaryocytic cell clumps outnumbered free megakaryocytes 3 : 1 (6 per 1000 cells vs 2 per 1000, respectively). Clumps also contributed a prominent 6C peak that contained few megakaryocytes, as determined by cell sorting. The most critical factors affecting multivariate analysis were (a) the degree of megakaryocyte specificity of APS vs the NRS control, (b) the preferential ability of megakaryocytes to survive hypotonic PI staining conditions with sufficient cytoplasm to stain with APS for dual color analysis and sorting and (c) the degree of cell clumping induced by the above staining conditions. The utility of this approach was shown by the close correspondence between the ploidy distributions obtained by flow cytometry vs microspectrophotometry as well as the up-regulation of ploidy following acute immune-induced thrombocytopenia and down-regulation following platelet transfusion-induced thrombocytosis.

B. DNA Content

Correlative multivariate analysis of DNA has been performed using a variety of probes in rat, mouse and human marrow (Table I).

Regardless of whether megakaryocytes were isolated from marrow that was either fixed or viable prior to staining or analyzed from unfractionated marrow, ploidy distributions displayed a remarkable similarity, even among species. Clearly, the most prominent ploidy peak in unperturbed megakaryocytopoiesis was 16C comprising about half of the megakaryocytes with the other 40% split about evenly in the 8C and 32C peaks and the remaining 10% in the 2C and 4C peaks. This

Table I

PLOIDY DISTRIBUTION OF DNA CONTENT BY MULTIVARIATE ANALYSIS

Species	Sample	Stain: DNA	Stain: Other	Instrument	DNA distribution (%): 2C	4C	8C	16C	32C	64C	Reference
Rat (Long–Evans)	Marrow (unfix, unfract.)	PI (7 mOsm)	APS + FITC-GAR	Coulter Epics V	18	9	17	50	7	—	(*24*)
Rat (S–D)	Marrow <1.060 g/ml	PI (7 mOsm)	APS + FITC-GAR	Cytomutt	—	—	24	60	16	—	(*29*)
Rat (S–D)	Marrow <1.060 g/ml	PI (7 mOsm)	APS + FITC-GAR	Cytomutt	—	—	16	63	21	0.7	(*30*)
Rat (S–D)	Marrow (CE)	Hoechst 33342	FITC-APS	BD FACS IV	—	—	17	60	23	—	(*73*)
Mouse (C57B1)	Marrow (unfract.)	PI (100 mOsm)	Biotin APS + FITC-avidin	BD FACScan	5	4	11 (estimate)	55	18	2	(*12*)
Mouse (C57B1)	Marrow (unfract.)	PI Tween-20/ 1% PFA	Biotin APS + FITC-avidin	BD FACScan	10	11	16	43	12	3	(*13*)
Human	Marrow <1.050 g/ml	PI (7 mOsm)	FITC-GPIIb/ IIIa MoAb	BD FACStar	4	8	12	41	30	5	(*63*)
Human	Marrow <1.050 g/ml	PI 1% PFA/ Trit X-100 (0.05%)	FITC-GPIIb/ IIIa MoAb	BD FACS 440	10	11	16	40	20	4	(*64*)
Human	Marrow (unfract.)	PI (hypo) Trit X-100 (0.05%)	FITC-GPIIb/ IIIa MoAb	Coulter Epics V	7	4	11	55	22	1	(*65*)
Human	Marrow (unfract.) + <1.050 g/ml	PI (295 mOsm) Trit X-100 (0.075%)	GPIIb/ IIIa MoAb + FITC-GAM	Coulter Epics V	10	13	15 (unfract.)	39	18	5	(*56*)
					6	8	15 (<1.050 g/ml)	42	27	3	

type of correlative approach clearly indicates that previous univariate analyses reporting 8C to be the predominant ploidy peak were incorrect (*4,34*). Other univariate approaches correctly reported 16C to be the majority ploidy peak (*45*) and correlated these data with the extensive literature on DNA ploidy by microdensitometry which also indicated 16C to be the predominant ploidy peak (*48*). Cells in progressively higher ploidy classes are larger in volume as measured by FLS and a linear relationship between DNA content and FLS has been reported by several groups (*34,45,56,64,73*). There are several advantages inherent in this approach:

(1) The most significant may be the ability to obtain DNA measurements on several thousand megakaryocytes from unseparated

marrow cell suspensions. The approach saves time since only double staining of the cells is needed. This circumvents potential problems with the selective loss of ploidy classes that may arise from the less than optimal combination of physical techniques to enrich for megakaryocytes prior to analysis.

(2) The efficiency of cell sorting is increased by the ability to use dual parameter gates to increase sort purity and sort rate, particularly when supravital DNA stains such as Hoechst 33342 are used in conjunction with SMI to obtain "pure" ploidy populations for subsequent biochemical analysis.

(3) Potential megakaryocyte progenitor cell populations with DNA content less than 8C are detected.

(4) Shifts in ploidy class distributions can be monitored in response to altered platelet demand, both experimentally induced and in certain clinical conditions of abnormal platelet cytokinetics (*65*).

There are several caveats to the above advantages:

(1) The specificity exhibited by the secondary probe presents a problem. Even though mature megakaryocytes may express enough of a unique protein to be detected, more immature megakaryocytes may not express enough of it to be identified; alternatively, the type and length of fixation and permeabilization used may destroy sufficient target molecule expression that cells with inherently low expression may drop below the level of detection, and thus escape quantitation (*56*). Independent means of quantitating expression on an individual cell basis may need to be performed for each probe in order to evaluate to what extent these limitations are relevant.

(2) Ethanol fixation and permeabilization with Triton X-100 degrades cell integrity sufficiently that subsequent ultrastructural analysis (by scanning or transmission electron microscopy) is not possible. However, supravital DNA stains together with optimized fixation conditions utilizing combinations of paraformaldehyde and glutaraldehyde have been used successfully to describe surface ultrastructure (*70*) and immunogold labeling of sorted megakaryocytes (*18,19*). Furthermore, ethanol fixation significantly degrades FLS and SLS measurements, a limitation observed to a lesser degree with PFA fixation.

(3) Optimization of the electronic gate for the second (non-DNA) probe affects the DNA distribution of the cells above the gate by increasing the proportion of cells with 2C, 4C and 8C content at the expense of those with more than 8C content. Clearly, the degree to which this becomes a serious limitation reflects the degree of specificity of the second probe as measured against its appropriate "control". Even with polyclonal anti-platelet serum, gate optimization occurs over a reasonable range of settings (*13,24*). This is less critical when monoclonals to GPIIb/IIIa in human megakaryocytes are applied where over a 50-fold difference exists between "specific" and "nonspecific" labeling (*56,63,64*). This difference may be further enhanced by using $F(ab')_2$ antibody fragments and high fluorescein to protein ratios ($F/P = 3$) (*63*) although it has been suggested that F/P ratios over 3 may contribute to higher backgrounds (*56*). Biotinylation of the primary antibody followed by labeled streptavidin may produce even more of a gain factor, as well as trivariate analysis and gating to take advantage of the high SLS of megakaryocytic cells. A word of caution in the comparable use of polyclonal vs monoclonal antibodies is in order. Although the latter may be more "specific" with regard to background fluorescence, it necessarily detects a single epitope; this contrasts with polyclonals which detect a wider epitope spectrum, albeit at the expense of positive signal gain over background.

(4) The physiologic importance of the 2C and 4C "megakaryocytes" that appear from correlated bivariate (DNA vs anti-platelet antigen expression) analysis remains unclear. Corash *et al.* (*13*) indicate that although 15% of 2C cells are present in the "dim" APS^+ fraction (indicative of non-megakaryocytic cells?), about 4% of these cells also are present in the "bright" fraction that comprises mostly

morphologically recognized megakaryocytes. Jackson *et al.* (*24*) report the presence of small cells not otherwise morphologically identified as megakaryocytes. They exhibit APS binding that is greater than that displayed by fully mature megakaryocytes and they also are $SAChE^+$. Kuter *et al.* (*29*) and Kuter and Rosenberg (*30*) showed that sorted rat marrow devoid of APS positive megakaryocytes (and yet containing small, APS positive progenitors?) produced megakaryocytes after 3 days in culture whose degree of polyploidization could be up- or down-regulated by an undefined factor present in added thrombocytopenic and thrombocytotic serum, respectively. Similarly, Rabellino *et al.* (*56*) showed in unfractionated human marrow that about 1% of 2C and 4C cells expressed GPIIb/IIIa (with 31% of 8C cells and 71–98% of cells with ploidy > 8C). Although these data are suggestive of progenitor cell populations, positive cell sorting for these target cells from unfractionated marrow still presents a technically formidable problem (*13*). The fraction of these cells is extremely low (< 1%) and supravital staining for DNA using Hoechst 33342 cannot adequately be substituted for hypotonic PI in unfixed, unfractionated marrow to obtain comparable CVs.

(5) The "catch-22" situation of decreased expression of specific markers for megakaryocytic progenitor cells as a function of degree of immaturity looms again. Clearly, new approaches are needed in addition to probing for platelet-specific marker expression.

C. Platelet-associated Protein (PAP) Expression

Although megakaryocytes maturing to end-stage platelet formation should contain increasing levels of PAP, the relationship is complicated by the fact that each class of polyploid megakaryocytes with > 8C DNA content undergoes this process for reasons that are not clearly understood. Recent studies have attempted to determine the relationship between degree of polyploidization and platelet formation (as quantitated by PAP cell content) using correlative, multivariate analysis on a single-cell basis.

Worthington *et al.* (73) used this approach on rat megakaryocytes enriched by CE that were supravitally stained with Hoechst 33342 for DNA content and FITC conjugated rabbit APS for SMI. DNA content varied linearly with SMI, supporting the conclusion of a proportional increase in PAP expression with increased polyploidization. This interpretation was complicated by the findings that subpopulations of different sized megakaryocytes (by FLS) with DNA contents > 4C were detected within individual ploidy peaks and that there was little correlation between APS-SMI and FLS for 8C, 16C and 32C cells. Some smaller megakaryocytes expressed more APS-SMI than larger (more mature?) cells, an observation confirmed independently (*13,24*). A similar observation of smaller megakaryocytes frequently fluorescing more intensely than larger ones was reported in human megakaryoctes expressing GPIIb/IIIa (*56,64*).

Tomer *et al.* (64) showed a linear relationship between increased GPIIb/IIIa expression and increase in FLS among > 4C ploidy classes of 1% PFA-fixed, PI-stained human megakaryocytes. Interestingly, derived estimates of the density (i.e. SMI/FLS) of GPIIb/IIIa expression were inversely correlated to FLS, suggesting that smaller cells exhibit a higher surface epitope density than larger ones, despite the total SMI of the larger megakaryocytes being greater. However, large statistical variations obtained in these correlative comparisons caution against too simplistic an interpretation, particularly since cell size measurements by FLS of cells as large and granular as megakaryocytes are complex and yet to be fully characterized. Nevertheless, it was encouraging that microscopic measurements using the ACAS analyzer on FACS-sorted megakaryocytes provided some independent confirmation.

Since many PAP are granule bound, the SMI

of unfixed cells (*73*) would be expected to detect only that fraction secreted to, or expressed at, the megakaryocyte outer surface, although the demarcation system in more mature cells with its inherent invaginations complicates this assumption. The use of very low hypotonicity (< 10 mOsm) (*24,29,30,63, 65*), fixation with 1% PFA (*13,64*), and the permeabilizing agents Tween-20 and Triton X-100 (*56,64,65*) should permit the cytoplasmic entry of antibodies; the latter then would presumably detect granule-bound PAP. Differences in the fraction of detected PAP as well as in the epitopes detected prevent direct comparisons among these data.

In another approach to the analysis of PAP expression, Hegyi *et al.* (*19*) applied quantitative immunogold probing (with a new monoclonal, 2E7) to the localization of PF-4 in 8C, 16C and 32C FACS-sorted megakaryocytes that were optimally perfusion fixed *in situ* prior to their near-homogeneous purification following combined CE and discontinuous Percoll density gradient centrifugation. The proportion of PF-4 labeled alpha-granules (AG) and labeled alpha-granule-related small vesicular structures (AGR-SVS, a possible, transport form linking the Golgi to the AG (*18*)), were quantitated in random cross sections of ploidy-sorted megakaryocytes that were subdivided into three stages of cytoplasmic maturation on the basis of ultrastructural criteria. Although there was no significant interaction effect between maturation and ploidy level for labeled AGR-SVS, the effect of polyploidization alone (controlling for maturation) was significant for the mean percentage of labeled AGR-SVS (16C > 32C > 8C). This was also true for the effect of maturation alone (controlling for ploidy level) where a higher mean percentage of labeled AGR-SVS was obtained as a function of increasing maturation (stage 1 < stage 2 < stage 3). The significant differences in the maturation effect were observed only for 8C vs 16C and 16C vs 32C cells. When the analyses were repeated for labeled AG, a significant interaction effect of maturation and ploidy was observed. Within each ploidy class, the more mature cells had proportionately more labeled AG, with the single exception of the 8C class. These ultrastructural data obtained on pure FACS-sorted ploidy classes of megakaryocytes provide evidence of the complexity of megakaryocytic differentiation involving both cytoplasmic maturation and nuclear endoreduplication as reflected in PF-4 expression (*46,49*).

D. Cell Size by Light Scatter (FLS vs SLS)

Cells as large as megakaryocytes and containing complex internal cell structures (demarcation membranes with multiple reflective interfaces and lobulated nuclei) present special problems for accurate flow measurements of cell size (FLS) and "structuredness" (SLS) (*47,48*). The relationship between FLS and megakaryocyte size is complex (*64*) while the contribution of the SLS signal to the FLS signal and what it is truly measuring remain poorly characterized. These limitations take on an even greater importance following the pretreatment of megakaryocytes prior to multivariate analysis by extended exposure to low hypotonicity (< 10 mOsm) (*24*), permeabilizing agents (*13,56,64,65*) and somewhat more subtle changes induced by low concentrations of PFA (*13,64*). Although microscopic controls have been used to compare FLS measurements on the same treated sample (*64*), no studies to date have systematically compared FLS to volume (impedance) and microscopic (diameter) measurements using unfixed, unfractionated marrow in suspension. This is essential to determine the accuracy with which FLS measures true volume over the extensive size range reflective of all polyploid megakaryocytes (*48*), particularly those whose ploidy is up-regulated (to >64C) following platelet demand. The optimal situation would be to substitute direct impedance measurements of cell volume for FLS in a laser-based flow cytometer. Although such sophisticated machines do exist (Los Alamos), electronic interference during sorting may remain a problem. Large statistical variations have been obtained in carefully performed studies using FLS and SLS (*64*). From the above considerations (not including species differences and preflow megakaryocyte isolation techniques), it is not unexpected that comparisons among different studies utilizing correlative multivariate analyses of FLS and SLS are not particularly useful. Nevertheless, the magnitude of the differences in these two measurements as compared to those defining the non-megakaryocytic nucleated cells in marrow (4- to 7-fold) (*64*) argues for their use as electronic gates in correlated multivariate analyses with DNA content and PAP expression. It would appear that SLS with a greater megakaryocyte to non-megakaryocyte differential than FLS may be a more discriminating gate parameter, despite our general ignorance of the former's significance (*64*). Since cytoplasmic maturation reflects to a major extent the proliferation of demarcation membranes and partitioning into platelet fields, the SLS signal may prove especially useful for dissecting differentiation in this cell lineage.

VII. CONCLUSIONS

Clearly, flow analysis of megakaryocytopoiesis has come of age. Despite limitations reflecting the uniqueness of this cell lineage, flow cytometry and cell sorting have been utilized in new and sophisticated ways. This information has enriched the cytokinetic literature and helped to explain how this system operates in the unperturbed state and in response to platelet and megakaryocyte demand. There is great potential in future applications that promise to shed more light on megakaryocytopoiesis in ways that can be obtained only by the judicious application of this evolving technology.

ACKNOWLEDGEMENTS

This work was supported by funds from the Wayne State University Ben Kasle Trust for Cancer Research, the Comprehensive Cancer Center of Metropolitan Detroit (CA 22453) and the Allie Fund. Special thanks to Kay Gounaris for her expertise and invaluable secretarial help.

REFERENCES

1. Bauman, J. G. J., and Chen, M. G. (1987). Light scatter and cell surface properties of murine megakaryocyte progenitor cells. *Exp. Hematol.* **15**, 1074–1079.

2. Berkow, R. L., Straneva, J. E., Bruno, E., Beyer, G. S., Burgess, J. S., and Hoffman, R. (1984). Isolation of human megakaryocytes by density gradient centrifugation and counterflow centrifugal elutriation. *J. Lab. Clin. Med.* **103**, 811–818.

3. Berthier, R., Marin, M., Duperray, A., Prenant, M., Newton, I., Schweitzer, A., Hollard, D., and Marguerie, G. (1987). Megakaryocytic development in liquid cultures of cryopreserved leukocyte stem cell concentrates from chronic myelogenous leukemia patients. *Exp. Hematol.* **198**, 50–58.

4. Bessman, J. D. (1984). The relation of megakaryocyte ploidy to platelet volume. *Am. J. Hematol.* **16**, 161–170.

5. Bodger, M. P., Izaguirre, C. A., Blacklock, H. A., and Hoffbrand, A. V. (1983). Surface antigenic determinants on human pluripotent and unipotent hematopoietic progenitor cells. *Blood* **61**, 1006–1010.

6. Breton-Gorius, J., and Vainchenker, W. (1986). Expression of platelet proteins during the *in vivo* and *in vitro* differentiation of megakaryocytes and morphological aspects of their maturation. *Semin. Hematol.* **23**, 43–67.

7. Briddell, R. A., Brandt, J. E., Straneva, J. E., Srour, E. F., and Hoffman, R. (1989). Characterization of the human burst-forming unit—megakaryocyte. *Blood* **74**, 145–151.

8. Brown, W. M., and Axelrad, A. A. (1976). Effect of Friend leukemia virus on megakaryocytes and platelets in mice. *Int. J. Cancer* **18**, 764–773.

9. Bruno, E., Miller, M. E., and Hoffman, R. (1989). Interacting cytokines regulate *in vitro* human megakaryocytopoiesis. *Blood* **73**, 671–677.

10. Chatelain, C., DeBast, M., and Symann, M. (1988). Identification of a light density murine megakaryocyte progenitor (LD-CFU-M). *Blood* **72**, 1187–1192.

11. Chatelain, C., Hamood, M., DeBast, M., and Symann, M. (1989). Cholinergic enhancement of megakaryocytopoiesis and granulocytopoiesis in culture: Mediation via T-lymphocytes. *Exp. Hematol.* **17**, 1067–1071.

12. Corash, L., Chen, H. Y., Levin, J., Baker, G., Lu,

H., and Mok, Y. (1987). Regulation of thrombopoiesis: Effects of the degree of thrombocytopenia on megakaryocyte ploidy and platelet volume. *Blood* **70**, 177–185.

13. Corash, L., Levin, J., Mok, Y., Baker, G., and McDowell, J. (1989). Measurement of megakaryocyte frequency and ploidy distribution in unfractionated murine bone marrow. *Exp. Hematol.* **17**, 278–286.
14. Ebbe, S., Yee, T., Carpenter, D., and Phalen, E. (1988). Megakaryocytes increase in size within ploidy groups in response to the stimulus of thrombocytopenia. *Exp. Hematol.* **16**, 55–61.
15. Gewirtz, A. M. (1986). Human megakaryocytopoiesis. *Semin. Hematol.* **23**, 27–42.
16. Gewirtz, A. M., Yu Xu, W., and Mangan, K. (1987). Role of natural killer cells in comparison with T-lymphocytes and monocytes in the regulation of normal human megakaryocytopoiesis *in vitro*. *J. Immunol.* **139**, 2915–2924.
17. Hagen-Aukamp, C., Wesemann, W., and Aumuller, G. (1980). Intracellular distribution of adenine and 5-hydroxytryptamine in megakaryocytes isolated by density gradient and velocity sedimentation from bone marrow. *Eur. J. Cell Biol.* **23**, 149–156.
18. Hegyi, E. and Nakeff, A. (1989). Ultrastructural localization of platelet factor 4 in rat megakaryocytes and platelets by gold-labeled antibody detection. *Exp. Hematol.* **17**, 223–228.
19. Hegyi, E., Heilbrun, L., and Nakeff, A. (1990). Immunogold probing of platelet factor 4 in different ploidy classes of rat megakaryocytes sorted by flow cytometry. *Exp. Hematol.* **18**, 789–793.
20. Hoffman, R. (1989). Regulation of megakaryocytopoiesis. *Blood* **74**, 1196–1212.
21. Irons, R. D., and Stillman, W. S. (1985). Flow cytofluorometric analysis of blood and bone marrow. *In* "Toxicology of the Blood and Bone Marrow" (R. D. Irons, ed.), pp. 101–118. Raven Press, New York.
22. Ishibashi, T., and Burstein, S. A. (1985). Separation of murine megakaryocytes and their progenitors on continuous gradients of Percoll. *J. Cell Physiol.* **125**, 559–566.
23. Ishibashi, T., Ruggeri, Z. M., Harker, L. A., and Burstein, S. A. (1986). Separation of human megakaryocytes by state of differentiation on continuous gradients of Percoll. Size and ploidy analysis of cells identified by monoclonal antibody to GPIIb/IIIa. *Blood* **67**, 1286–1292.
24. Jackson, C. W., Brown, L. K., Somerville, B. C., Lyles, S. A., and Look, A. T. (1984). 2-Color flow cytometric measurement of DNA distributions of rat megakaryocytes in unfixed unfractionated marrow cell suspensions. *Blood* **63**, 768–778.
25. Jackson, C. W., Steward, S. A., Brown, L. K., and Look, A. T. (1986). Inverse relationship between megakaryocyte buoyant density and maturity. *Br. J. Haematol.* **64**, 33–44.
26. Jackson, C. W., Steward, S. A., Hutson, N. K., and Ashmun, R. A. (1987). Interaction of ristocetin and bovine plasma with guinea pig megakaryocytes: A means to enrich megakaryocytes based on membrane rather than physical properties. *Blood* **69**, 173–179.
27. Kanz, L., Bross, K. J., Mielke, R., Lohr, G. W., and Fauser, A. A. (1986). Fluorescence-activated sorting of individual cells onto poly-L-lysine-coated slide areas. *Cytometry* **5**, 491–494.
28. Kanz, L., Mielke, R., and Fauser, A. A. (1988). Analysis of human hemopoietic progenitor cells for the expression of glycoprotein IIIa. *Exp. Hematol.* **16**, 741–747.
29. Kuter, D. J., Greenberg, S. M., and Rosenberg, R. D. (1989). Analysis of megakaryocyte ploidy in rat bone marrow cultures. *Blood* **74**, 1952–1962.
30. Kuter, D. J., and Rosenberg, R. D. (1990). Regulation of megakaryocyte ploidy *in vivo* in the rat. *Blood* **75**, 74–81.
31. Lawry, J., Kristensen, S. D., and Martin, J. F. (1989). Flow cytometric analysis of the megakaryocyte DNA distribution in whole and partially purified human and rabbit bone marrow. *J. Clin. Lab. Hematol.* **11**, 361–368.
32. Levene, R. B., Lamaziere, J. M. D., Broxmeyer, H. L., Lu, L., and Rabellino, E. M. (1985). Human megakaryocytes. V. Changes in the phenotypic profile of differentiating megakaryocytes. *J. Exp. Med.* **161**, 457–474.
33. Levine, R. F., and Fedorko, M. E. (1976). Isolation of intact megakaryocytes from guinea pig femoral marrow. Successful harvest made possible with inhibitors of platelet aggregation: enrichment achieved with a two-step separation technique. *J. Cell Biol.* **69**, 159–172.
34. Levine, R. F., Bunn, P. A. Jr., Hazzard, K. C., and Schlam, M. L. (1980). Flow cytometric analysis of megakaryocyte ploidy. Comparison with feulgen microdensitometry and discovery that 8N is the predominant ploidy class in guinea-pig and monkey marrow. *Blood* **56**, 210–217.
35. Levine, R. F., Hazzard, K. C., and Lamberg, J. D. (1982). The significance of megakaryocyte size. *Blood* **60**, 1122–1131.
36. Long, M. W., and Shapiro, D. N. (1985a) Immune regulation of *in vitro* murine megakaryocyte development: the role of T-lymphocyte and Ia antigen expression. *J. Exp. Med.* **162**, 2053–2067.
37. Long, M. W., Gragowski, L. L., Heffner, C. H., and Boxer, L. A. (1985b). Phorbol diesters stimulate the development of an early murine progenitor cell: The burst forming unit-megakaryocyte. *J. Clin. Invest.* **76**, 431–438.
38. Lu, L., Walker, D., Broxmeyer, H. E., Hoffman, R., Hu, W., and Walker, E. (1987). Characteriz-

ation of adult human marrow hematopoietic progenitors highly enriched by two-color cell sorting with MY10 and major histocompatibility class II monoclonal antibodies. *J. Immunol.* **139**, 1823–1829.

39. Lu, L., Briddell, R. A., Graham, C. D., Brandt, J. E., Bruno, E., and Hoffman, R. (1988). Effect of recombinant and purified haematopoietic growth factors on *in vitro* colony formation by enriched populations of human megakaryocyte progenitor cells. *Br. J. Haematol.* **70**, 149–156.
40. Mayer, M., Schaefer, J., and Queisser, W. (1978). Identification of young megakaryocytes by immunofluorescence and cytophotometry. *Blut* **37**, 265–270.
41. Mazur, E. M. (1987). Megakaryocytopoiesis and platelet production: A review. *Exp. Hematol.* **15**, 340–350.
42. McDonald, T. P. (1989). The regulation of megakaryocyte and platelet production. *Int. J. Cell Cloning* **7**, 139–155.
43. Nakeff, A., and Maat, B. (1974). Separation of megakaryocytes from mouse bone marrow by velocity sedimentation. *Blood* **43**, 591–595.
44. Nakeff, A., and Floeh, D. P. (1976). Separation of megakaryocytes from mouse bone marrow by density gradient centrifugation. *Blood* **48**, 133–138.
45. Nakeff, A., Valeriote, F., Gray, J. W., and Grabske, R. J. (1979). Application of flow cytometry and cell sorting to megakaryocytopoiesis. *Blood* **53**, 732–745.
46. Nakeff, A. (1980). Flow cytometric analysis of megakaryocytopoiesis: Strategy for defining mitotic and endoreduplicative events in the committed progenitor compartment. *Blood Cells* **6**, 665–678.
47. Nakeff, A. (1982). Megakaryocytopoiesis: progenitor pool. *In* "The Human Bone Marrow" (S. Trubowitz and S. Davis, eds), Vol. 1, p. 209. CRC Press, Boca Raton, FL.
48. Nakeff, A. (1984). Megakaryocytic cells. *Bibl. Haematol.* **48**, 131–209.
49. Nakeff, A. (1987). Multivectorial analysis of megakaryocytic differentiation. *Ann. NY Acd. Sci.* **509**, 41–58.
50. Nakeff, A., Loeffler, M., Wichmann, E., and Mackey, M. (1989). A re-examination of mathematical modeling in megakaryocytopoiesis. *Exp. Hematol.* **17**, 528.
51. Pallavicini, M. G., Levin, J., Summers, L., and Levin F. (1987). Multivariate flow cytometric characterization and enrichment of murine megakaryocyte progenitors CFU-Meg. *Exp. Hematol.* **15**, 704–709.
52. Pantel, K., and Nakeff, A. (1989). Differential effect of natural killer cells on modulating CFU-Meg and BFU-E proliferation *in situ*. *Exp. Hematol.* **17**, 1017–1021.
53. Pretlow, T. G., and Stinson, A. J. (1976). Separation of megakaryocytes from rat bone marrow cells using velocity sedimentation in an isokinetic gradient of ficoll in tissue culture medium. *J. Cell Physiol.* **88**, 317–322.
54. Rabellino, E. M., Nachman, R. L., Williams, N., Winchester, R. J., and Ross, G. D. (1979). Human megakaryocytes. I. Characterization of the membrane and cytoplasmic components of isolated marrow megakaryocytes. *J. Exp. Med.* **149**, 1272–1287.
55. Rabellino, E. M., Levene, R. B., Leung, L. L. K., and Nachman, R. L. (1981). Human megakaryocytes. II. Expression of platelet proteins in early marrow megakaryocytes. *J. Exp. Med.* **154**, 88–100.
56. Rabellino, E. M., and Bussel, J. B. (1990). Human megakaryocytes. VII. Analysis of megakaryocytes for nuclear DNA content distribution in whole marrow cell suspensions by flow cytometry. *Exp. Hematol.* **18**, 167–173.
57. Schick, P. K., Schick, B. P., and Williams-Gartner, K. (1989). Characterization of guinea pig megakaryocyte subpopulations at different phases of maturation prepared with a Celsep separation system. *Blood* **73**, 1801–1808.
58. Seitz, R., and Wesemann, W. (1980). Studies on megakaryoctyes: Isolation from rat and guinea pig and incorporation of 5-hydroxytryptamine. *Eur. J. Cell Biol.* **21**, 183–187.
59. Sims, R. B., and Gewirtz, A. M. (1989). Human megakaryocytopoiesis. *Annu. Rev. Med.* **40**, 213–224.
60. Sitar, G., Marini, G., Pellicciari, C., Casirola, G., and Ascari, E. (1977). Separation and enrichment of human megakaryocytes by velocity sedimentation on continuous albumin gradient. *Haematologica* **62**, 119–126.
61. Smith, C. M. II, Burris, S. M., and White, J. G. (1989). Micropipette aspiration of guinea pig megakaryocytes: Absence of fragmentation and dependence on maturation stage. *Blood* **73**, 1570–1575.
62. Stenberg, P. E., and Levin, J. (1989). Ultrastructural analysis of acute immune thrombocytopenia in mice-dissociation between alterations in megakaryocytes and platelets. *J. Cell Physiol.* **141**, 160–169.
63. Tomer, A., Harker, L. A., and Burstein, S. A. (1987). Purification of human megakaryocytes by fluorescence-activated cell sorting. *Blood* **70**, 1735–1742.
64. Tomer, A., Harker, L. A., and Burstein, S. A. (1988). Flow cytometric analysis of normal human megkaryocytes. *Blood* **71**, 1244–1252.
65. Tomer, A., Friese, P., Conklin, R., Bales, W., Archer, L., Harker, L. A., and Burstein, S. A.

(1989). Flow cytometric analysis of megakaryocytes from patients with abnormal platelet counts. *Blood* **74**, 594–601.

66. Van Zant, G. and Fry, C. G. (1983). Hoechst 33342 staining of mouse bone marrow: Effects on colony-forming cells. *Cytometry* **4**, 40–46.

67. Watson, J. V., and Chambers, S. H. (1977). Fluorescence discrimination between populations of diploid cells on their RNA content: A possible distinction between clonogenic and non-clonogenic cells. *Br. J. Cancer* **36**, 592–600.

68. Watson, J. V. (1980). Enzyme kinetic studies in cell populations using fluorogenic substrates and flow cytometric techniques. *Cytometry* **1**, 143–151.

69. Watt, S. M., Katz, F. E., Davis L., Capellaro, D., Gordon, M. Y., Tindle, R. W., and Greaves, M. F. (1987) Expression of HPCA-1 and HLA-DR antigens on growth factor- and stroma-dependent colony forming cells. *Br. J. Haematol.* **66**, 153–159.

70. Weller, M. A., Szela, J., Barnhart, M. I., and Nakeff, A. (1986). Surface ultrastructure of human megakaryocytes sorted on the basis of DNA content. *Scan Electron Microsc.* **4**, 1437–1443.

71. Worthington, R. E., and Nakeff, A. (1981). Thromboxane synthesis in megakaryocytes isolated by centrifugal elutriation. *Blood* **58**, 175–178.

72. Worthington, R. E., and Nakeff, A. (1982). Aspirin inhibits rat megakaryocyte thromboxane synthesis. *Prostaglandins* **23**, 841–854.

73. Worthington, R. E., Nakeff, A., and Micko, S. (1984). Flow cytometric analysis of megakaryocyte differentiation. *Cytometry* **5**, 501–508.

74. Young, K. M., and Weiss, L. (1987). Megakaryocytopoiesis: Incorporation of tritiated thymidine by small acetylcholinesterase-positive cells in murine bone marrow during antibody-induced thrombocytopenia. *Blood* **69**, 290–295.

1.5

Cell Kinetics of Hemopoiesis

O. D. LAERUM,[1] F. LUND-JOHANSEN,[1] O. SLETVOLD,[2] and R. SMAALAND[1]

[1] *The Gade Institute*
Department of Pathology, University of Bergen
Haukeland Hospital, N-5021 Bergen, Norway
[2] *Department of Geriatrics*
The Deaconess Hospital, Haraldsplass
N-5009 Bergen, Norway

I. GENERAL ASPECTS

A. Hemopoietic Cell Kinetics in Man and Animals

As already mentioned in the introductory chapter, hemopoietic tissues have a rapid turnover of cells. However, there are large variations in the proliferative activity between the various maturation stages. Approximately 90% of the multipotent stem cells are considered to be in a resting phase, while the remaining 10% are actively proliferating. This serves two purposes. One is to produce new stem cells. The other is to provide new cells to enter the maturing compartments and compensate for the continuous cell loss. The multipotent stem cell pool is considered to be a rather heterogeneous cell population with somewhat differing cycling properties (see also Chapter 1.1). Since these cells are so rare, i.e. 1 per 2000

ISBN 0–12–432940–3

nucleated bone marrow cells, their cell kinetic properties have largely been measured by indirect biological methods.

At any time a certain fraction of the multipotent stem cells will become committed for entering the different maturing compartments, giving rise to a progeny of either erythrocytes, granulocytes, megakaryocytes or lymphocytes. Once they become committed, they enter the next stage, the progenitor cell compartment. The progenitor cell will undergo a variable number of divisions to amplify their numbers, after which they enter the defined maturing stages called the precursor compartment. The progenitor cells are relatively rare in the bone marrow, 1 per 500 to 1 per 1000, depending on the assay procedure.

It is estimated that more than half of the progenitor cells are actively proliferating. The next step consists of the early precursor cells, proerythroblasts, myeloblasts, or lymphoblasts, all considered to be in the cell cycle. For each cell division they increase their degree of maturation, i.e. they transcribe the genetic program for their specialized functions. For myelopoiesis this leads to production of granules, lysosomal enzymes and specialization of their oxidative metabolism, and in erythroblasts hemoglobin is produced. After a number of cell divisions, they reach the end stage as non-proliferative cells. In granulocytic cells the decision is taken at the myelocyte stage, where a myelocyte may either divide once or twice, or further mature to a non-proliferative metamyelocyte and thereafter to a mature granulocyte. For erythropoiesis the same occurs at the normoblast stage after which the nucleus is expelled and mature erythrocytes result (for reviews, see *5,6,34*).

Monocytes are developed in parallel with granulocytes, but will leave the bone marrow with an intact capacity for cell proliferation. After a certain time of circulation in peripheral blood, they enter the tissues and with a few divisions develop into mature macrophages. Lymphocytes originating in the bone marrow will also retain the capacity to proliferate after they enter peripheral blood and lymphatic tissues.

In essence, the multipotent stem cells have a small growth fraction, the progenitor cells have a larger growth fraction, while almost all of the precursor cells are proliferating, and are followed by a non-proliferative end stage.

Studies on cell cycle kinetics of the bone marrow have been done in combination with precise microscopic identification of the different cell compartments and maturating stages. During the last 20 years a large amount of kinetic information on hemopoietic cell proliferation in man and animals has been collected. In general, the cell cycle time is approximately the same for all compartments in a given species. In mice t_c of myelopoietic and erythropoietic cells varies between 8 and 12 h, the S phase is 3–5 h, G_2 phase 0.5–1 h, and mitosis around 0.5 h. In rats and higher animals the corresponding cell cycle periods are somewhat longer, and in man the cell cycle time is 15–20 h, with 10 h for the S phase, 2–3 h for the G_2 phase and 0.5–1 h for mitosis (*10*).

It should be remembered that approximately one third of all nucleated bone marrow cells are mature, non-proliferative cells. This is because bone marrow functions as storage for mature cells ready for export to the peripheral blood, and is an important cell reserve for actual needs. Most of these are granulocytes, since 50–60% of the bone marrow consists of myelopoiesis.

Since the migration of mature granulocytes into peripheral blood is both dependent on the actual needs in the body and on several hormonal and regulatory factors, the non-proliferative pool in the marrow may vary considerably from time to time (*8*). For example, factors such as physical exercise, cortisone and acute bacterial infections are strong mobilizers of granulocytes. In addition, there may be strong rhythmic variations in these processes, which further complicate the picture (see below).

The above-mentioned data are largely based on conventional cell kinetic methods such as mitotic counts and the stathmokinetic method, radioactive labeling with [^{3}H-]Tdr and other isotopes, microspectrophotometry and various biological and culture methods. Some kinetic data have been obtained by studying the regenerating bone marrow following different types of perturbations, including irradiation and cytostatic damage. The basis for these studies has been morphological identification of the various cell types and maturation stages based on light microscopy of stained smears. In addition, some simple histochemical methods have been used for identification of, for example, myelopoietic cells.

With the development of flow cytometric techniques in combination with different biological markers, such as monoclonal antibodies to various specific antigens, a more biologically orientated classification of hemopoietic cell populations has developed (Chapters 1.2 and 3.4). Since these two approaches are not entirely overlapping, comparisons of "old" and "new" data should be done with reservation. However, the basic knowledge of kinetic data still seems to be applicable.

B. Sampling Techniques

Although hemopoiesis should be considered as an organ system where the different locations of bone marrow are regulated systematically and in concert (*14*), there may be some regional variations within the bone marrow (*35*). As already mentioned, there are also variations with age and sex, and this may apply to the density of stem and progenitor cells as well as the amount of hemopoietic cells in relation to the amount of fatty tissue (*20*).

In addition, the sampling technique may greatly affect the resulting cell suspension. This is mainly because of the high vascularity of hemopoietic tissue with corresponding high risk of bleeding during the puncture procedure. A bleeding will lead to admixture of non-proliferaing granulocytes from peripheral blood and thus dilute the sample of real hemopoietic cells. There are several methods to overcome this problem, and some of them have been specifically designed for flow cytometric analyses (*14,49*).

II. APPLICATIONS OF FLOW CYTOMETRY

A. DNA Distribution

The most commonly used method for estimating the kinetic properties of bone marrow cells has been by use of fluorochrome staining of DNA and measurement of cellular DNA distribution. This gives a rough estimate of the proliferative activity of the bone marrow as a total, but does not take into consideration the complexity of hemopoietic tissues. Although the dominating cell population will be myelopoiesis, 50–60%, the proliferative compartments are diluted by the store of non-proliferative mature cells. An increase in the total numbers of mature cells will give the impression of a reduction of proliferating cells. Conversely, a loss of the mature cells to the bloodstream will lead to a corresponding relative increase in the numbers of S and G_2 phase cells. Still a DNA distribution curve gives a rough estimate of the total activity in the same way as incorporation of ^{3}H-Tdr into DNA with subsequent liquid scintillation.

A DNA distribution curve can be useful for studying time-dependent variations (see below), as well as different perturbations with regenerative cell proliferation. The most widely used methods have been fluorochrome staining DNA with ethidium bromide, either alone or in combination with mitramycine, propidium iodide or DAPI, after RNA has been enzymatically removed (*12,25,42*).

B. Other Proliferative Estimates

Since a DNA distribution curve gives a static estimate of the cell cycle distribution, there is a need for alternative methods which measure the real kinetics. One marker which has been widely used in different tissues, including bone marrow, is the monoclonal antibody to the Ki-67 antigen. This is a nuclear protein which is only expressed in actively proliferating cells, mainly S phase, but also in late G_1/ and G_2/M phases (*11,41*). In addition the BrdU technique has been applied for hemopoietic cells. BrdU is a nucleotide analog which is incorporated into cells in the DNA synthetic phase. Because of the lower fluorochrome stainability of BrdU-DNA than ordinary DNA, the rate of incorporation of the substance can be estimated. The same can be achieved by use of a monoclonal antibody to BrdU. This gives an estimate of the rate of DNA synthesis (*17*). However, these methods will only give a quantitation of the total activity unless other parameters are induced in order to discriminate between the various subpopulations (see below).

C. Multiparameter Kinetic Measurements

A one-parameter method implies very crude estimates of the total proliferative activity and does not take into consideration the complexity of hemopoiesis. One of the greatest progresses in the field of flow cytometry has been that multiple parameters can be measured on the same cells. Cell kinetic properties can specifically be measured on various subpopulations using additional parameters and electronic gating. The measured cell populations can be sorted and verified by conventional cytological methods. Although such studies on hemopoiesis are only in their infancy, there are already some promising data which have good potential for practical kinetic studies (*38,48*).

Whereas analysis of one-dimensional DNA histograms provides information about the fractions of the total cell population in the G_0/G_1, S and G_2/M phases, multiparameter analysis differentiates between the cell cycle phases in subpopulations. Flow cytometric methods have been developed for simultaneous measurement of proliferation or cell cycle phase-associated nuclear antigens such as Ki-67, PCNA, P53 or c-myc, and DNA content using fluorescent conjugated monoclonal antibodies (*16,24*). Combined analysis of Ki-67 expression and DNA content allows discrimination between G_0 and G_1 cells (*16*), whereas PCNA appears to be more specific for the S phase. A more detailed picture of cell cycle distribution may also be achieved using acridine orange as a nucleic acid stain (*24*) or by simultaneous measurement of nucleic DNA content and light scatter (*13*). DNA synthesis rate can also be estimated by measuring the incorporation of BrdU, either by fluorochrome conjugated specific monoclonal antibodies, or BrdU quenching of the DNA dye Hoechst 33258 in hemopoietic cells (reviewed in *18*).

A major application of cell kinetic measurement is the study of regulation of cell proliferation. As bone marrow represents a mixture of different cell types known to be regulated independently, the necessity of studying each cell population separately is obvious. During recent years a number of monoclonal antibodies have been developed that allow discrimination between bone marrow cells of different lineages and maturity (*9,23,32,33*). The combined measurement of cell surface antigens and DNA content gives valuable information about lineage and maturation-specific differences in cell kinetics. In addition, contamination of peripheral blood cells may be identified and excluded from analysis.

Although informative, the use of combined analyses of surface markers and DNA content analysis have been limited by technical difficulties. DNA dyes usually require substantial cell membrane permeabilization in order to obtain the low coefficient of variation of the G_0/G_1 peak that is necessary for evaluation of DNA histograms. These permeabilization methods may result in loss or redistribution of antibodies bound to cell surfaces (*3,7,9,23,33*). Recently, methods have been developed to overcome these problems. Sequential fixation with paraformaldehyde (PFA) and ethanol has proved to be useful in some applications (*31*) and this approach has been successfully applied to bone marrow cells (*36*). Another technique has been developed by Hollander and Loken (*21*). Using the relatively membrane permeable DNA dye, Hoechst 33342, in combination with PFA fixation and a mild detergent it was possible to measure cellular light scatter as well as DNA content and immunofluorescence.

Cellular light scatter is a particularly important parameter when antibodies that bind to more than one cell lineage are used, as several cell types can be distinguished by differences in light scattering properties (*21,33*). This procedure may require a two-laser instrument. Recently, an assay for simultaneous dual color immunofluorescence and cellular DNA content using a single 488 nm excitation source has been published (*40*). By introducing a new DNA stain, 7-amino actinomycin D, with an emission maximum further into the red part of the spectrum than propidium and ethidium bromide, the investigators were able to measure DNA content in cells stained with two different monoclonal antibodies conjugated to FITC or phycoerythrin using a 488 nm laser. Applied to bone marrow this method may allow further differentiation between cell types and maturation stages. Other recent methodological advances include procedures for combined measurement of cell surface antigens, light scatter and Ki-67 expression (*15*) or oncogene expression (*32*).

Recently, a method has been designed to detect β-globin mRNA in murine hemopoietic cells by use of flow cytometry (*4*). This was done by measuring strongly positive fluorescent signals which are emitted from fluorochrome-labeled antisense β-globin mRNA. The identification of specific mRNAs in hemopoietic cells may become another important approach to the study of cell kinetics, either as an indirect parameter, or for discrimination of subpopulations.

An example of such multiparameter kinetic analysis is shown in Fig. 1. This illustrates combined measurement of cell surface antigens and DNA content of human bone marrow cells (*36*). A panel of monoclonal antibodies to myeloid differentiation antigens appearing at distinct levels of maturation was used to estimate proliferative activity of cell subsets. VIM2 (CDw65) defines the whole myeloid population and stains mature and immature cells with approximately the same intensity. CD11b (mAb Bear 1) is present in increasing amounts on all myeloid cells approximately

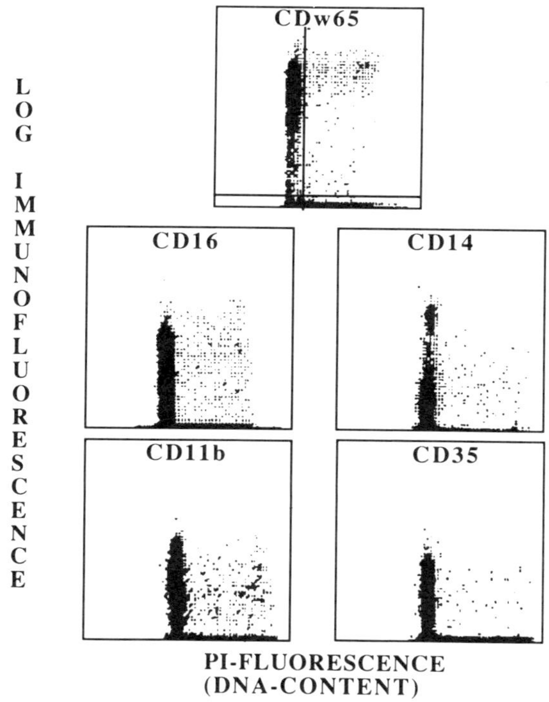

Fig. 1. Combined measurements of DNA content and cell surface antigen expression in human bone marrow cells. Nucleated bone marrow cells were stained with monoclonal antibodies to CDw65, CD16, CD14, CD11b and CD35, fixed and stained with propidium iodide after RNAse treatment (for details see *31*).

from the promyelcyte/myelocyte stage (see earlier chapters). The mAb 3G8 (CD16, FcγRIII) and mAb AML-2-23 (CD14) are relatively specific for the granulocytic and monocytic populations, respectively, and expression of these antigens is also increased during maturation (Chap. 3.4). Finally the mAb E11 reacts with the complement receptor 3 (CD35) present mainly on mature granulocytic and monocytic cells.

As shown in Table I, the percentage of proliferating cells (DNA content > 2n) was significantly lower in the CD35 positive population than in the myeloid population as a whole (CDw65 positive). Cells positive for CD11b, CD14 ad CD16, showed intermediate proliferative activity (Table I). In addition to

Table I

COMBINED MEASUREMENT OF DNA CONTENT AND CELL SURFACE ANTIGEN EXPRESSION ON HUMAN BONE MARROW CELLS[a]

	Cell population				
	CDw65+	CD16+	CD14+	CD11b+	CD35+
S and G_2/M (%)[b]	10.3 ± 2.0	7.7 ± 1.0	6.6 ± 2.0	3.7 ± 2.0	0.7 ± 0.5
Relative fluorescence (proliferating to resting)[c]	1.8 ± 0.2	1.0 ± 0.1	0.4 ± 0.2	0.7 ± 0.2	n.d.

[a] Nucleated bone marrow cells were stained with monoclonal antibodies to cell surface antigens and FITC anti-mouse immunoglobulins, fixed sequentially in paraformaldehyde and ethanol, treated with RNAse, and stained with propidium iodide (for details see *36*).

[b] The percentage of cells in S and G_2/M were determined by quadrant analysis (for details see *36*).

[c] The relative fluorescence of proliferating to resting cells was defined as the mean immunofluorescence intensity of cells in S and G_2/M divided by that of cells in G_0/G_1.

allowing cell cycle analysis of subpopulations of cells, multiparameter analysis may reveal interesting correlations between different cell characteristics. In Fig. 1 and Table I the cells in S and G_2/M phases are shown to have high expression of CDw65, and the mean VIM2 fluorescence intensity of S, G_2/M cells was higher than that of G_0/G_1 cells. In contrast proliferating cells had lower expression of CD11b and CD14 than resting cells (Fig. 1, Table I). The CDw65 antigen is a glycolipid which has been associated with malignancy and cell activation (*23*). Thus, increased expression of the antigen during the S and G_2/M phases of the cell cycle may reflect a role for the antigen in regulation of cell growth.

Using a similar approach, Brons *et al.* (*7*) recently presented results from simultaneous measurements of DNA content and cell surface immunofluorescence of human bone marrow cells by use of light excitation with a single laser. They employed a "cocktail" of different unconjugated monoclonal antibodies which were bound to the surface of unfixed human bone marrow cells, followed by FITC labeling with goat anti-mouse antibodies as the second step. This was followed by staining for 24 h with a hypotonic solution of propidium iodide. For lymphoid cells they used antibodies against CD2 and CD19, for erythroid cells anti-glycophorin-A, for myelomonocytic cells CD13 and CD33, and for monocytic cells CD14. A correction was made for the cross-over of the propidium iodide emission in the FITC channel. With these antibodies they measured specifically the fraction of cells with S phase DNA for main subpopulations in the marrow. The erythroid cells showed the highest proportion of S phase cells with 26.4%, the myelomonocytic cells had 8.5%, the monocytic cells 6.1% and the lymphoid cells only 2.5%. For cells within the lymphoid lineage, the percentage of S phase cells was significantly lower in the T cells (2%), as compared to the B cells (5.1%). They also showed that this method could be used for the selective study of proliferative parameters in erythroleukemia cells.

At present such methods constitute a potent tool for selective measurements of proliferative properties of the different maturation lines among the precursor cells of the bone marrow. In addition, using maturation-dependent antibodies, it is technically possible to discriminate between proliferative and non-proliferative stages of hemopoiesis. Therefore, the cell cycle distribution of practically all subclasses of the bone marrow cell population can specifically be measured by use of electronic gating. The best opportunities are offered for measurements of human cells, while the panels of monoclonal antibodies to differentiation markers are not so abundant for mouse and rat hemopoietic cells.

D. Response to Growth Regulators

As a method to monitor rapidly occurring kinetic changes, flow cytometry has already been used in the study of responses to various growth regulators in hemopoietic cells. Using one-parameter DNA measurements cell cycle effects of the hemoregulatory peptide, an acidic pentapeptide associated with mature granulocytes, were studied in the mouse bone marrow after *in vivo* injections. The reaction to this inhibitor was biphasic, showing an initial reduction of S phase cells (and ^{3}H-Tdr uptake) after 24 h. This was followed by a similar reduction occurring after 6–12 days. The latter reaction was probably secondary to a reduction of myelopoietic progenitor cells (*26*).

Similarly, Pennline *et al.* (*39*) studied the sequence of reactions in the bone marrow of mice injected with recombinant GM-CSF. Using a multiparameter approach with dual fluorescence and various antibodies, they found an increase in the myeloid cell population, as well as other changes in the hemopoietic cell population.

At present, flow cytometry seems to be a promising tool for the study of kinetic reactions to growth factors.

III. CIRCADIAN AND OTHER TIME-DEPENDENT VARIATIONS OF BONE MARROW AND PERIPHERAL BLOOD

A. Bone Marrow in Normal and Aging Mice

It has been a general rule that studies on cell kinetics should always be performed at the same time of the day, since proliferative parameters may change with time. It has long been known that hemopoietic activity may vary considerably throughout 24 h periods, and often with a regular circadian pattern (*29*). This is seen in myelopoietic cells, and to some extent in the other cell compartments. In our laboratory we have addressed this question in more detail. The background was that the rapidity of flow cytometric techniques enables sequence sampling and analysis over prolonged periods. As an estimate of the total proliferative activity we investigated the cell cycle distribution in the femur marrow of C3H mice using DNA measurements. In line with other workers, it came out that there were strong circadian variations, where both the fraction of S phase as well as $G_2 + M$ phase cells varied with up to ± 50% of the 24 h mean value (*28*). The same applied to the stem and progenitor cell compartments as studied by spleen colony and agar colony technique (*1*,*2*). These fluctuations showed an unstable pattern, where the period of maximal activity of the day varied with the time of the year. Although the maximum in most cases was found throughout the day, there were several exceptions.

With increasing age of mice, these rhythms became more unstable and often had a different phasing than young mice. If the fractions of cells with S phase DNA content in young and old mice were compared at only one time point the result of the comparison was totally dependent on their phasing. At some points young mice had higher activity than the old ones, and at other time points it was the opposite (*44*).

On this basis we concluded that the conflicting results in the literature on studies of quantitative changes with aging in the bone marrow could largely be due to such variations. This stresses the importance of taking time-dependent variations into consideration when quantitative studies on proliferative activity in hemopoiesis are performed. The time of the day and the season are equally important (*43*,*45*).

B. Bone Marrow in Man

There are some studies from the 1960s indicating that there are also circadian variations in bone marrow proliferation of humans with a higher activity during the day (*22*,*37*). Based on our studies on mice, we designed a systematic investigation of the same parameters in humans, including CFU-GM, differential counts, total cell cycle distribution by flow cytometry and total glutathione content. By sequential sampling of bone marrow from the iliac crest and sternum every 4 h for 24 h periods on 21 healthy male volunteers, we found similar circadian variations as in mice (*46,47*). In humans, however, there was a more regular pattern, where the maximum of the fraction of cells in S and $G_2 + M$ phases roughly followed the fluctuations of the core body temperature. Maximum was found around 16:00 h, and minimum between 00:00 and 04:00 h. The differences were of the same order of magnitude as in mice, although there were wide individual variations. The study indicates that quantitative estimates of hemopoietic cell proliferation are insufficient when only taken at one time point of the day.

The variations of proliferative activity were verified for the myelopoietic and erythropoietic compartments when studied selectively by use of monoclonal antibodies and electronic gating (unpublished results).

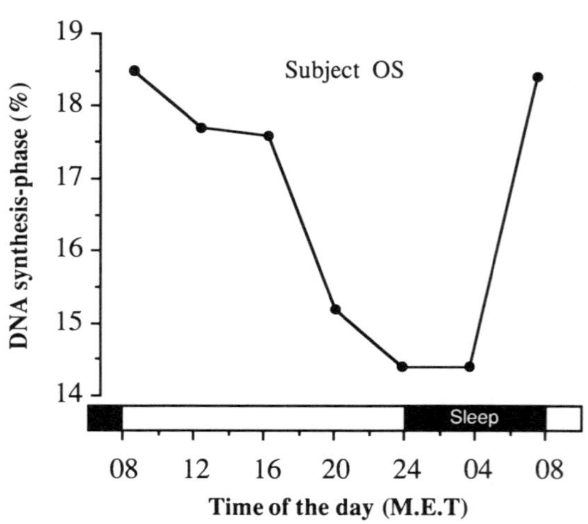

Fig. 2. Example of circadian variation of the fraction of DNA synthesizing cells in the whole bone marrow cell population obtained by serial aspiration from a healthy male volunteer.

From a practical point of view flow cytometry turned out to be a simple and straightforward method for studying sequence samples from the bone marrow in man, where the results were obtained and evaluated within a few hours after sampling. We hold the opinion that such studies can be valuable for evaluating chemotherapeutic side effects in cancer patients, as well as finding periods of the day where patients have the lowest proliferative hemopoietic activity, and hence a lower risk of bone marrow damage.

C. Peripheral Blood

For completeness, we shall briefly mention circadian variations in peripheral blood parameters of mice and humans. Similar circadian patterns are present as in the bone marrow, although the phasing can be somewhat different. In aging mice the amplitude is larger than in young mice, and occasionally with different phasing (*30*). The highest leukocyte numbers are usually found in the late afternoon, mainly because of a stable rhythm of the lymphocytes (*19*).

In humans both granulocyte and lymphocyte numbers roughly follow the rhythm of the core body temperature with maximal numbers between 16:00 and 20:00 h. Flow cytometric analysis of phagocytic capacity follows the temperature, and in addition seems to have an early morning peak coinciding with the peak of the cortisol excretion (*27*).

IV. CONCLUSIONS

When all the different methodologies of flow cytometry are considered together, they offer unique possibilities as compared with conventional, more laborious methods for studying cell kinetics in complex cell populations:

(1) Cell cycle distribution can be measured specifically on different compartments and maturation stages of hemopoiesis using multiple simultaneous parameters and electronic gating.

(2) Cell proliferation and maturation, both morphological and functional, can be studied specifically in conjunction with each other.

(3) The rapidity and simplicity of flow cytometric preparation and analysis techniques enable multiple sequential samples for studying kinetic reactions and time-dependent variations of hemopoiesis.

(4) DNA distribution curves on whole bone marrow give a rough estimate of the total proliferative activity, but should be replaced by multiple parameter analysis on selected subpopulations.

REFERENCES

1. Aardal, N. P. (1984). Circannual variation of circadian periodicity in murine colony-forming cells. *Exp. Hematol.* **112**, 61–67.
2. Aardal, N. P., and Laerum, O. D. (1983). Circadian variations in mouse bone marrow. *Exp. Hematol.* **11**, 792–801.
3. Bauer, K. D. (1988). Application of multiparameter DNA content and immunofluorescence analysis in the investigation of human neoplasia. *Pathol. Immunopathol. Res.* **7**, 371–380.
4. Bayer, J. A., and Bauman, J. G. J. (1990). Flow cytometric detection of β-globin mRNA in murine haemopoietic tissues using fluorescent in situ hybridization. *Cytometry* **11**, 132–143.
5. Benestad, H. B., and Laerum. O. D. (1989). The neutrophilic granulocyte. *In* "Current Topics in Pathology. Vol. 79: Cell Kinetics of the Inflammatory Reaction" (O. H. Iversen, ed.), pp. 7–36. Springer-Verlag, Berlin.
6. Boggs, D. R. (1967). The kinetics of neutrophilic leukocytes in health and disease. *Seminars Hemat.* **4**, 359–386.
7. Brons, P. P. T., Pennings, A. H. M., Haanen, C., Wessels, H. M. C., and Boezeman, B. M. (1990). Simultaneous measurements of DNA content and cell-surface immunofluorescence of human bone marrow cells using a single laser flow cytometer. *Cytometry* **11**, 837–844.
8. Butcher, E. C. (1990). Cellular and molecular mechanisms that direct leukocyte traffic. *Am. J. Pathol.* **136**, 3–11.
9. Civin, C. I. (1990). Human monomyeloid cell membrane antigens. *Exp. Hematol.* **18**, 461–467.
10. Dancey, J. T., Deubelbeiss, K. A., Harker, L. A., and Finch, C. A. (1976). Neutrophil kinetics in man. *J. Clin. Invest.* **58**, 705–715.
11. Danova, M., Giordano, M., Mazzini, G., and Riccardi, A. (1990). Proliferation-associated antigens and oncogene products in hematological malignancies: A flow cytometric study. Abstract. *In* "Abstracts for the XIV International Meeting of the Society for Analytical Cytology," Asheville, North Carolina, March 18–23, 1990. *Cytometry*, Suppl. 4, 29.
12. Darzynkiewicz, Z., Crissman, F. H., Traganos, F., and Steinkamp, J. (1982). Propidium iodide. *J. Cell Physiol.* **113**, 465–474.
13. Dent, G. A., Leglise, M. C., Pryzwansky, K. B., and Ross, D. W. (1989). Simultaneous paired analysis by flow cytometry of surface markers, cytoplasmic antigens, or oncogene expression with DNA content. *Cytometry* **10**, 192–198.
14. Dosik, G. M., Barlogie, B., Göhde, W., Johnston, D., Tekell, J. L., and Drewinko, B. (1980). Flow cytometry of DNA content in human bone marrow: A critical reappraisal. *Blood* **55**, 734–740.
15. Drach, J., Gattringer, C., Gassl, H., Schwarting, R., and Huber, H. (1989). Simultaneous flow cytometric analysis of surface markers and nuclear Ki-67 antigen in leukemia and lymphoma. *Cytometry* **10**, 743–749.
16. Gerdes, J., Lemke, H., Baisch, H., Wacker, H. H., Schwab, U., and Stein, H. (1984). Cell cycle analysis of a cell proliferation-associated human nuclear antigen defined by the monoclonal antibody Ki-67. *J. Immunol.* **133**, 1710–1715.
17. Gray, J. W., ed. (1985). "Monoclonal Antibodies Against Bromodeoxyuridine." *Cytometry* 6(6). Alan R. Liss, New York.
18. Gray, J. W., and Darzynkiewiz, Z., eds (1987). "Techniques in Cell Analysis." Humana Press Clifton, NJ.
19. Haus, E., Lakatua, D. J., Swoyer, J., and Sackett-Lundeen, E. (1983). Chronobiology in hematology and immunology. *Am. J. Anat.* **168**, 647.
20. Hiddemann, W., Büchner, T., Andreeff, M., Wörmann, B., Melamed, M. R., and Clarkson, B. D. (1982). Cell kinetics in acute leukemia. A critical reevaluation based on new data. *Cancer* **50**, 250–258.
21. Hollander, Z., and Loken, M. R. (1988). Simultaneous analysis of DNA content and surface antigens in human bone marrow. *Cytometry* **9**, 485–490.
22. Killmann, S. A., Cronkite, E. P., Fliedner, T. M., and Bond, V. P. (1962). Mitotic indices of human bone marrow cells. I. Number and cytologic distribution of mitoses. *Blood* **19**, 743–750.
23. Knapp, W., ed. (1989). "Leukocyte Typing IV". Oxford University Press, New York.
24. Kurki, P., Ogata, K., and Tan, E. M. (1988). Monoclonal antibodies to proliferating cell nuclear antigen (PCNA)/cyclin as probes for proliferating cells by immunofluorescence microscopy and flow cytometry. *J. Immunol. Met.* **109**, 49–59.
25. Laerum, O. D., and Aardal, N. P. (1981). Chronobiological aspects of bone marrow and blood cells. *In*

"11th International Congress of Anatomy, Part C, Biological Rhythms in Structure and Function" (H. von Mayersbach, L. E. Scheving, and J. E. Pauly, eds), pp. 87–97. Alan R. Liss, New York.

26. Laerum, O. D., and Paukovits, W. R. (1984). Modification of mouse hemopoietic cell proliferation *in vivo* by a hemoregulatory pentapeptide (HP5b). Relation to circadian rhythms. *Virchows. Arch. (Cell Pathol.)* **46**, 333–348.

27. Laerum, O. D., Smaaland, R., Bjerknes, R., Sletvold, O., and Lote, K. (1987). Flow cytometric measurements of circadian variations in human hemopoiesis and phagocytosis. XII Annual Meeting of International Society of Cytology, Cambridge.

28. Laerum, O. D., Sletvold, O., and Riise, T. (1988). Circadian and circannual variations of the cell cycle distribution in the mouse bone marrow. *Chronobiol. Int.* **5**, 19–35.

29. Laerum, O. D., Smaaland, R., and Sletvold, O. (1989). Rhythms in blood and bone marrow: Potential therapeutic implications. *In* "Chronopharmacology" (B. Lemmer, ed.), pp. 371–393. Marcel Dekker, New York.

30. Laerum, O. D., Smaaland, R., and Sletvold, O. (1989). Rhythms in blood and bone marrow: Potential therapeutic implications. *In* "Chronopharmacology. Cellular and Biochemical Interactions" (B. Lemmer, ed.), pp. 371–393. Marcel Dekker, New York.

31. Lakhanpal, S., Gonchroff, N. J., Katzmann, J. A., and Handwerger, B. S. (1987). A cytofluorometric double staining method for determination of human mononuclear cell surface phenotype and cell cycle phase. *J. Immunol. Met.* **96**, 35–40.

32. Larsen, J. K., Munch-Pettersen, B., Christiansen, J., and Jørgensen, K. (1986). Flow cytometric discrimination of mitotic cells: resolution of m as well as G1, S, and G2 phase nuclei with mitramycin, propidium iodide and ethidium bromide after fixation with formaldehyde. *Cytometry* **7**, 54–63.

33. Loken, M., Shah, V. O., Dattilio, K. L., and Civin, C. (1987). Flow cytometric analysis of human bone marrow: I. Normal erythroid development. *Blood* **69**, 225–263.

34. Lord, B. I. (1983). Haemopoietic stem cells. *In* "Stem Cells. Their Identification and Characterization" (C. S. Potten, ed.), pp. 118–154. Churchill Livingstone, Edinburgh.

35. Lord, B. I. (1990). The architecture of bone marrow cell populations. *Int. J. Cell Cloning* **8**, 317–331.

36. Lund-Johansen, F., Bjerknes, R., and Laerum, O. D. (1990). Flow cytometric assay for the measurement of human bone marrow phenotype, function and cell cycle. *Cytometry* **11**, 610–617.

37. Mauer, A. M. (1965). Diurnal variation of proliferative activity in the human bone marrow. *Blood* **26**, 1–7.

38. Pallavicini, M. G., Summers, L. J., Giroud, F. J., Dean, P. N., and Gray, J. W. (1985). Multivariate analysis and list mode processing of murine hemopoietic subpopulations for cytokinetic studies. *Cytometry* **6**, 539–549.

39. Pennline, K. J., Pellerito, F., DaFonseca, M., Monahan, P., Siegel, M. I., and Smith, S. R. (1990). Flow cytometric analysis of recombinant murine GM-CSF (rmuGM-CSF) induced changes in the distribution of specific cell populations *in vivo*. *Cytometry* **11**, 283–291.

40. Rabinovitch, P. S., Torres, R. M., and Engel, D. (1986). Simultaneous cell cycle analysis and two color surface immunofluorescence during 7-amino-actinomycin D and single laser excitation: applications to study of cell activation and the cell cycle of muring Ly-1 B cells. *J. Immunol.* **136**, 2769–2774.

41. Sasaki, K., Matsumura, K., Tsuji, T., Shinozaki, F., and Takahashi, M. (1988). Relationship between labeling indices of Ki-67 and BrdUrd in human malignant tumours. *Cancer* **62**, 989–993.

42. Shapiro, H. M. (1986). The little laser that could: Applications of low power lasers in clinical flow cytometry. *Ann. NY Acad. Sci.* **468**, 18–27.

43. Sletvold, O. (1987). Circadian rhythms of peripheral blood leukocytes in aging mice. *Mech. Ageing Dev.* **39**, 251–261.

44. Sletvold, O., and Laerum, O. D. (1988). Alterations of cell cycle distribution in the bone marrow of aging mice measured by flow cytometry. *Exp. Gerontol.* **23**, 43–58.

45. Sletvold, O., Laerum, O. D., and Riise, T. (1988). Age-related differences and circadian and seasonal variations of myelopoietic progenitor cell (CFU-GM) numbers in mice. *Eur. J. Haematol.* **40**, 42–49.

46. Smaaland, R., Lote, K., Sletvold, O., Bjerknes, R., and Laerum, O. D. (1989). Circadian stage dependent variations in the DNA synthesis-phase and G2/M-phase of human bone marrow. *Proceedings AACR* 30:35.

47. Smaaland, R., Laerum, O. D., Lote, K., Sletvold, O., Sothern, R. B., and Bjerknes, R. (1991). DNA synthesis in human bone marrow is circadian stage dependent. *Blood* **77**, 2603–2611.

48. Terstappen, L. W. M. M., and Loken, M. R. (1988). Five-dimensional flow cytometry as a new approach for blood and bone marrow differentials. *Cytometry* **9**, 548–556.

49. Zbroja, R. A., Wass, J., Vincent, P. C., and Young, G. A. R. (1986). Fragment filtration: a method for the accurate determination of flow cytometric kinetic data from bone marrow aspirates. *Exp. Hematol.* **14**, 85–89.

2

Normal Peripheral Blood Cells

2.1

Reticulocytes and Mature Erythrocytes

HANS J. TANKE

Department of Cytochemistry and Cytometry
Sylvius Laboratories, Wassenaarseweg 72
2333 Al Leiden, The Netherlands

I. INTRODUCTION

A. RED BLOOD CELLS: THEIR MORPHOLOGY, METABOLISM AND FUNCTION

Erythrocytes in peripheral blood were described as early as 1673 by the Dutch microscopist van Leeuwenhoek: he called them "small round globules" and attributed the colour of blood to them. In 1747 Menghini demonstrated the presence of iron in blood, and Funke isolated hemoglobin in crystalline form somewhat later. The main function however of the erythrocyte was discovered in 1867 by Hoppe-Seyler by demonstrating that hemoglobin had the property to bind and discharge oxygen. At the same time it was found that erythrocytes are formed in the bone marrow, and morphological studies of erythrocytes and their precursors have been carried out since, on specimens that were stained according to procedures originally proposed by May, Grünwald and Giemsa, and by Romanovski using conventional light microscopy. Much later information about the morphology of the (living) erythrocyte was obtained by studies using phase contrast microscopy (stimulated by the work of Bessis and colleagues (*2*)), and by scanning electron microscopy. The morphological definition of

FLOW CYTOMETRY IN HEMATOLOGY
ISBN 0-12-432940-3

the erythrocyte seems relatively easy. Erythrocytes in the circulation are approximately 8 μm in size, and they have the shape of a biconcave disc. Upon visualization with phase contrast microscopy they appear dark grey, with a clear center and a surrounding halo; in Giemsa smears they look pink. When examined with violet light (Soret band absorption of hemoglobin at 410 nm) differences in hemoglobin content may be observed, although the human eye is not very sensitive to light of this wavelength.

Mature red cells circulate for approximately 120 days in peripheral blood before they die and are removed from the circulation. Morphological differences with respect to age cannot be detected. However, precursor erythrocytes can be distinguished from mature red cells. They are called reticulocytes, since upon staining with supravital dyes such as methylene blue or brilliant cresyl blue a typical "reticulum-like" precipitation pattern of the rRNA is observed. The beginning of the reticulocyte stage is defined as the moment that the nucleus is extruded from the normoblast, the nucleated cell that originates from the early red cell precursors. Determining the end of the reticulocyte stage is more difficult (*16*), since it is defined as the stage where mitochondria and ribosomes, necessary for synthesis of the remaining 20% of hemoglobin, are lost. At this time the erythrocyte also obtains its disc shape. Reticulocytes are larger than mature red cells and contain stainable amounts of rRNA, by which they have been classified according to age by Heilmeyer (*17*) in stages I to IV (I being immature with relatively large amounts of RNA, IV being almost mature with minor amount of RNA left). According to the definitions used there is no practical parameter that can be used to determine where the reticulocyte ends and the mature red cells begins, since it largely depends on the sensitivity for detecting RNA, on demonstrating size increments or on detecting certain membrane antigens (transferrin receptors, glycophorin).

The metabolism in reticulocytes and mature red cells is, as in other cells, complex. In this chapter the metabolism of hemoglobin is specially emphasized, since most flow cytometric studies of red cells concern the measurement of parameters that are somehow related to hemoglobin, such as the total amount of hemoglobin per cell, the amount of rRNA that is still present in the maturing reticulocyte, and the presence of transferrin receptor on the membrane.

The rate of hemoglobin synthesis reaches its maximum in the polychromatophilic stages of the erythroid precursors and continues through the orthochromatic stages. After exclusion of the nucleus in the normoblast approximately 20% of necessary hemoglobin is still to be synthesized during the various stages of reticulocyte maturation.

Hemoglobin basically consists of a heme group to which globin molecules have been attached. The heme group contains one molecule of iron, that plays a key role in the binding and release of oxygen. For the synthesis of hemoglobin iron has to be transported from the plasma to the erythrocytes. For this purpose plasma Fe^{3+} ions are bound by a glycoprotein, called transferrin, to form a complex, that subsequently binds to the transferrin receptor that is present on the membrane of all hemoglobin synthesizing cells.

This very basic background to metabolism thus shows that first, reticulocytes have less hemoglobin than mature red cells, second that reticulocytes have transferrin receptors (unlike mature red cells), and finally that reticulocytes have significant amounts of RNA (unlike mature red cells).

B. Reticulocytes, Erythrocytes and Erythropoiesis

It is evident that the number of mature red blood cells needs to be controlled and maintained by the body within certain limits under both normal and abnormal conditions. The mechanisms involved in this control process are very complex, however the major steps are described as follows.

A decrease in hemoglobin concentration leads to alterations in the oxygen tension in the tissues, also in the kidney. Here, a factor is secreted that gives rise to the production of erythropoietin, a hormone that stimulates primitive precursor cells in the bone marrow to differentiate into erythroid precursors, leading to an increased production of red cells. This, in turn, leads to an increase in oxygen level at the tissue, and subsequently to a decrease in erythropoietin production, and a feedback mechanism is established.

Clinical situations often require the study of this complex control mechanism. This can be achieved by investigating the erythropoietic activity of the bone marrow, for which several methods are presently available. The rate of production of red cells can be calculated from independent measurements of erythrocyte life span and red cell volume using chromium isotopes (*36*) or by measuring the turnover rate of red cell iron labeled with radioactive ^{59}Fe (*1*). A disadvantage of these methods is that they are relatively complex, slow, and moreover require steady state conditions, which often cannot be reached.

A relatively simple and commonly used method to estimate the rate of erythropoiesis is the visual reticulocyte count. Reticulocytes vitally stained with dyes such as brilliant cresyl blue and methylene blue (*3*) can be classified according to the RNA precipitation pattern in stages I to IV, according to Heilmeyer (*17*). These studies have shown that under conditions of stimulated erythropoiesis, reticulocytes of stage I and II, which normally occur only in the bone marrow, appear in the peripheral blood (reticulocyte shift) (*30*). This implies that the number of reticulocytes in the peripheral blood no longer reflects the situation of the bone marrow and cannot be used as an adequate parameter of the activity of the red cell production system. Hillman and Finch have attempted to correct the absolute count for this shift (*18*). Theoretically, counting of morphologically distinguishable subclasses of reticulocytes could be applied to determine the degree of anemia. However, problems and shortcomings of the manual counting method were analyzed in a joint study of the College of American Pathologists (*33*). In this report the authors concluded that "the degree of imprecision is severe and that inaccuracy is impossible to assess."

This, but also the immediate clinical need for a simple and rapid method to count the total number of reticuloctyes with respect to their maturation stage, has stimulated the use of flow cytometric techniques for this purpose.

II. FLOW CYTOMETRY OF RETICULOCYTES

A. Requirements for Reticulocyte Counting

The visual methods for counting reticulocytes on the basis of vital staining of rRNA with brilliant cresyl blue or new methylene blue have a number of serious disadvantages. First, since only a few hundred red cells are counted they are statistically very unreliable for determining percentages of reticulocytes lower than 1% or detecting small but significant changes in reticulocyte populations as a consequence of blood loss, irradiation or chemotherapy. Moreover, routinely they do not provide information on the reticulocyte maturation stage (according to Heilmeyer), since subclassification on the basis of stain precipitation patterns is not only time consuming but also very subjective.

Flow cytometry provides a useful alternative for quantification of reticulocytes. By using fluorescent dyes that bind to RNA, reticulocytes cannot only be counted flow cytometrically, but also information on their maturation stage can be obtained as far as expressed in the RNA fluorescence intensity (*9,37,38*). It has been shown that flow cytometry is statistically much more reliable than manual counts, since 10 000–100 000 cells are readily counted per test. First attempts for flow cytometric counting of reticulocytes were carried out using acridine orange (*27*). Later several other dyes were used for this purpose, such as pyronin Y (*37*), DiOC-1-[3] (*20*), thioflavin T (*32*), thiazole orange (*25*), and auramin O (*23*).

In addition reticulocyte staining has been proposed with monoclonal antibodies such as OKT 9 and L5.1 against the transferrin receptor (*24*). Reticulocyte counting on the basis of staining with monoclonal antibodies is only incidently used for experimental purposes, and therefore not extensively discussed here.

B. Flow Cytometric Counting of Reticulocytes

a. *The Acridine Orange Method*

Acridine orange (AO) is a metachromatic basic dye (Fig. 1) that shows complex binding mechanisms to nucleic acid, but also to other anions because of its positive charge. Under strictly controlled conditions, extensively studied by Darzynkiewicz and colleagues (*7*), AO fluorescence relates almost exclusively to nucleic acids, in most of the cell types. The binding to double stranded nucleic acid is by intercalation into the double helix leading to green fluorescence emission (530 nm) at excitation with blue light. AO interacts with single stranded nucleic acids by a co-operative association resulting in condensation and precipitation of the product. This binding mechanism causes metachromasia and a shift in the emission spectrum towards the red (640 nm). Defined staining results can however only be obtained if important parameters such as dye concentration, ionic composition and strength and pH are strictly controlled. This may be hampered by undefined impurities and other acridine compounds in the commercial dye product. Chromatographically pure dyes, which can be purchased from various companies, are recommended.

Acridine orange stains unfixed reticulocytes since it passes the intact cellular membrane. The AO staining solution originally developed for the Ortho Spectrum III flow cytometer, however, contains low amounts of formaldehyde (*29,41*). AO staining of reticulocytes

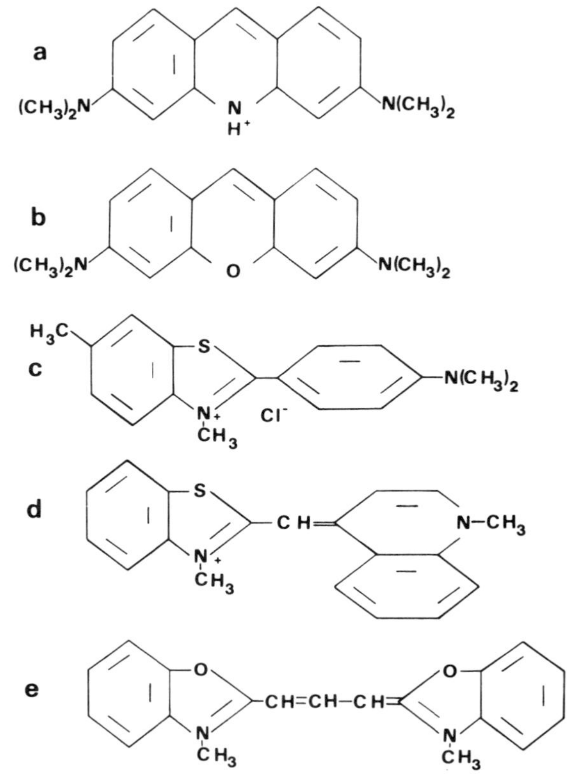

Fig. 1. Chemical structure of fluorescent RNA dyes: (a) acridine orange; (b) pyronin Y; (c) thioflavin T; (d) thiazole orange; (e) DiOC1 [3].

for flow cytometry is rapid and easy. Ten microliters of whole EDTA blood are added to 2 ml commercially available AO reagent, whirlmixed for approximately 5 sec and incubated for 4–6 min at room temperature.

Flow cytometry of AO stained red cells is carried out using an argon ion laser emitting at 488 nm. Two-parameter cytograms of forward angle light scatter and red fluorescence of AO are recorded. Non-rectangular gate setting is applied to separate the platelets from the erythrocyte/reticulocyte population. For the red cell population a one-dimensional fluorescence histogram is recorded on a linear scale, and curve fitting by a computer algorithm is applied to calculate the percentage of reticulocytes (*41*). A modification of this staining and analysis procedure using an AO solution without fixative, and discrimination of mature red cells and reticulocytes on the basis of logarithmic detection and pulse processing, has been proposed as well (*34*). It should be noted that various manufacturers of flow cytometers have their own AO method, reagents and software to perform reticulocyte analysis. Examples are the procedures developed by Ortho Diagnostic Systems for the Spectrum, and by Coulter Electronics for the EPICS C and the Profile.

b. The Pyronin Y Method

Pyronin Y (PY) is the xanthene homolog of AO (Fig. 1). The exact binding mechanism of PY to nucleic acids is still not clear. The current concept is that PY binds to both single and double stranded nucleic acids. However, PY only fluoresces with sufficient quantum efficiency when it is intercalated into double stranded nucleic acid. Cells have to be fixed prior to PY staining, since the dye does not pass the membrane of live cells, although vital staining of mitochondria has been observed. Staining of double stranded RNA is concentration dependent. At concentrations lower than 2 μM PY fluorescence is almost RNA conformation independent and roughly proportional to double stranded RNA content (*8*).

PY staining of reticulocytes requires fixation in 7% formaldehyde in PBS for at least 2 hours, or better overnight at 4°C. After washing with PBS the red cells are stained for 30 min in 0.01% pyronin Y in sodium acetate buffer (1.0 *M*, pH 4.7), washed once with PBS and directly analyzed (*37*).

Flow cytometry is performed with an argon ion laser emitting at least 500 mW of 514 nm light. Two-parameter cytograms of forward angle light scatter and red fluorescence of pyronin Y (LP 570) are recorded. Platelets and leukocytes are distinguished on the basis of their light scatter properties and gated out, and the fluorescence distribution of red cells is recorded on a linear scale. Discrimination between reticulocytes and red cells is carried out by computer using a discrimination level defined as the modal value of the mature red cell peak (channel number) plus the width of the peak at half maximum (FWHM) (in channels). Furthermore, the integrated RNA content and the median value (channel number) of the reticulocyte population is calculated. The latter is referred to as the reticulocyte maturation index (*38*).

c. The Thioflavin T, Thiazole Orange and DiOC1 [*3*] *Methods*

All these dyes have a thiazolium or oxazolium moiety conjugated to an aromatic amine (Fig. 1). They pass the cellular membrane under physiological conditions and can be used for vital staining. Unlike other vital dyes such as brilliant cresyl blue and new methylene blue, they do not precipitate the rRNA, but stain the reticulocyte homogeneously. Their fluorescence increases upon binding to RNA: 3,3′-dimethyloxacarbocyanine DiOC1 [3] by a factor of 3, thioflavin T by 100, and thiazole orange by 3000. The quantum efficiency of the latter dye is relatively high (0.2) (*25*). The excitation and emission maxima for RNA–thioflavin T are 422 nm and 487 nm, respectively; for DiOC1 [3]–RNA these values are 482 nm versus 510 nm, and for thiazole orange–RNA 509 nm versus 533 nm (*25*). The fact that thiazole orange can be well excited at 488 nm (argon ion laser), i.e. below the optimal wavelength, and still fluoresces brightly, reflects the increased quantum efficiency of this dye (X 3000) when bound to RNA.

Thioflavin T (TT) (*32*) and thiazole orange (TO) (*25*) staining are very similar. TT staining consists of adding 10 μl whole blood to 10 μl of a 0.01% thioflavin T solution in isotonic buffered saline; incubation continues for 7 min followed by dilution of the solution with 1 ml isotonic saline.

TO staining is carried out by preparing a stock solution of TO (1 mg/ml) in methanol. This stock solution is diluted 1 : 10 000 in PBS containing 0.002 *M* EDTA and 0.02% sodium azide. Whole blood (5 μl) is then added to 1 ml diluted stain solution and incubated at room temperature for 1 h.

Flow cytometry of TT stained samples requires excitation with violet light (457 nm from an argon ion laser or 436 nm from a mercury arc lamp). Depending on the type of flow cytometer two-parameter correlated cytograms of size (either forward angle light scatter or cell volume) in combination with green fluorescence (500 nm interference filter) are then recorded. The gate setting for red cells is determined on the basis of size and the fluorescence distribution is recorded on a logarithmic scale. At first an unstained diluted blood sample is analyzed to determine the channel position above which no more than 0.1% of the total number of recorded events is found. This channel position is then used as a fixed threshold for the stained sample to determine the percentage of reticulocytes (*32*).

Flow cytometric analysis of TO stained samples and calculation of the percentage of reticulocytes is very similar to that described for TT staining. However, excitation of TO stained cells is generally carried out using 488 nm excitation light from an argon ion laser. Gating of red cells is performed on two-parameter cytograms of forward angle light scatter and green thiazole orange fluorescence (interference band pass filter at 530 nm).

Staining with cyanin dye DiOC1 [3] (*20*) is carried out by adding 5 μl of whole blood to 1 ml dye solution (0.5–15 μ*M* DiOC1 [3] in PBS containing 5 m*M* EDTA). After staining for at least 20 min, cells are analyzed by excitation with 488 nm light from an argon ion laser, and two-parameter cytograms of forward angle light scatter and green fluorescence (LP 515 nm filter) are recorded. Cells other than red cells are gated out on the basis of size, and an interactive computer algorithm is used to fit the fluorescence distribution and determine the population of reticulocytes (*20*).

The staining methods described above refer to the originally published procedure. Modifications and adaptions of these techniques for routine clinical application have been proposed by others (*4,5,19*). A thiazole orange staining kit can be obtained from Becton Dickinson and allows flow cytometric counting of reticulocytes on the FACStar and FACScan using special software.

In most cases flow cytometric analysis of RNA stained red cells involves the interpretation of correlated histograms of size (forward scatter or Coulter volume) versus fluorescence. These two-parameter size–fluorescence histograms depend on the staining method and on the instrument that is used, but basically

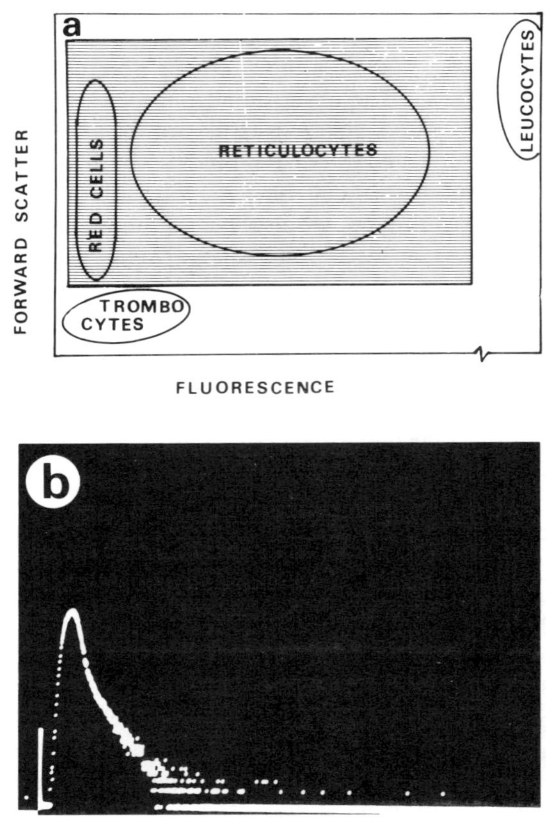

Fig. 2. (a) Schematic representation of a size–fluorescence cytogram observed for formalin fixed and pyronin Y stained blood cells as used for flow cytometric counting of reticulocytes. The window for gated analysis is shown. (b) Fluorescence distribution of red cells obtained by gated flow cytometric analysis. Mature red cells are found in the major population at the left, whereas the reticulocyte population extends to the right with increasing RNA content. (Ordinate: log number of cells.)

contain mature red cells, reticulocytes, nucleated cells (mainly leukocytes), and platelets. Artefacts such as cell fragments and debris may be observed as well. A schematic representation is given in Fig. 2(a). Proper identification of the red cell population including the reticulocytes is of utmost importance. A gated analysis by which the influence of all other objects is eliminated is then carried out, and a fluorescence distribution is recorded either on a linear or logarithmic scale. The major peak at the left in the fluorescence histogram represents the mature red cells, whereas the population extending to the right represents reticulocytes with increasing amounts of RNA (Fig. 2(b)).

C. Comparison of Methods for Flow Cytometric Counting of Reticulocytes

Methods for reticulocyte staining and flow cytometric counting should fulfil certain requirements. First of all the staining procedure should be fast, reproducible and simple, especially for routine applications. Flow cytometric enumeration is expected to be accurate, sensitive and reproducible. Obviously, the information that is obtained must reflect the red cell population and should not be influenced by interfering artefacts such as platelets, leukocytes or cell debris under normal and abnormal hematological conditions.

Pyronin Y does not fulfil the requirement for staining, since it requires fixation with formaldehyde, preferably overnight, which is not acceptable in a routine clinical environment. For experimental purposes, however, the pyronin Y method may have some advantages. Samples taken at various time intervals can be stored and analyzed at the same time. Moreover, it has been shown that pyronin Y fluorescence of reticulocytes is linearly related to RNA content (*37*) and thus to the degree of maturation of the reticulocyte, something that has not been shown (yet) for other staining methods.

Most of the vital dyes have the advantage that the quantum efficiency is significantly enhanced when they bind to RNA. Especially, thiazole orange has a large fluorescence enhancement (X 3000), and therefore can be excited with medium power light sources. This implies that cheaper flow cytometers can be used. This also holds for staining with acridine orange, that fluoresces brightly red. The specificity of acridine orange staining, however, is known to be sensitive to external factors such as dye concentration, staining time and temperature. Under strictly controlled conditions with well-defined staining kits as provided for use with commercially available flow cytometers reproducible results are obtained that are comparable to conventional manual counts.

A critical step in the flow cytometric analysis is the discrimination between mature red cells and the reticulocytes. Basically, there are two approaches: (1) the reticulocyte population is calculated from the total fluorescence histogram by mathematical curve analysis and fitting procedures (*20,37,41*), or (2) by determining a discrimination level either quite arbitrarily based on experience, or better, on the basis of an unstained control sample (*32*).

For routine counting of reticulocytes using thiazole orange some groups have reported reproducible and reliable results by using one fixed discrimination level, that is determined for a few control samples but not for each sample individually (van Hove *et al.* personal communication). This obviously reduces analysis time and therefore costs. However, such an approach is only advisable if proper control and reference samples are analyzed in each series to assure accuracy and reproducibility.

A general problem in flow cytometric reticulocyte counting is the interference of objects other than reticulocytes. Examples are large platelets that often take up fluorescent dye, Howell–Jolly bodies, malaria parasites, nucleated red blood cells and (fragments of) leukocytes. Gating of the red cell population on the basis of light scatter or cell volume is sufficiently effective in most cases to eliminate these objects. However, under abnormal hematological conditions events other than red cells may occur in the red cells window. Careful examination of the correlated two-parameter fluorescence–scatter histograms is therefore absolutely necessary to detect these cases. In the case of thiazole orange staining contaminating objects may be eliminated on the basis of their slightly different emission characteristics after staining (*19*). In practice this is achieved by analyzing green versus red fluorescence, and subsequent gating of the various subpopulations that deviate from the theoretically expected position on the diagonal (Fig. 3). Interference of (large) platelets as seen in patients with myelofibrosis or idiopathic thrombocytopenia with the gate setting for reticulocytes analysis has been observed as well (M. Bins, personal communications). The various steps of the thiazole orange method are shown in Fig. 4.

Lastly, DiCO1 [3] staining of reticulocytes is of special interest in research studies in experimental hematology. The discrimination of DiCO1 [3] stained red cells and reticulocytes is probably a function of RNA content, membrane potential and size. The fluorescence properties of DiOC1 [3] allow combination with Hoechst 33342, thereby offering a procedure to perform combined DNA analysis of nucleated precursor cells. However, the iterative and interactive data analysis procedure that is required to calculate reticulocyte percentages is rather complicated. The DiCO1 [3] method is not often used for routine counting of reticulocytes, since the population of immature red cells that is stained by DiOC1 [3] is not strictly equivalent to reticulocytes determined by brilliant cresyl blue. Therefore, as stated by the authors, "any routine analysis with DiOC1 [3] would require re-education of the community receiving the information" (*20*).

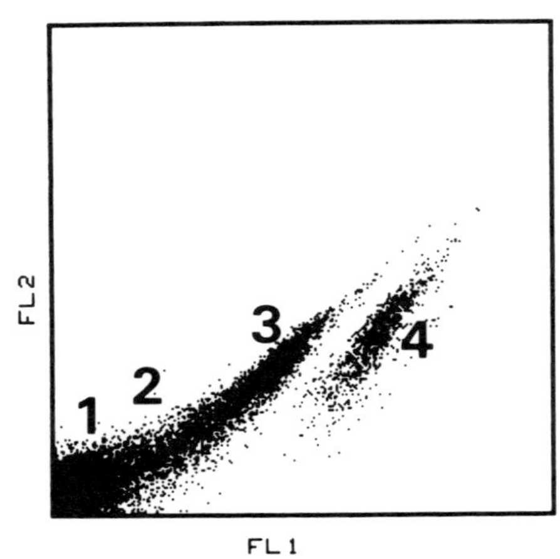

Fig. 3. Cytogram of FL1 (green) versus FL2 (red) fluorescence of thiazole orange stained blood cells. Cells differ slightly in fluorescence emission and autofluorescence properties, which can be used for proper gating of red cells only. (1) Red cells; (2) region where red cells containing Howell–Jolly bodies can be seen (especially at somewhat higher FL2 values); (3) reticulocytes; (4) nucleated red blood cells and leukocytes (data provided by Dr L. van Hove, Leuven, Belgium).

An aspect that has not been sufficiently addressed so far is the standardization of the staining and the flow cytometric analysis. At present, a well-defined internal standard that is accepted for routine counting by clinical hematologists is missing. Such a standard may consist of an artificial mixture of cells, which have scatter properties that differ from red cells, and which contain different but known amounts of RNA. Ideally, the standard cells could be mixed with the blood sample and simultaneously stained for RNA. Such a procedure would highly facilitate the interpretation of RNA distributions and the setting of discrimination levels, especially if a standard could be developed that marks the discrimination level.

A working party of the International Committee for Standardization in Haematology and of the National Committee for Clinical Laboratory Standards is considering the use of a mixture of microspheres with defined amounts of RNA for this purpose. Their suitability is currently being evaluated.

D. Applications

a. *Routine Clinical Hematology*

The methods that have been developed for flow cytometric counting of reticulocytes have

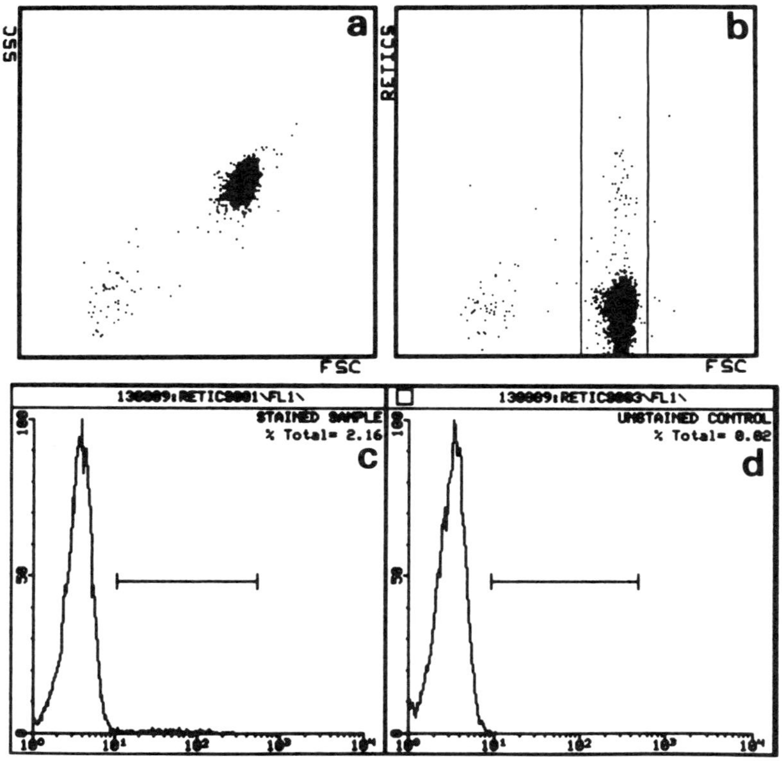

Fig. 4. Flow cytometric counting of reticulocytes using thiazole orange staining: (a) cytogram of forward and perpendicular light scatter used to identify red cells (gate setting); (b) forward light scatter versus fluorescence allows setting of the region in which the percentage of reticulocytes is determined; (c) fluorescence histogram and calculated percentage of reticulocytes (2.16%); (d) unstained control sample used for determination of the reticulocyte threshold (for exact procedure see *25* and *32*).

been adapted with respect to staining conditions and data analysis in such a way that percentages are obtained that correlate well with conventional manual counts using brilliant cresyl blue or new methylene blue.

One of the first techniques that fulfilled this requirement was the acridine orange method. It was used initially for simultaneous staining of reticulocytes and platelets and extensively tested under clinical conditions. An excellent correlation with conventional counts has been reported (e.g. $r = 0.989$ and $r = 0.923$), depending on the type of blood samples (*41*).

The same holds for the more recently introduced staining methods. As an example, the thioflavine T method correlates well with new methylene blue manual count ($r = 0.97$) (*32*); the thiazole orange method behaves very similarly ($r = 0.97$). However, these data have been obtained under non-routine conditions. Both methods, although slightly adapted, were also extensively tested in a routine clinical laboratory and normal values were established in a group of healthy blood bank donors (*4,5*). A mean absolute number of reticulocytes was found of 84.6 and 86.8 ($\times 1000/\mu l$ of whole blood) for thioflavin T and thiazole orange, respectively. Comparison of the thioflavin T and thiazole orange method with conventional NMB counts showed a coefficient of correlation of 0.74 and 0.85, respectively. Most investigators report slightly higher values for the fluorescence methods in comparison with manual counts, which is explained by the fact that the more sensitive fluorescence techniques detect more mature reticulocytes with minor amounts of RNA. The precision of the flow cytometric methods, as evaluated by performing 30 independent replicate determinations, varied from a coefficient of variation (CV) of 28% for decreased reticulocytes (e.g. 0.5%), and a CV of 9% for normal levels (e.g. 1.5%) to a CV of 6% for reticulocyte percentages higher than 5% (*4,5*).

Recently, the performance of a new optoelectronic reticulocyte counter, e.g. the R-1000 system (Sysmex-TOA), has been evaluated (*23*). A correlation with hand counts of $r = 0.941$ was found. The reproducibility was examined for 15 specimens each tested five times. Its CV was 5.84%.

Another recent study using a modified thiazole orange method showed an excellent correlation with manual counts ($r = 0.967$), a similar precision as discussed above, and moreover independence with respect to the type of flow cytometer used (*19*).

In conclusion, the kits for staining of reticulocytes for flow cytometric enumeration provided by Becton Dickinson (thiazole orange), Coulter Electronics and Ortho Diagnostic Systems (acridine orange), Sysmex-TOA (auramine O), and other systems and methods not discussed here, allow reticulocyte counting with much greater accuracy and reproducibility than possible with hand counting methods (*33*).

b. *Monitoring in Bone Marrow Transplantation*

Patients suffering from aplastic anemia (AA), severe combined immunodeficiency (SCID) or acute myeloid leukemia (AML) are often treated with heterologous bone marrow transplantation (BMT). Also solid tumours of non-hematological origin are increasingly treated by autologous BMT to circumvent the problem of bone marrow toxicity, thereby enabling therapy with higher doses of chemotherapy or radiation. Such patients achieve a state of reversible aplasia, the period of which has to be carefully monitored with regard to the prophylactic and therapeutic use of antimicrobial agents and blood products. Hence, laboratory methods that provide early and reliable information about engraftment of a bone marrow transplant contribute to the better clinical management of BMT patients.

Flow cytometric counting of reticulocytes has been used to monitor children suffering from AA, SCID or AML who were treated with BMT (*37,38*). Very low reticulocyte values were observed after the ablative regimen prior to BMT, as expected. Significantly increased numbers of reticulocytes were seen between day 7 and 14. These early appearing reticulocytes contained relatively high amounts of RNA, as expressed in the maturation index (*9,38*), and in all cases indicated engraftment as confirmed several days later (*38*). A typical example is shown in Fig. 5.

Similar studies of autologous bone marrow transplantation in adult patients with AML have been carried out recently using the thiazole orange method. In the majority of patients (12/20) the reticulocyte maturation index was found to be the earliest indicator of bone marrow engraftment (*10*).

c. *Monitoring of Patients under Chemotherapy*

Cytotoxic side effects of the bone marrow represent an important problem during chemotherapeutic treatment of cancer patients. Bone marrow damage is usually evaluated by regular counting of leukocytes and platelets, and by measuring the peripheral blood hemoglobin content during therapy. Reliable information on changes in erythropoietic activity is generally not obtained, because of the insensitivity of hemoglobin measurements due to the relatively long life time of erythrocytes (120 days), and because of the imperfection of manual reticulocytes counts discussed above. Effects of drugs that preferentially affect

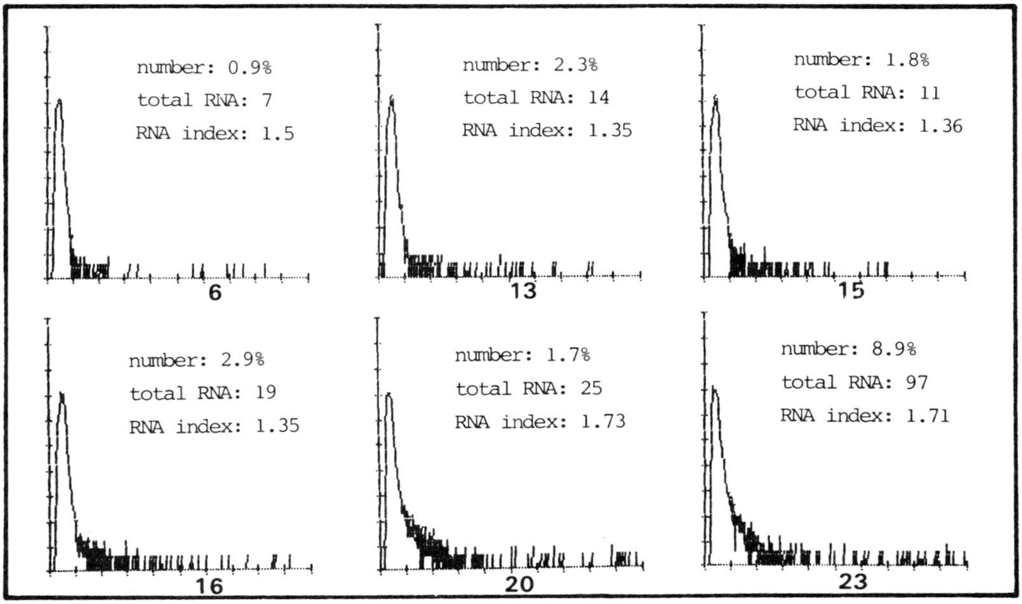

Fig. 5. RNA histograms of fluorescently stained human reticulocytes. Pyronin Y stained blood samples from a patient with aplastic anemia taken at various days (6, 13, 15, 16, 20, 23) after successful bone marrow transplantation. Note large numbers of immature reticulocytes (Heilmeyer stage 1 and 2). For calculations of percentage and maturation parameters see *38*.

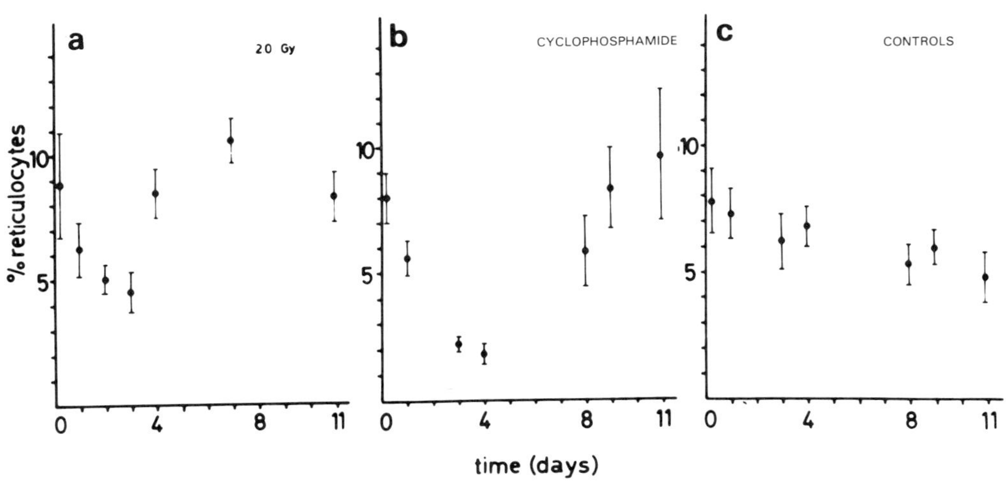

Fig. 6. Changes of reticulocyte values determined by flow cytometry with time (pyronin Y method) as a consequence of (a) 2.0 Gy irradiation and (b) cyclophosphamide intraveneously (70 mg/kg), both given to rats; (c) untreated control rats. Note the decrease in reticulocyte numbers after irradiation or drug treatment, followed by an increase after day 4. All data points represent average reticulocyte values of five rats.

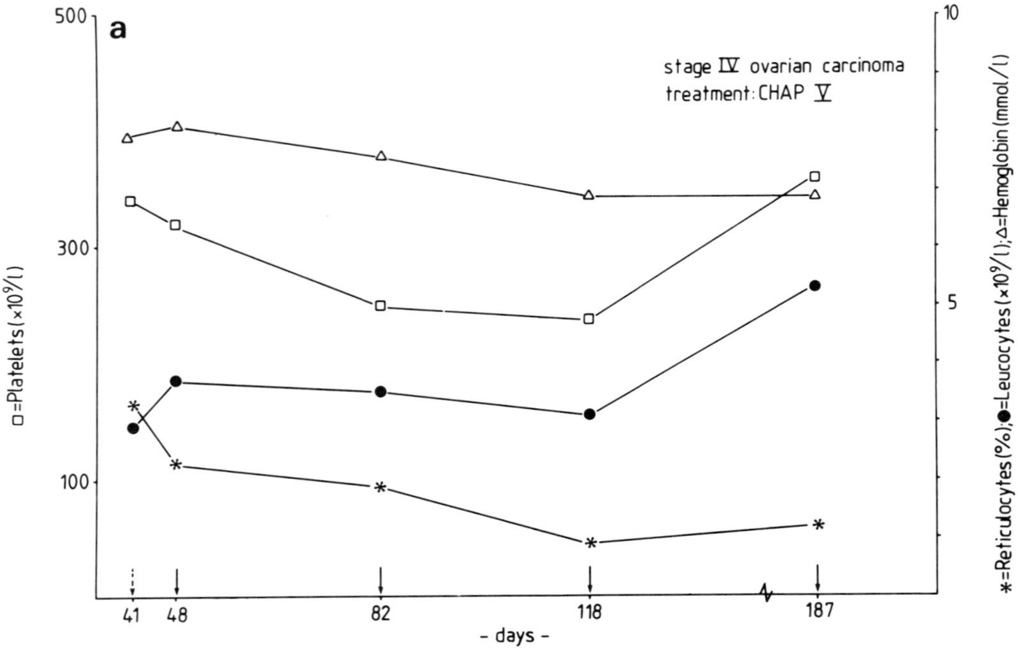

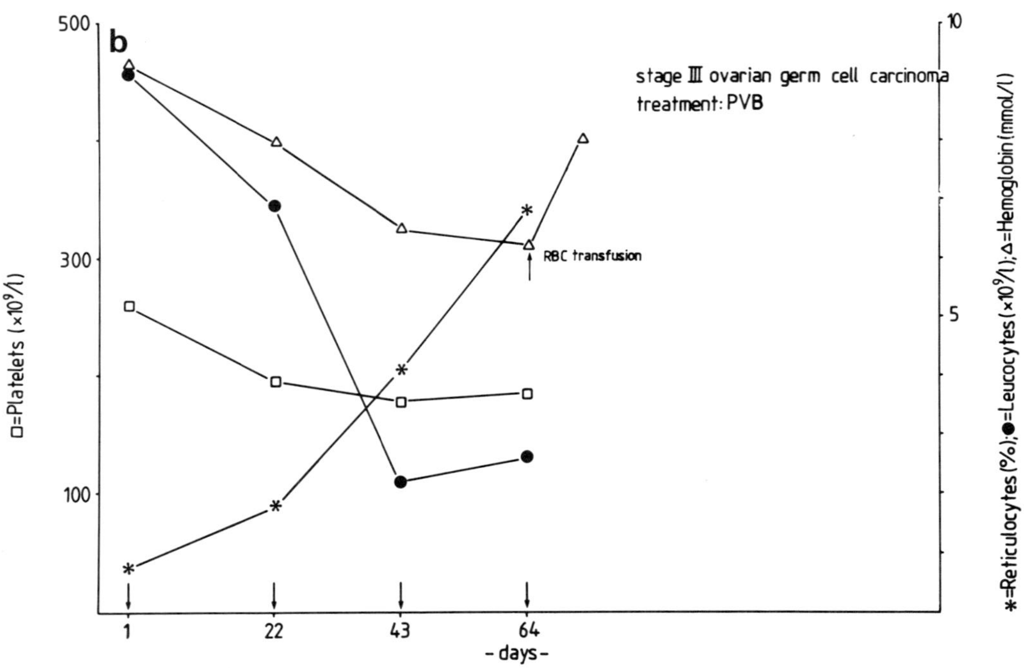

Fig. 7. Changes in number of peripheral blood cells in patients with ovarian carcinoma treated with (a) CHAP 5 (cyclophosphamide, hexamethylmelanine, adriamycin, *cis*-platinum) and (b) PVB (*cis*-platinum, vinblastin, bleomycin). Arrows on abscissa indicate the first day of each chemotherapy period; blood cell counts were performed on these days.

erythropoiesis can therefore not be studied adequately.

Flow cytometric counting of reticulocytes in principle allows the reproducible detection of minor but significant changes in even low numbers of reticulocytes. Model studies in rats using the pyronin Y method have shown that the effect of irradiation doses as low as 0.5 Gy on the bone marrow can be detected (*39*). Also the cytotoxic effect of cyclophosphamide given to rats (70 mg/kg) has been observed by studying changes in the reticulocyte population. Irradiation or drug treatment results in an initial decrease of the percentage of reticulocytes (day 3–4) followed by compensation and an overshoot approximately on day 8 (Fig. 6). At this time relatively immature reticulocytes appear in peripheral blood, as can be concluded from their high average fluorescence (RNA content). Similar profiles have been obtained for rats treated with other drugs (e.g. vincristin (0.2 mg/kg) and mitomycin C (1.0 mg/ml)), although the effect was less prominent (*39*). Comparable results on changes induced by actinomycin-D in rats have been reported using the acridine orange method (*35*).

The model studies in rats indicate that reticulocyte counting may be of use in monitoring patients under chemotherapy. A systematic study to establish the clinical value of such a monitoring procedure has not been reported yet. Some initial attempts indicate that reticulocyte counting may be of use, especially in those cases where drugs are used such as *cis*-platinum, that are considered as specific affectors of erythropoiesis. Figure 7 shows examples of the long-term monitoring of patients with ovarian carcinoma treated with CHAP V (cyclophosphamide, hexamethylmelanine, adriamycin, *cis*-platinum) or PVB (*cis*-platinum, vinblastin, bleomycin). The first example (Fig. 7(a)) shows more or less constant cell counts. The second example (Fig. 7(b)) however shows strongly increasing reticulocyte percentages, whereas the hemoglobin concentration only slowly declines. This situation may indicate a high erythropoietic activity to compensate for specific red cell loss. Red blood cell transfusions were found to be necessary at day 64.

Increasing reticulocyte counts with decreasing leukocytes and platelet counts are difficult to explain in terms of general damage of early precursor cells in the bone marrow. An explanation may be the fact that *cis*-platinum interacts with mature red cells and shortens their life span (*15*). The subsequent increase of reticulocytes would then be a reaction of the (although damaged) bone marrow to compensate for this red cell loss. By interpretation of these reticulocyte counts it should be noted that the normal values found with the pyronin Y method are somewhat higher than with conventional procedures.

III. FLOW CYTOMETRY OF MATURE ERYTHROCYTES

A. Methods for Analysis of Unstained Erythrocytes

a. *Measurements of Electrical Resistance*

Electrical resistance measurements of biological objects in suspension were introduced by Coulter (*6*). Basically, cells suspended in a conducting fluid are passed through an orifice with a small diameter. Across the orifice a constant current is maintained by two electrodes, positioned at both sides. Each cell, being a relatively poor conductor due to its membrane composition, that passes between the electrodes changes the electrical resistance and therefore, according to Ohm's law, generates a voltage pulse which is easily amplified and measured. This signal is proportional to the volume of the cell, and a volume distribution curve of objects thus can be derived from a recorded pulse height distribution. Measurements of electrical resistance obviously require a difference between the suspension medium and the object. This is readily achieved for live cells. However, it should be realized that cell death or fixation significantly reduces the electrical resistance of cellular membranes. Coulter-type cell counters have an orifice that is adaptable to the size of the objects to achieve optimal resolution of the measurements.

b. Measurements of Light Scatter

When cells in suspension are passed through a beam of light, part of the light is reflected due to the difference in refractive index that generally occurs between the cells and the suspension medium. This light scatter pattern contains important information about the cells. The theoretical phenomena that play a role in the light scatter of small particles are complex. Factors such as particle size, shape, structure, orientation and refractive index all influence the scatter signals, which makes it very difficult to predict the scatter intensity (*21*). In practice, however, light scatter measurements are extremely useful in classifying different cell types. Light scatter is generally measured in the forward direction, giving information about cell size, and in the perpendicular direction, as a parameter related to the number of internal reflection surfaces and thus to nuclear shape or the presence of granules (*11*). An example of a two-angle light scatter histogram is shown in Fig. 8. It is shown that red cells, leukocytes and platelets can be well separated on the basis of their scatter intensity in the forward and perpendicular direction.

The discoid shape of live erythrocytes causes the forward angle light scatter signal to be rather variable due to the various orientation possibilities of the cells in the laser beam. In cases where exact definition of the red cell population is necessary and all interference of for instance large platelets have to be avoided, sphering of live red cells may be performed by swelling under controlled hypotonic conditions (*22*). On the basis of the improved resolution in the two-parameter scatter histogram better discrimination of cell populations is possible. Sphering of red blood cells is currently applied in commercial red cell analyzing systems such was the Technicon H1.

Finally, as for electrical resistance, light scatter intensity changes in the case of cell death or upon treatment of the cells with fixatives.

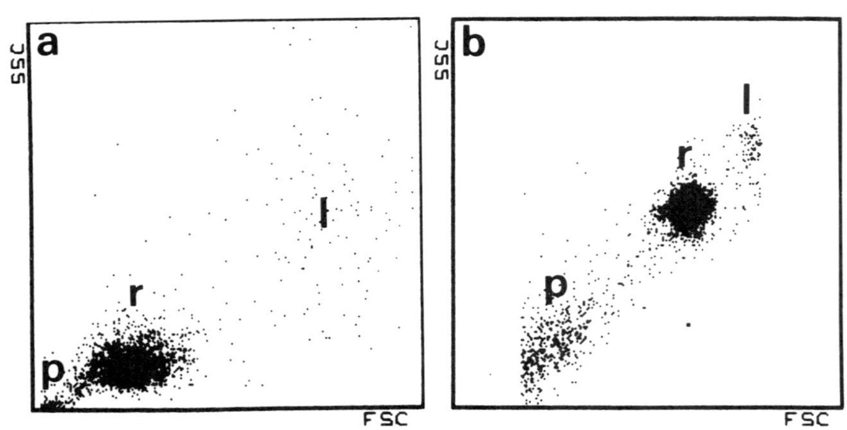

Fig. 8. Separation of the major cell types (p = platelets; r = red blood cells; l = leukocytes) in human peripheral blood on the basis of light scatter. Ascissa: forward angle light scatter (FSC); ordinate: perpendicular light scatter (SSC). For demonstration purposes the sample was enriched for platelets and leukocytes: (a) FSC and SSC linear scale; (b) FSC and SSC logarithmic scale.

B. Clinical Applications

a. *Analysis of Red Cells: the MCV and MCHC Parameters*

Counting of the number of erythrocytes in peripheral blood is one of the oldest and certainly the most frequently performed cytometric test in routine hematology laboratories. It is based in most cases on electrical pulse sizing, as originally developed by W. H. Coulter (*6*). The simplest instruments only allow the discrimination of red cells and other objects on the basis of the change in electrical resistance, and subsequent counting. More advanced instruments generate a distribution of signals that give information about the heterogeneity in size of the population (see below). The principle of cell counting on the basis of electrical pulse resistance is so commonly applied that the technique is only mentioned here for reasons of completeness and not further discussed in detail.

Besides counting erythrocytes, red cell indices are determined to characterize the red cell population. These indices are the mean cell volume (MCV), the mean cell hemoglobin concentration (MCHC) and the mean cell hemoglobin mass (MCH) (*42*). Flow cytometric techniques have been employed to determine the cell volume distribution on the basis of light scatter measurements or by electronic resistive pulse sizing, as performed in the Coulter-type instruments, and then subsequently derive the MCV. The accuracy of individual cell volume measurements is limited by factors that are related to the unknown and variable hemoglobin content of the individual cell. A varying hemoglobin changes the refractive index of the cell and thereby the light scatter intensity. Measurements based on electrical sizing depend on the deformation of the cells as they pass through the sensor orifice. The deformability is determined by the cell fluid viscosity and therefore also affected by the variable hemoglobin content.

The interference of hemoglobin can be overcome by simultaneous measurements of both volume and hemoglobin content of each cell (*28*). This is performed by measuring the light scatter intensity in two non-overlapping directions (2.5–3.5° and 5–15°) (*21,40*). To avoid the effect of cell shape and orientation the red blood cells are sphered as mentioned above (*22*). Derivation of the red cell volume and the hemoglobin content is achieved by applying the Mie theory of scattering, which allows the calculation of size and internal density based on light scatter measurements at two angles (*21*). A good correlation between this method and conventional techniques for determining red cell volume and mean hemoglobin content has been shown and an increasing number of laboratories are utilizing this method (*31,40*). An example of hematological parameters that can be derived from flow cytometric scatter cytograms is shown in Fig. 9.

The real clinical value of independent measurements of cell volume and hemoglobin content by flow cytometry has not been determined yet, since the method has only been introduced recently. With these new parameters, however, anemias can be classified on the basis of the width of the red cell distribution (RDW), describing the heterogeneity of the population, and on the width of the hemoglobin distribution (HDW), which reflects the degree of anisochromia, an indicator of modifications in hemoglobin synthesis. This information is of particular interest in the case of anemia due to iron deficiency.

For a long time a low MCV value has been considered as the earliest and most sensitive indicator of iron deficiency. However, it has been shown that a low MCV value is preceded by changes in red cell size (*12*). Therefore the early detection of anisocytosis and microcytosis has become increasingly important in this respect. In the case of anemia due to iron deficiency, elevated values for RDW and HDW are found. Often a population of hypochromatic cells can be observed in the two-dimensional histogram before decreased indices or microcytosis are noted. Upon treatment of anemia due to iron deficiency a normocytic cell population appears, whereas the microcytic and hypochromatic cells become less frequent (*13*).

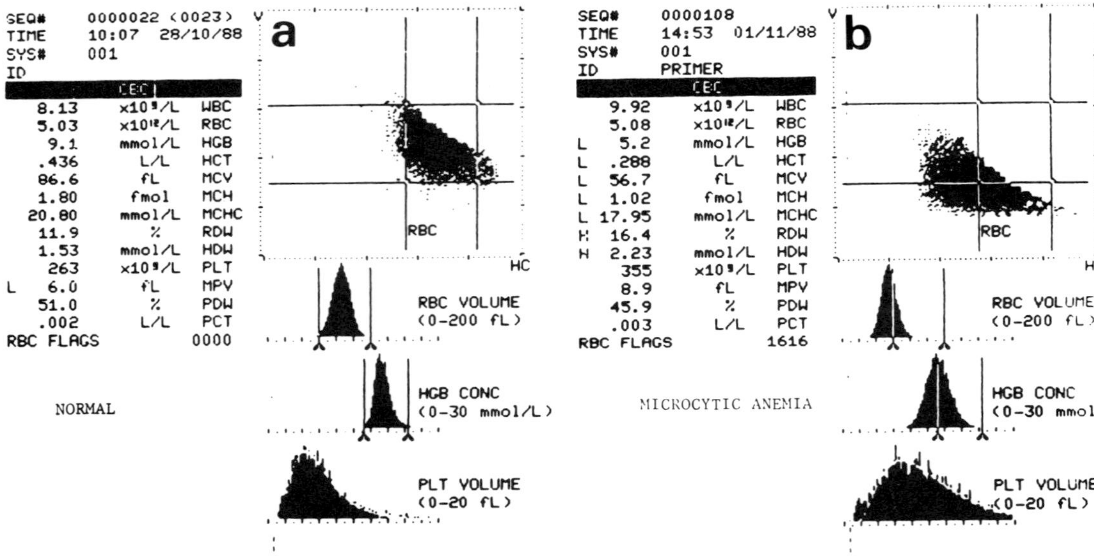

Fig. 9. Examples of flow cytometric analysis of red blood cells with a Technicon H1. Parameters such as HC (hemoglobin content) and V (cell volume) are calculated from two-parameter light scatter histograms and plotted. Other hematological parameters are listed as numerical data or plotted as single parameter histograms. (a) Normal blood sample; (b) blood sample of a patient with microcytic anemia due to iron deficiency; note the shift of the red cell population in the HC–V histogram, resulting in higher values for RDW and HDW in the microcytic anemia (data provided by Dr G. J. den Ottolander, Leiden, Netherlands).

Generally, small changes in the two-dimensional histogram reflect early changes in the blood cell population, that are not detected by the reference methods for determining MCV and MCHC. An example is a non-symmetrical erythrocyte size distribution exhibiting a shoulder on the left which is indicative of the presence of schistocytes as found in patients with cardiac valve protheses or microangiopathies, as well in patients who were transplanted (*13*).

In hemoglobinopathies (heterozygous beta-thalassemias) a normal RDW parameter is observed, whereas the HDW value indicates anisochromia. This has been claimed to be useful in distinguishing hypochromatic microcytic anemia from hemoglobinopathy, although cases involving anemia caused by severe iron deficiency that have become very microcytic and homogeneous may give misleading results (*13*).

In general, measurements of HDW and RDW allow the monitoring of the physiopathologic course of anemia: an abnormal HDW value is indicative for a disorder that stems from hemoglobin production, regardless of its origin, whereas a modified RDW parameter indicates a disorder from other forms of erythropoiesis (deficiency in DNA synthesis), or medullary regeneration (*13*).

C. Membrane Antigens and Receptors

During maturation from reticulocyte to mature red cell the expression of the transferrin receptor (TFR) (*24*) decreases, while glycophorin A (GPA) expression (*14*), that starts at the erythroblast stage, is mostly constant throughout further maturation. Reticulocytes are therefore TFR^+/GPA^+, whereas mature red cells are TFR^-/GPA^+ (*26*). Antibodies against the TRF have occasionally been used to determine the percentage of reticulocytes, in a more experimental context rather than for routine clinical studies. A comparison between the acridine orange method and TFR staining revealed that AO stained all RNA containing cells (including relatively mature reticulocytes), whereas antibodies against TFR showed positive staining with the immature reticulocytes only, which possess relatively high density of cell surface receptors (*35*).

A detailed study of the typical antigens and receptors on the erythrocyte membrane is given elsewhere in this book (Chapters 1.3 and 3.4), and therefore only briefly mentioned here.

IV. CONCLUSIONS AND PROSPECTS

In clinical hematology two major applications of flow cytometry, as far as red cell analysis is concerned, have been introduced in the last few years: the counting of reticulocytes and the analysis of the red cell indices MCV and MCHC as studied by measurements of RDW and HDW. In experimental hematology the situation is less focused, particularly since many monoclonal antibodies have become available that are directed against defined red cell membrane proteins, and are used to study the fundamental aspects of normal and abnormal erythropoiesis (see Chapters 1.3 and 3.4).

The methods introduced for counting of reticulocytes by flow cytometry so far all aimed at reticulocyte numbers that were in agreement with the data provided by manual counting methods. Obviously, the flow cytometric methods are faster, statistically more accurate, and in some cases more sensitive. To order to achieve a good correlation with manual counts the conditions for staining and data analysis were adapted.

It should be realized however that flow cytometric analysis of reticulocyte provides data that have been hardly used so far, and of which the clinical value has only been incidently demonstrated. An example is the RNA content, or maturation index of the reticulocyte population (*9,38*). Examples of this parameter in transplantation have been discussed. An advantage of such a test is that the RNA distribution of the entire red cell population (mature red cells and reticulocytes) is addressed, avoiding the issue of mathematical analysis of the data, or setting the discrimination level between red cells and reticulocytes. The definition of a reticulocyte "as a red cell with no DNA and stainable amounts of RNA" depends on the sensitivity of the entire cytochemical and cytometric procedure. In other words, a method with a theoretical detection sensitivity of only one RNA molecule most likely would result in reticulocyte percentages that approach 100%, since there is a gradual loss of RNA during maturation towards mature red cell.

Nevertheless, there is no doubt that the methods developed so far have proven their value in practice, and that their routine use in clinical hematology will become further established, especially if proper standards become available. It is likely that the clinical need to count reticulocytes will change. Especially, the reproducible counting of percentages of reticulocyte below 1% opens new possibilities for follow-up of patients under chemotherapy or in cases of transplantation. The real clinical value of such information is unknown, since means for counting very low percentages of reticulocytes were practically non-existent.

A more or less similar situation holds for the analysis of two-parameter scatter histograms of red blood cells. The current concept of the clinical hematologist is that conventional parameters such as hemoglobin content, hematocrit and MCV will remain of great importance in the diagnosis and treatment of anemia or polycythemia. New parameters such as the red cell distribution width (RDW) and the hemoglobin distribution width (HDW) are not likely to change this concept. However, especially for research studies, they provide additional information about normal and abnormal erythropoiesis and may reveal more of the fundamental aspects of this complex mechanism. Sufficient data, obtained with commercial-type flow cytometers, will become available in the near future. Then the real clinical value of these tests can be evaluated.

REFERENCES

1. Alpen, E. L., and Granmore, D. (1959). Observations on the regulation of erythropoiesis and on cellular dynamics by Fe59 autoradiography. *In* "The Kinetics of Cellular Proliferation" (F. Stohlman Jr, ed.), pp. 290–305. Grune and Stratton, New York.
2. Bessis, M. (1977). "Blood Smears Reinterpreted." Springer, Berlin.
3. Brecher, G. (1949). New methylene blue as a reticulocyte stain. *Am. J. Clin. Pathol.* **19**, 895–896.
4. Corash, L., Rheinschmidt, M., Lieu, S., Meers, P., and Brew, E. (1988). Fluorescence-activated flow cytometry in the hematology clinical laboratory. *Cytometry* (Suppl.) **3**, 60–64.
5. Corash, L., Rheinschmidt, M., Lieu, S., Meers, P., and Brew, E. (1988). Enumeration of reticulocytes using fluorescence-activated flow cytometry. *Pathol. Immunopathol. Res.* **7**, 381–394.
6. Coulter, W. H. (1956). High speed automatic blood cell counter and analyser. *Proc. Natl. El. Conf.* **12**, 1034–1040.
7. Darzynkiewicz, Z., and Kapuscinski, J. (1990). *In* "Flow Cytometry and Sorting" (M. R. Melamed, T. Lindmo and M. L. Mendelsohn, eds). Alan R. Liss, New York.
8. Darzynkiewicz, Z., Kapuscinski, J., Traganos, F., and Crissman, H. A. (1987). Application of pyronin Y(G) in cytochemistry of nucleic acids. *Cytometry* **8**, 138–145.
9. Davis, B. H., and Bigelow, N. C. (1989). Flow cytometric reticulocyte quantification using thiazole orange provides a clinically useful reticulocyte maturity index. *Arch. Pathol. Lab. Med.* **113**, 684–689.
10. Davis, B. H., Bigelow, N., Ball, E. D., Mills, L., and Cornwell, G. G. (1989). Utility of flow cytometric reticulocyte quantification as predictor of engraftment in autologous bone marrow transplantation. *Am. J. Haematol.* **32**(2), 81–87.
11. Engh, G. van den, and Visser, J. W. M. (1984). Flow cytometry in experimental hematology. *Bibl. Haematol.* **48**, 42–62.
12. England, J. M., Ward, S. M., and Down, M. C. (1976). Microcytes, anisocytosis and the red cell indices in iron deficiency. *Br. J. Haematol.* **34**, 589–597.
13. Fossat, C., David, M., Harle, J. R., Sainty, D., Horschowski, N., Verdot, J. J., and Mongin, M. (1987). New parameters in erythrocyte counting. *Arch. Pathol. Lab. Med.* **111**, 1150–1154.
14. Gahmberg, C. G., Jokinen, M., and Andersson, L. C. (1978). Expression of the major sialoglycoprotein (glycophorin) on erythroid cells in human marrow. *Blood* **52**, 379–387.
15. Getaz, E. P., Beckley, S., Fitzpatrick, J., and Dozier, A. (1980). Cisplatin-induced haemolysis. *N. Engl. J. Med.* **302**, 334–335.
16. Gilmer, P. R., and Koepke, J. A. (1976). The reticulocyte: an approach to definition. *Am. J. Clin. Pathol.* **66**, 262–267.
17. Heilmeyer, L., and Westhaeuser, R. (1932). Reifungsstadien an ueberlebenden Reticulozyten *in vitro* und ihre Bedeutung fuer die Schaetzung der taeglichen Haemoglobin-Produktion *in vivo*. *Z. Klin. Med.* **121**, 361–655.
18. Hillman, R. S., and Finch, C. A. (1967). Erythropoiesis: normal and abnormal. *Semin. Hematol.* **4**, 327–348.
19. Hove, L. van, Goossens, W., van Duppen, V., and Verwilghen, R. L. (1990). Reticulocyte count using thiazole orange. *Clin. Lab. Haematol.* **12**, 287–299.
20. Jacobberger, J. W., Horan, P. K., and Hare, D. J. (1984). Flow cytometric analysis of blood cells stained with cyanine dye DiOC1[3]. *Cytometry* **5**, 589–600.
21. Jovin, T. M., Morris, S. J., Striker, G., *et al.* (1976). Automatic sizing and separation of particles by ratios of light scattering intensities. *J. Histochem. Cytochem.* **24**, 269–283.
22. Kim, Y. R., and Ornstein, L. (1983). Isovolumetric sphering of erythrocytes for more accurate and precise cell volume measurement by flow cytometry. *Cytometry* **3**, 419–427.
23. Kojima, K., Niri, M., Setoguchi, K., Tsuda, I., and Tatsumi, N. (1989). An automated optoelectronic reticulocyte counter. *Am. J. Clin. Pathol.* **92**, 57–61.
24. Lebman, D., Trucco, M., Bottero, L., Lange, B., Pessano, S., and Rovera, G. (1982). A monoclonal antibody that detects expression of transferrin receptor in human erythroid precursor cells. *Blood* **59**, 671–678.
25. Lee, L. G., Chen, C. H., and Chiu, L. A. (1986). Thiazole orange: a new dye for reticulocyte analysis. *Cytometry* **7**, 508–517.
26. Loken, M. R., Shah, V. O., Dattilio, K. L., and Civin, C. I. (1987). Flow cytometric analysis of human bone marrow: I. Normal erythroid development. *Blood* **69**, 255–263.
27. Melamed, M. R., and Kamentski, L. A. (1975). Automated Cytology. *In* "International Reviews of Experimental Pathology," Vol. 14, pp. 206–219. Academic Press, New York.
28. Mohandas, N., Kim, Y. R., and Tycko, D. (1986). Accurate and independent measurement of volume and hemoglobin concentration of individual cells by laser light scattering. *Blood* **68**, 506–513.

29. Natale, P. J. (1982). Method and reagents for quantitative determination of reticulocytes and platelets in whole blood. US Patent No. 4336029.
30. Perotta, E. L., and Finch, C. A. (1972). The polychromatic erythrocyte. *Am. J. Clin. Pathol.* **57**, 471–474.
31. Ross, D. W., and Bentley, S. A. (1986). Evaluation of an automated hematology system (Technicon H-1). *Arch. Pathol. Lab. Med.* **110**, 803–808.
32. Sage, Jr, B. H., O'Connell, J. P., and Mercolino, T. J. (1983). A rapid vital staining procedure for flow cytometric analysis of human reticulocytes. *Cytometry* **4**, 222–227.
33. Savage, R. A., Skoog, D. P., and Rabinovitch, A. (1985). Analytical inaccuracy and imprecision in reticulocyte counting: a preliminary report from the College of American Pathologists reticulocyte project. *Blood Cells* **11**, 97–112.
34. Schmitz, F. J., and Werner, E. (1986). Optimization of flow cytometric discrimination between reticulocytes and erythrocytes. *Cytometry* **7**, 439–444.
35. Seligman, P. A., Allen, R. H., Kirchanski, S. J., and Natale, P. J. (1983). Automated analysis of reticulocytes using fluorescent staining with both acridine orange and an immunofluorescence technique. *Am. J. Hematol.* **14**, 57–66.
36. Sterling, K., and Gray, S. J. (1950). Determination of the circulating red cell volume in man by radioactive chromium. *J. Clin. Invest.* **29**, 1614–1620.
37. Tanke, H. J., Nieuwenhuis, I. A. B., Koper, G. J. M., Slats, J. C. M., and Ploem, J. S. (1980). Flow cytometry of human reticulocytes based on RNA fluorescence. *Cytometry* **1**, 313–320.
38. Tanke, H. J., Rothbarth, P. H., Vossen, J. M. J. J., Koper, G. J. M., and Ploem, J. S. (1983). Flow cytometry of reticulocytes applied to clinical hematology. *Blood* **61**, 191–197.
39. Tanke, H. J., van Vianen, P. H., Emiliani, F. M. F., Neuteboom, I., de Vogel, N., Tates, A. D., de Bruijn, E. A., and van Oosterom, A. T. (1986). Changes in erythropoiesis due to radiation or chemotherapy as studied by flow cytometric determination of peripheral blood reticulocytes. *Histochemistry* **84**, 544–548.
40. Tycko, D. H., Metz, M. H., Epstein, E. A., and Grinbaum, A. (1985). Flow-cytometric light scattering measurements of red blood cell volume and hemoglobin concentration. *Appl. Opt.* **24**, 1355–1365.
41. Vaughan, W. P., Hall, J., Johnson, K., Dougherty, C., and Peebles, D. (1985). Simultaneous reticulocyte and platelet counting on a clinical flow cytometer. *Am. J. Hematol.* **18**, 385–391.
42. Wintrobe, M. (1974). "Clinical Hematology," 7th edn. Lea and Febiger, Philadelphia.

2.2

Leukocyte Differential Counting

LEON W. M. M. TERSTAPPEN and MICHAEL R. LOKEN

Becton Dickinson Immunocytometry Systems
San Jose
CA, USA

1. INTRODUCTION

This article will review the techniques used in leukocyte differential counting focusing primarily on those employing flow cytometry. We will attempt to define the advantages and limitations of original and current technology in light of new techniques which can be used to address these same problems.

The goal of differential leukocyte counting is to be able to identify and to monitor abnormalities which are reflected in the blood or bone marrow of the patient. The process of differential counting can be divided in three major steps.

(1) To enumerate cells of different lineages and to compare these numbers with normal range values. The lineages of interest include the lymphocytes, monocytes, neutrophils, eosinophils, basophils and erythroid cells.

(2) To discriminate between different maturational stages within cell lineages. The relative proportion of immature and mature cells can be used to identify abnormalities of production or destruction of cells. The difficulty in studying immature cells is that they have lost the characteristics which are used to discriminate between lineages in the mature stage.

(3) To distinguish between normal and neoplastic (leukemic) cells. The identification and characterization of abnormal leukemic cells is required to assign a specific treatment protocol. The discrimination between normal and abnormal cells is important for moni-

FLOW CYTOMETRY IN HEMATOLOGY
ISBN 0–12–432940–3

toring these cells during the treatment of a patient.

There are many problems involved in the differential analysis of leukocytes. Differential counting by morphology is reliable for qualitative analysis of the end stages of maturation of each of the cell lineages. However, reduced precision in classification of cell lineage and/or maturational stage occurs when immature maturational stages are analyzed. Quantification of the analysis is virtually impossible, since only 100 leukocytes are evaluated and the different leukocytes are found in dramatically different frequencies in normal peripheral blood. For example, neutrophils comprise approximately 60% of all leukocytes while basophils contribute less than 1% to the leukocyte count. Automated differential leukocyte counting by flow cytometry has the significant advantage over the morphological differential in that it is quantitative since a large number of leukocytes are counted. The disadvantage of current hematology analyzers is that the discriminators used to distinguish between leukocytes are not precise enough to properly classify each event. As with morphology the precision in classifying cells to various lineages decreases dramatically when immature cells appear in the analysis.

The development of new methods will solve some of the problems involved in current leukocyte differentiation. New methods to discriminate between cells have been described which extend the analysis of leukocytes to highly abnormal states and to the automated assessment of differential bone marrow counting (*63–65,67*). The two significant departures from current technology in the analysis of leukocytes in peripheral blood and bone marrow include: (1) simultaneous multidimensional analysis of cells as they flow through the instrument and (2) the application of monoclonal antibodies to identify normal and abnormal appearance of cell surface antigens. These two improvements can extend the analysis of white blood cells and can provide a more quantitative description of both normal and abnormal hematopoiesis.

II. LEUKOCYTE DIFFERENTIAL COUNTING BY MORPHOLOGY

Although the study of leukocyte differentiation could be traced back to the invention of the light microscope by Antoni van Leeuwenhoek in 1674, it was Paul Ehrlich who introduced the staining of blood cells on a glass slide two hundred years later (*19*). Mixtures of acidic and basic dyes were used to distinguish between the acido(eosino)philic, neutrophilic and basophilic components of the cells (*19,58,71*). This fundamental concept is currently still used for almost all morphologic examination of blood smears.

In the years following Ehrlich's discovery a variety of cytochemical stains were introduced for studying blood cells. The most commonly used cytochemical dyes are the Romanowsky dyes (Wrights, Giemsa, May Grunwald) which aid in discriminating the different cell lineages (i.e. neutrophils, eosinophils, basophils, monocytes, lymphocytes and nucleated red cells). The criteria used for cell classification are nuclear to cytoplasmic ratio, cell and nuclear shape, size and number of granules, differences in color of all the cell components and context in which the cell is seen. The visual evaluation of the more immature stages of maturation introduces more uncertainty with substantial interobserver variations in the assignation of cells (*29*). This is inevitable since morphologies of cells gradually change during the maturational process. The assignment of a discrete maturational stage (e.g. myeloblasts, promyelocytes, etc.) for individual cells along a continuous process is somewhat arbitrary with the criteria for classification differing from observer to observer.

The Romanowsky dyes have proved insufficient to discriminate between the most immature cells of the different lineages (*23,71,75*). As a result, cytochemical dyes staining enzymes specific for a cell lineage were introduced (myeloperoxidase, periodic acid-schiff (PAS), Sudan black, acid phosphatase, α-naphtyl butyrate esterase, naphtol ASD chloroacetate esterase) (*23,58,71,75*). These dyes are especially useful for lineage classification within the acute leukemias. The cytochemical stains are useful for lineage assignment but provide little information with respect to maturational differences between cells of a single lineage (*23,58,71,75*).

The lineage assignment of individual cells can be adequately made when the most mature stages of maturation are concerned. However, accurate quantitation of these cells in blood smears is difficult because of the large differences in frequencies of the cell populations. Quantitative visual differentials are routinely performed on 100–200 nucleated cells (the approximate number of nucleated cells found on a single smear prepared from normal blood). The error in counting the most frequent leukocytes (i.e. lymphocytes and neutrophils) is larger than 25%, while the counting error for cells occurring in a frequency less than 10% (i.e. eosinophils and monocytes) can go up to 100% (*28,48*). Obtaining quantitative results for cells occurring in frequencies below 1% (i.e. nucleated red cells or basophils) is impossible since only 100–200 cells are evaluated in the first place (*17,40,41,47,49,50,51*). Even with these limitations, the visual differential is the standard for comparing results obtained with automated hematology analyzers (*18,45*). The role of visual examination of blood smears remains important in the clinical laboratory since up to 25% of the blood samples run on automated hematology analyzers are flagged as abnormal and must be manually assessed (*4,47*).

III. AUTOMATED LEUKOCYTE DIFFERENTIAL COUNTING

A. Principles

Two alternative technologies, flow cytometry and image analysis, are used for automated leukocyte differential counting. Image analysis, in principle, automates or replicates the visual differential. Blood smears are analyzed by high speed optics on a computer-controlled stage. Algorithms apply similar criteria as the visual differentials on the cells thereby enumerating cell subpopulations in a more reproducible and quantitative fashion (*2*). In the 1970s and early 1980s several commercial image-based leukocyte differential counters were available (Lark from Corning Medical Instruments, Hematrak from Geometric Data, Diff 4 from Coulter Electronics, ADC-500 from Abott Diagnostics) (*3,42*). Currently no commercial leukocyte differential counters based upon image analysis are sold. They were withdrawn from the market place as a result of competition from the flow cytometer-based leukocyte differential counters which were more rapid and more complete (J. Bacus, personal communications).

Two basic principles have been used for detection of cells in flow-based hematology analyzers, electric impedance measurements (Coulter principle) and optical measurements. Both techniques use the hydrodynamic focusing introduced by Crosland-Taylor by which a narrow core fluid containing the cells is passed through the center region of a laminar wider sheath flow. This results in the passage of physically separate individual cells through an observation point.

For electric impedance measurements, introduced by Coulter in 1953, cells flow through a small orifice across which a low current electric field is applied (*13,14*). A cell passing through this orifice will momentarily change the electric impedance which is proportional to the volume of the cell. In 1970 an oscillating, radio frequency (RF) current was introduced as a parameter in addition to the constant current fields (*15*). While the constant current fields go around a blood cell, some radio frequency current also goes through the cell. The ease at which the radio frequency goes through a cell, "electrical transparency", is a measure of density and composition of material within the cell. The combination of both parameters improves the distinction between different leukocytes. Terms, such as opacity, are mathematically derived from these two parameters. Currently the major focus using this technology is in the preparation of cells in order to maximize the separation between different leukocyte populations identified by both electrical parameters.

For optical measurements a light source is focused on the observation point and the resulting light scattered or absorbed by the cells passing the observation point is measured. Absorbed light is measured directly from the undiffracted light beam. This is dependent on differences in absorbing materials at the illuminating wavelength. For instance, erythrocytes can be distinguished from leukocytes based on the differences in hemoglobin when an arc lamp is used as a light source and hemoglobin absorption at 420 nm is measured (*12*,*57*). Cell discrimination based upon differences in absorption of cells stained with cytochemical dyes is an approach used in several hematology analyzers. Alcian blue was used in the Technicon Hemalog D and H-6000 to distinguish between basophilic granulocytes (whose granules stain with Alcian blue) and other leukocytes which did not stain with this dye (*24*,*39*). Peroxidase is used in the Hemalog D, H-6000 and H-1 (Technicon) to distinguish between lymphocytes (unstained), monocytes (dimly stained) and neutrophils and eosinophils (brightly stained) (*7*,*10*,*24*,*39*,*59*).

Light scattering can also be used to discriminate between different leukocyte populations (*53*,*62*,*72*). In this technique the undiffracted light beam is blocked and light is collected at different angles with respect to the incident light beam. Three angles have been traditionally used, forward light scattering, orthogonal light scattering and back scattering (*9*,*36*,*37*, *52*,*53*,*60*,*62*,*66*,*72*). Forward light scattering is collected in the direction of the incident light beam at angles usually between 0.5° and 20°. The amplitude of the signal at narrow angles is dependent predominantly on cell size (*9*,*44*,*52*). Forward light scattering can also be used to distinguish between cells with a damaged membrane and living cells (*27*,*32*,*36*). Small differences in the angle at which the forward light scattering is detected can result in large differences of signals in the leukocyte populations (*52*,*62*). Distinction of the major leukocyte populations can be obtained solely by collection of two small angle light scattering parameters (*62*).

Orthogonal light scattering is obtained from light scattered within a cone centered at 90° with respect to the incoming light beam. It provides information of the granularity of the cells (*53*,*64*,*65*,*66*,*72*). This parameter is useful for defining leukocyte differentiation as described in detail in Section IV.

Backward light scattering is obtained when scattered light is collected within a cone of 180° with respect to the direction of the incoming beam. The information provided for leukocyte differentiation has only been exploited in research settings (*60*).

Light scattering measurements have been extended by assessing the degree of polarization relative to the direction of the incident laser light (*21*). A polarizing filter is placed in front of a detector such that all scattered light polarized in the direction of the incident beam is absorbed. The scattered light measured by this detector is referred to as depolarized light scattering. Depolarized orthogonal light scattering measurements appeared to be useful for leukocyte differentiation since eosinophilic granulocytes have distinctly larger depolarized orthogonal light scattering signals as compared to other leukocytes (*21*,*62*).

B. Commercial Instruments

Different combinations of the principles just discussed are applied in the commercially available automated leukocyte differential counters. All these instruments provide an enumeration of red blood cells and platelets and also measure hemoglobin in addition to performing leukocyte differentials. In fact, all instruments providing leukocyte differentials contain at least three measurement channels, a fact not often appreciated by the user confronted by only displays and numbers. We will focus only on the segments of the instruments which provide information regarding leukocyte differentials.

In order to be able to differentiate between leukocytes, current hematology analyzers must eliminate the overwhelming number of erythrocytes which interfere with the measurements. Instruments based on electronic sizing principles require more rigorous lytic procedures to remove the red blood cells as compared with instruments using optical measurements. Lytic procedures such as hypotonic lysis, ammonium chloride lysis and ethylene glycol lysis produce red blood cells which are transparent with respect to optical measurements but not towards impedance measurements (*33*). Erythrocyte lysis and subsequent measurement and analysis are performed automatically and quantitatively in all hematological analyzers. This feature provides a significant advantage over flow cytometers used for immunofluorescence measurements which lack an online automatic cell preparation and do not give absolute cell numbers.

Many of the hematology instruments report a three-part differential, i.e. granulocytes, monocytes and lymphocytes. The instruments using the electronic sizing principle include Coulter T-series, Toa E-series, and Unipath Cell-Dyn 1600. Instruments using optical measurements include Ortho ELT 800 and 1500. The recent challenge for companies producing hematology analyzers has been to build instruments reporting a five-part differential: i.e. neutrophils, eosinophils, basophils, monocytes and lymphocytes.

The first company to achieve this was Technicon with the Hemalog D (*39*). The instrument combined light scattering and absorption measurements. The separation of the eosinophils, neutrophils, monocytes and lymphocytes was achieved by measurement of differences in absorption and light scattering of cells stained with peroxidase. The identification of basophils was achieved by differences in light scattering and absorption of cells stained with alcian blue. Since the construction of a five-part differential counter is difficult (because of the infrequent occurrence of eosinophils and basophils) companies released instruments with the five-part "interpretation" differential. These instruments (Unipath Cell-Dyn 2000, Coulter S Plus, Toa E2500) were essentially the same instruments as the three-part differential counters made by these companies, with the addition of mathematical algorithms written to detect variations in the normal distributions of the measured histograms which were indicative of the occurrence of other cell populations (*46*).

A second generation of the five-part differential counters are now available including the Technicon H-1, Toa NE 8000, Coulter VCS and Unipath Cell-Dyn 3000. The methodology used in these instruments to accomplish a distinction between neutrophils, eosinophils, basophils, monocytes and lymphocytes is illustrated in Fig. 1. The same blood sample was run on all instruments. The dotplots and histograms generated by the instruments are illustrated for the H-1 in Fig. 1(A) and (B), for the NE 8000 in Fig. 1(C), (D), and (E), for the VCS in Fig. 1(F), (G) and (H) and for the Cell-Dyn 3000 in Fig. 1(I) and (J). The parameters measured are indicated on the axes and the locations of the leukocyte populations are indicated with N for neutrophils, E for eosinophils, B for basophils, M for monocytes and L for lymphocytes.

The five-part differential of the H-1 is generated by two separate flow channels. In one channel, the light scattering and absorption of the cells stained with peroxidase are measured. Four populations are identified in the correlation of those parameters: lymphocytes, monocytes, neutrophils and eosinophils (Fig. 1(A)). In the second flow channel, two narrow angle light scattering signals are measured of a specially lysed blood sample. This lysing procedure results in a distinction of basophils from the other leukocytes as is shown in Fig. 1(B).

The five-part differential of the Toa NE8000 is measured in three separate flow channels. This instrument measures electric volume and radio frequency of cells obtained by three different preparation procedures. In the first channel, red cells are lysed and leukocytes are differentiated into lymphocytes, monocytes and granulocytes as is shown in Fig. 1(C).

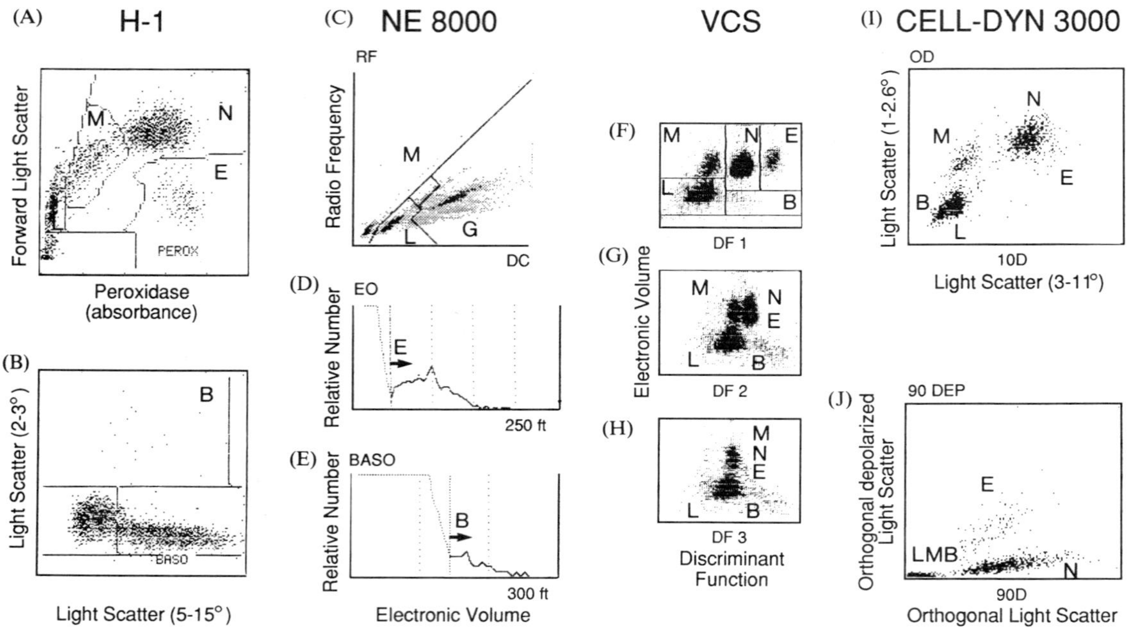

Fig. 1. A single peripheral blood sample was analyzed on four hematology analyzers. The displays illustrating the different cell clusters identified by the instruments are shown in (A) and (B) for the H-1 (Technicon, Tarrytown, NY), (C), (D) and (E) for the NE8000 (TOA, Medical Electronics, Carson, CA), (F), (G) and (H) for the VCS (Coulter Electronics, Hialeah, FL) and (I) and (J) for the Celldyn-3000 (Unipath, Mountain View, CA). The locations of the leukocyte populations are indicated with G for granulocytes, N for neutrophils, E for eosinophils, B for basophils, M for monocytes and L for lymphocytes.

Differences in sensitivity of basophils and eosinophils to pH extremes are used in the second and third flow channels. The histogram in Fig. 1(D) shows the electric volume distribution of the cell preparation where all cells except eosinophils are lysed. The histogram in Fig. 1(E) shows the electric volume distribution of the cell preparation in which all cells but the basophils are lysed.

In Coulter's VCS, three parameters are measured in a single channel from an erythrocyte lysed cell preparation. A flow cell is constructed such that electric volume and conductivity as well as light scattering are measured simultaneously. Three different displays are generated by this instrument in which the electric volume is plotted versus three different discriminating functions. These discriminating functions are derived from the combination of conductivity versus light scattering measurements. The three displays are illustrated in Fig. 1(F), (G) and (H), respectively. The five-part differential is best illustrated in Fig. 1(F). Different regions are indicated in which lymphocytes, monocytes, neutrophils, eosinophils and basophils can be distinguished.

In Fig. 1(I) and (J) the leukocyte differentiation obtained with the Unipath Cell-Dyn 3000 is displayed. The principles used in this instrument are entirely based upon optical measurements of unstained erythrocyte lysed cell preparations. Four independent light scattering parameters are measured in one passage of the cells. Light scattered orthogonally, depolarized orthogonally and light scattered between 1–2.6° and 3–11° are measured simultaneously with the data acquired in listmode. The position of the different leukocytes in the two displays of the four-dimensional data are indicated in both figures.

One approach for comparing the ability of instruments to discriminate between leukocyte populations is to determine the separation of each of the cell clusters from the others. This can be achieved by determination of the mean and standard deviation of each cell cluster with an assessment of the interdonor variation for each of the parameters. Such data are not usually provided by the manufacturers of the analyzers.

As an alternative, the separations of the cell clusters obtained on the same blood sample can provide a way of identifying differences in resolution between the various instruments (Fig. 1). The two major cell populations, i.e. lymphocytes and neutrophils, are clearly separated from each other in all instruments (Fig. 1). The cluster of cells formed by eosinophils is better resolved in all instruments as compared with the monocytes (Fig. 1). Lack of complete resolution of the less frequent monocytes from the other cell populations brings the accuracy of the instruments to reliably count the claimed cell population into question. The distinction of monocytes from neutrophils and lympyhocytes in the H-1 (Fig. 1(A)) and NE8000 (Fig. 1(C)) is rather arbitrary. The monocytes in this particular illustration of the Toa instrument fall outside the gate indicated in Fig. 1(C). (This is due to the improperly calibrated demonstration instrument used for this analysis. After calibration the angle at which the cells are displayed would be changed and the monocyte population will fall in this gate. The actual separation of the monocyte cluster from the other cell clusters, however, will not improve dramatically.) In Coulter's VCS (Fig. 1(F), (G) and (H)) and Unipath Cell-Dyn3000 (Fig. 1(I) and (J)) the monocyte cluster is better separated from the two major leukocyte populations.

Identification of basophilic granulocytes is in question in all instruments since no apparent "basophil" cluster is visible in any of the four instruments (Fig. 1). The infrequency of the basophils and the number of cells actually measured do not permit accurate measurements of this cell population. When abnormal blood samples are measured, the enumeration of basophils is totally unreliable as illustrated in Fig. 2. In this example peripheral blood and bone marrow samples from the same patient were run on the same hematology analyzer. The manual differential of the peripheral blood smear did not show any basophilic granulocytes while the automated system counted 0.2%. Manual examination of the bone marrow smear also did not show any basophils. Analysis of the bone marrow sample on the automated instrument revealed a large population of cells falling within the basophil region (Fig. 2(D)). Such a major variation from normality indicates that cells other than basophils can be counted as basophils in this system.

The ability to resolve each leukocyte population with respect to each characteristic is related to the mean and the coefficient of variation of the populations to be resolved. For each characteristic the differences in mean peak position and coefficients of variation for the five leukocyte populations are not sufficiently different to permit resolution by a single parameter, whether it is electronic volume, light scattering or absorption. The addition of a correlated second parameter also does not provide for sufficient increased resolution to separate all five populations (see Fig. 1). Increased resolution between the populations can be obtained by correlating a third or fourth parameter with the first two. For example in the Cell-Dyn3000 the monocytes are not only separated by both small angle light scattering parameters but also by the orthogonally, and depolarized orthogonal light scattering signals (Fig. 1(I) and (J)) (*62*). The instruments capable of using the multidimensional approach are Coulter VCS (three dimensions) and Unipath Cell-Dyn3000 (four dimensions). A further advantage of a multidimensional approach is that the instruments are less complicated because they only use one

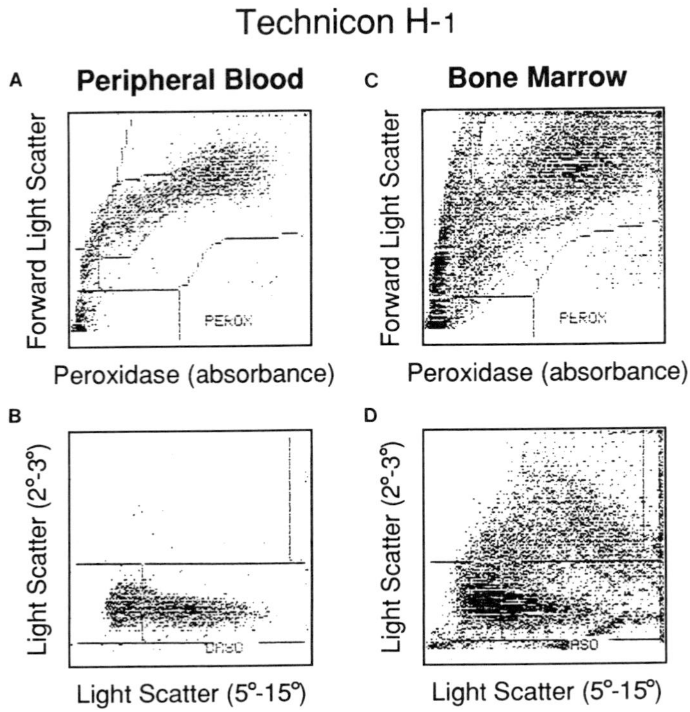

Fig. 2. A peripheral blood and bone marrow sample of the same donor was analyzed on an H-1 hematology analyzer. (A) and (C) show the forward light scattering and absorbance of peripheral blood and bone marrow stained with peroxidase respectively. (B) and (D) show two forward light scattering angles of the peripheral blood and bone marrow sample of which "all cells except the basophils are lysed" respectively. Note the large population of cells occurring in the region typical for basophils in (D) and the appearance of cell populations in (C) not present in the peripheral blood. No correlation between both observations can be obtained since they are generated in different flow channels.

flow channel and one cell preparation procedure.

There is an important distinction to be made between *multiparameter* and *multidimensional* analysis. Multiparameter analysis refers to the collection of data with regards to several characteristics on each sample. The information obtained by one detector on any individual cell may not necessarily be correlated with the data obtained in other detectors. For example the information obtained from the analysis of a bone marrow aspirate on a hematology analyzer illustrated in Fig. 2(C) and (D) is very interesting since a variety of cell populations are observed. One is not able to correlate the clusters of cells seen in Fig. 2(C) with those in Fig. 2(D) since they are measured in two different channels. Multidimensional analysis, on the other hand, infers that the multiparameter data are correlated so that for each and every event the data are combined. In this method of analysis the data in the multiple detectors are collected simultaneously as the cell passes through the sensing region. These data are reanalyzed to create a multidimensional space in which cell clusters can be more easily identified (as illustrated in Section IV). The increased information obtained by the addition of an extra detector is an additive process when the data are not directly correlated with all the other detectors. In contrast, a geometric increase in information content is obtained by correlating all the signals from each cell passing the detectors.

A major limitation of the current data is the validation of the identified leukocyte populations. Currently, no absolute standard is available for leukocyte differentiation. The reference method published by the National Committee for Clinical Laboratory Standards (NCCLS), the H20-T, requires an 800-cell manual count performed by four technologists on eight smears of each specimen (*45*). The amount of work required to execute the reference method is enormous and often is not performed. This method is flawed since the number of cells counted is still too low to quantify cells occurring in a frequency below 5% (*48–51*). In addition the appearance of immature cells on the slides causes a loss of specificity. Nonetheless current studies are performed comparing the performance of hematology analyzers with the manual reference method (*8,26,46*). More frequently, the data obtained with one hematology analyzer are compared with the results of other analyzers in order to validate the instrument performance (*1,7,26,31,46,54,73,74*). Both approaches do not prove the identity of the identified leukocyte populations since only correlative data are compared.

For detection of infrequent and abnormal cell populations, clearly another approach to leukocyte differentiation must be applied. Addition of immunofluorescence parameters in optical-based hematology analyzers can be used either for validation of the results or for improving the quality of the differential. Validation of leukocyte differentiation can also be made by physically sorting each cluster of cells. One can actually examine the identity and purity of the designated cell clusters by light microscopic examination of the cells. The feasibility of this approach is illustrated in Section IV.

IV. HIGH RESOLUTION LEUKOCYTE DIFFERENTIATION

While the light scattering measurements of human leukocytes provide the basis for hematology analyzers, the role of light scattering in immunoflow cytometry for distinguishing between the different components of peripheral blood is often under-utilized. The angle at which the forward angle light scattering is collected is critical for the discrimination of leukocytes but has not been well defined or controlled in commercial immunoflow cytometers (*52,62*). Combining high resolution light scattering with three colors of immunofluorescence yields a powerful approach in identifying the different cell components present in peripheral blood or bone marrow aspirates.

The analysis of multidimensional data is facilitated by a computer program which employs color to identify cell clusters in a multidimensional space (*61,63*). Three primary colors (green, red, dark blue) are utilized in different two-parameter displays to highlight (i.e. color) cells. When cells are overlapping with respect to two primary colors they become a secondary color (yellow, purple, light blue). The dots appear black when all three primary colors overlap. Once a cell cluster is identified, this cell population can be assigned a specific color allowing other colors to be used to distinguish other cell populations. In this manner populations of cells can be identified in a multidimensional space which could not be distinguished in any two-dimensional projection of the data.

This multidimensional approach can be illustrated following the analysis of a peripheral blood sample stained with CD14PE (*16,20*), CD45FITC (*32,55*) and anti-human IgE-Duochrome. The two light scattering parameters and three immunofluorescence parameters were acquired on the erythrocyte lysed sample and were stored in listmode. Different colors represent regions typical for each leukocyte population in the two-parameter projections of the data (Plate 9). Six two-parameter displays of the five-dimensional data are shown in which neutrophils are colored dark blue, eosinophils are purple, basophils are light blue, monocytes are red and lymphocytes are green. The gray dots are cells which are not clustered into one of the identified leukocyte populations.

A complete separation of the five leukocyte populations cannot be obtained by either immunofluorescence or light scattering alone (Plate 9). The requirement for using all five dimensions to identify a population is illustrated by focusing on the basophil population in Plate 9. In this example, basophils (light blue) have distinctive light scattering properties (Plate 9A), but a complete separation cannot be obtained since an overlap exists between the lymphocytes (green) and monocytes (red). In the projection of CD45 and CD14 fluorescence, (Plate 9(B)) the basophils are similar in staining to the neutrophils (dark blue). Although the presence of IgE in a high density is typical for basophils (light blue) (Plate 9(C) and (D)) (*43,64*), a few gray dots are present in this cell cluster. These (gray) cells are not basophils since they do not have light scattering characteristics of basophils (Plate 9(A)) and/or do not express the CD14 and CD45 antigens in a density typical for basophils (Plate 9(B)). This was confirmed by physically sorting the cells and examining their morphology by light microscope.

Microscope slides were prepared from the populations identified in Plate 9 and were stained with Wright–Giemsa. The morphologies of the sorted cell populations are illustrated in Plate 10. Although by morphology the cells illustrated in Plate 10(C) are homogeneous, their identity as basophils was confirmed by toluidine blue staining (*71*) (Plate 10(F)). Toluidine blue did not stain the slides prepared for the neutrophil, monocyte, eosinophil and lymphocyte fractions (data not shown). The sorted cell populations all had purities greater than 99%. The results clearly illustrate that a high resolution leukocyte differential can be obtained with multidimensional analysis by combining light scattering and immunofluorescence parameters.

The analysis of multidimensional data is facilitated when the cell populations are equally distributed over the acquired parameters (*65*). An example is illustrated for the orthogonal light scattering signals of leukocytes in Plate 11. Plate 11(A) shows the forward and orthogonal light scattering signals of the same leukocyte preparation shown previously (Plate 9). Light scattering regions typical for neutrophils were colored dark blue; for eosinophils, violet; for monocytes, red; for basophils, light blue; and for lymphocytes, green. It is clear that the five leukocyte populations are not equally distributed over the orthogonal light scattering dimension in Plate 11(A). The neutrophils and eosinophils occupy at least 60% of the orthogonal light scattering dimension whereas the lymphocytes are confined to the lower 5% of the orthogonal light scattering. The orthogonal light scattering signals can be transformed according to a polynomal function to obtain a more equal distribution of five leukocyte populations (*65*). The results of such a transformation are illustrated in Plate 11(B), where the leukocytes are colored identically as in Plate 11(A). The five leukocyte populations are now more evenly distributed in the two-parameter light scattering display (compare Plate 11(B) with Plate 11(A)). The advantage of the transformation of the orthogonal light scattering signals is a better visual separation of the five leukocyte populations. A large variety of cell populations can be identified in the cell cluster with relatively low orthogonal light scattering signals which can be more readily identified after transformation of the orthogonal light scattering signals (*6,11,34,35,66–69*). In the analysis of normal human bone marrow where the precursors of each of the cell lineages are found in addition to the end maturational stages the space added at the low orthogonal light scattering region aids in distinguishing between cell lineages (*11,34,35,38,55,64*). In addition the ability of cluster algorithms to identify clusters of cells in a multidimensional space improve by more homogeneous shape and size dispersion of the cell clusters.

By application of multidimensional analysis one actually can define maturation tracks for each of the cell lineages as described in Chapter 1.3 (*38*). In normal human bone marrow, the pathway leading from progenitor cells to the end stage of maturation of a certain leukocyte population is well defined with only small variations between donors. In abnormal marrow samples, these pathways can be quite different as compared with normal (*70*).

The potential of multidimensional analysis, in which high resolution light scattering is combined with immunofluorescence, is illustrated for a patient with chronic myeloid leukemia who was followed during a 4-month period. During this time the patient developed a blast transformation of the leukemia. The blood samples were stained with CD45 FITC and CD14 PE. Two two-parameter displays are illustrated in Plate 12(A) and (B) for $t = 0$, in Plate 12(C) and (D) for $t = 2$ months and in Plate 12(E) and (F) for $t = 4$ months. Cell clusters were identified using their unique characteristic of forward light scattering, transformed orthogonal light scattering, and CD45/CD14 immunofluorescence staining. Each population was assigned a particular color: neutrophils (dark blue) exhibited a much broader distribution in light scattering in comparison with peripheral blood (compare Plate 11(B) with Plate 12(A), (C) and (E)). This broader distribution in light scattering is correlated with a broader distribution of the neutrophils with respect to CD14 PE (compare Fig. 3(B) with Plate 12(B),(D) and (F)). This in fact indicates that the neutrophil population in these patients is not homogeneous. The loss of the CD14 antigen is indicative of a more immature differential stage of the neutrophils (*55*). Basophils and eosinophils do not occur in any significant numbers and therefore are not indicated in the figures. Normal lymphocytes are identified as green and are still detectable, although they occur in a frequency less than 1%. Note that the lymphocytes can only be recognized by multidimensional analysis. The monocytes, red, increase in frequency during the 4-month period (compare Plate 12(A) and (B) with 12(C) and (D) and 12(E) and (F)). Two cell populations colored yellow and black are identified in the peripheral blood of the CML patients and are not detected in normal peripheral blood. The frequency of both populations clearly increases during the 4-month period. In a separate experiment, the yellow populations were identified as $CD34^+CD33^-$ while the black population expressed both $CD34^+$ and $CD33^+$. The unique position of these population of cells allows a quantitation when using the multidimensional analysis which provides a tool to predict blastic transformation.

V. MULTIDIMENSIONAL ENUMERATION OF ALL FORMED ELEMENTS IN BLOOD AND MARROW

Blood cell differentials performed on current hematology analyzers are determined on red cell lysed cell preparations, which may introduce noise and artefacts. For example, not all red cells become transparent for the measurements and these will consequently interfere with leukocyte enumeration. In addition some of the leukocytes will be lysed or will be altered in a way that they fall out of the region set for the particular leukocyte population. In instruments using pH-dependent lysis of cells, some of the population of interest will always lyse while some of the cells which should be eliminated will not lyse. The noise introduced by these cell preparation procedures is almost negligible for normal blood samples. However, the artefacts introduced by the cell preparation procedure of abnormal samples can severely interfere with the accuracy of the measurements. Differences in the composition of the serum can influence the ability of the lysing reagents to effectively remove the cells, or the leukocytes are more vulnerable for the lysing reagents (as in certain leukemias).

These problems can be avoided by performing leukocyte differentials in the presence of red cells. Feasibility of identifying nucleated cells in the presence of the overwelming number of red blood cells by flow cytometry has been demonstrated more than two decades ago using the fluorescence intensity of acridine orange which stains the DNA and RNA present in the nucleated cells but does not stain the red blood cells (*22,25,30,56*). In a recent study a technique was introduced by which a peripheral blood or bone marrow sample was incubated with a combination of two nucleic acid dyes and a monoclonal antibody for leukocyte differential analysis in the presence of red blood cells (*63*). The nucleic acid dye thiazole orange preferentially binds to RNA, the nucleic acid dye LDS-751 preferentially binds to DNA and phycoerythrin labeled monoclonal antibody CD45 identifies a common leukocyte antigen (*5,55*). These three dyes can be excited at 488 nm and their emission can be spectrally separated as is illustrated in Fig. 3. After dilution with phosphate buffered saline (1 : 100) two light scattering parameters and three fluorescence parameters of 50 000 cells present in blood or bone marrow aspirates are acquired into listmode.

The combination of these five parameters define a five-dimensional space in which each of the cell populations forms unique clusters. This is illustrated for a peripheral blood sample in Plate 13 and for a bone marrow sample of the same donor in Plate 14. Six different two-parameter displays of the five-dimensional data are shown. The four major cell populations are

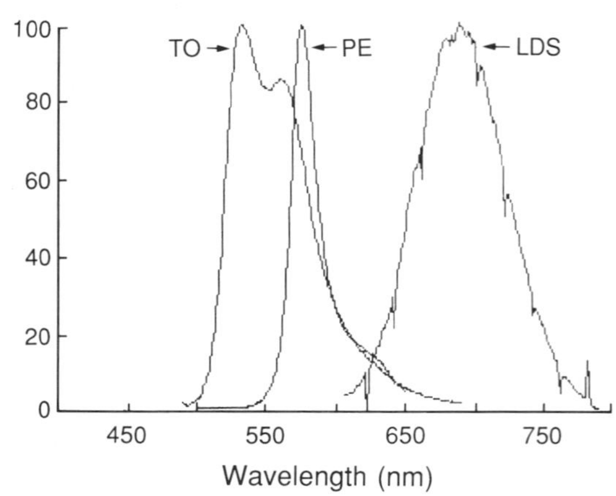

Fig. 3. Emission spectra of thiazole orange, R-phycoerythrine and LDS-751 excited at 488 nm. The emission of thiazole orange was measured in the presence of RNA; the emission of LDS-751 was measured in the presence of DNA.

each presented by a color. Red blood cells (colored red) have relatively large forward and orthogonal light scattering now displayed on a logarithmic scale and do not stain with thiazole orange, LDS-751 or CD45 PE. Reticulocytes (colored yellow) are distinguishable from red cells because of their staining with thiazole orange and from the nucleated cells colored black because of the lesser staining with thiazole orange and LDS-751. Note that the nucleated cells (colored black) have a relatively large forward light scattering and show a differential staining with CD45-PE. The platelets (depicted green) have relatively low forward and orthogonal light scattering, stain with LDS-751 but lack CD45. Cells or particles not belonging to one of the four cell clusters are depicted as gray. As illustrated in the previous section the cell populations cannot be separated in any of the two parameter projections, whereas they form unique clusters in the five-dimensional space in which they are identified.

The four identified cell clusters have identical positions in the five-dimensional space for both peripheral blood and bone marrow. The only obvious difference is the relative frequency of the cell populations, i.e. relatively more nucleated cells and reticulocytes are present in bone marrow as compared with peripheral blood. Another difference between peripheral blood and bone marrow is that a portion of the reticulocyte population in bone marrow stains more brightly with thiazole orange (more immature) as compared with peripheral blood. The nucleated red cells in the bone marrow appear as a part of the nucleated cells but do not express the CD45 antigen.

For an improvement of the distinction between the nucleated cells in peripheral blood and bone marrow 5000 cells were acquired into listmode with a gate set on cells with a relatively large fluorescent intensity for both thiazole orange and LDS-751 (Plates 15(B), and 16(B)). The amplifiers used for both light scattering parameters were then switched from logarithmic to linear. This approach results in a clear discrimination of the different nucleated cell populations as is shown in four two-parameter displays for peripheral blood in Plate 15 and for bone marrow in Plate 16. Neutrophils are depicted by a light blue color, eosinophils are depicted black, monocytes are depicted green, lymphocytes are depicted yellow, nucleated red cells are depicted red and immature nucleated cells are depicted dark blue. Cells not belonging to the identified cell clusters are depicted gray. The identity of these populations was confirmed after light microscopic examination of the sorted cell clusters as described previously (*63*).

The addition of the staining with CD45 of the nucleated cells to the forward and orthogonal light scattering introduces an important improvement for the discrimination within the nucleated cell population. Nucleated red cells, lacking the CD45 antigen (*55*), and immature nucleated cells, dimly expressing the CD45 antigen (*55*), cannot be distinguished from the lymphocyte population using light scattering only (Plate 16).

The multidimensional approach towards cell differentiation opens new prospectives for assessment of the hematologic status of patients. All parameters currently measured by a variety of hematology analyzers can be determined in one flow channel, while important additional information is obtained for a further discrimination of blood cells. A complete picture can be formed of the red blood cell lineage starting with the immature cells, nucleated red cells, reticulocytes gradually losing their RNA content until they reach the mature erythrocyte stage. Parameters such as red blood cell volume and optical density (hemoglobin concentration) are available in a much more detailed fashion since they can be determined with respect to five parameters. Similar information can be obtained for each of the leukocyte populations. The information present for each of the identified cell population is enormous with respect to their distribution in the five-dimensional space. Correlation of this data with the clinically useful information is currently being performed.

Actual introduction of the multidimensional approach for cell differentiation in clinical hematology analyzers in the form presented here will take some time. The microprocessors which are increasing in speed each year, more sophisticated cluster algorithms, optimalization of the optics and the hydrodynamic focusing promises that the actual acquisition and analysis of multidimensional data can be performed in the near future with the 120 samples/hour presently demanded by the larger hematology laboratories.

REFERENCES

1. Allen, J. K., and Batjer, J. D. (1985). Evaluation of an Automated Method for Leukocyte Differential Counts Based on Electronic Volume Analysis. *Arch. Pathol. Lab. Med.* **109**, 534–539.

2. Bacus, J. W. (1986). The development of automated differential systems. *In* "Differential Leukocyte Counting", Ch. 1, pp. 95–117, College of American Pathologists, New York.

3. Bacus, J. W. (1986). Design and performance of the LARC automated leukocyte classifier. *In* "Differential Leukocyte Counting," Ch. 1, pp. 119–134, College of American Pathologists, New York.

4. Bessman, J. D. (1986). "Automated Blood Counts and Differentials. A Practical Guide." The Johns Hopkins University Press, Baltimore.

5. Beverely, P. C. L., Linch, D., and Delia, D. (1980). Isolation of human hematopoietic progenitor cells using monoclonal antibodies. *Nature* **287**, 332–333.

6. Bockstaele, D. R. van, Berneman, Z. N., Peetermans, M. E. (1986). Flow cytometric analysis of hairy cell leukemia using right angle light scatter. *Cytometry* **7**, 217–221.

7. Bollinger, P. B., Drewinko, B., Brailas, C. D., Smeeton, N. A., and Trujillo, J. M. (1986). The Technicon H*1™: an automated hematology analyzer for today and tomorrow. *Am. J. Clin. Pathol.* **87**, 71–78.

8. Breakell, E. S., Marchand A., Marcus, R., and Simsom, E. (1985). Comparison of performance for leukocyte differential counting of the Technicon H6000™ system with a manual reference method using the NCCLS standard. *Blood Cells* **11**, 257–279.

9. Brunsting, A., and Mullaney, P. F. (1974). Differential lightscattering from spherical mammalian cells. *Biophys. J.* **5**, 439–453.

10. de Cresce, R. (1986). The Technicon H*1™: a discrete fully automated complete blood count and differential analyzer. *Lab. Med.* **17**, 17–21.

11. Civin, C. I., and Loken, M. R. (1987). Cell surface antigens on human marrow cells: dissection of hematopoietic development using monoclonal antibodies and multiparameter flow cytometry. *Int. J of Cell Cloning* **5**, 267–288.

12. Curbelo, R., Schildkraut, E. R., Hirschfield, T., Webb, R. H., Block, M. J., and Shapiro, H. M. (1976). A generalized machine for automated flow cytology system design. *J. Histochem. Cytochem.* **24**, 388–395.

13. Coulter, W. H. (1953). Means for counting particles suspended in fluid. USA patent PN 2,656,508.

14. Coulter, W. H. (1956). High speed automated blood cell counter and cell size analyzer. *Proc. Nath. Electr. Conf.* **12**, 1034.

15. Coulter, W. H. (1970). USA patent PN 3,502,974.

16. Dimitriu-Bona, A., Burmester, G. R., Waters, S. J., and Winchester, R. J. (1983). Human mononuclear phagocyte differentiation antigens. I. Patterns of antigenic expression on the surface of human and macrophages defined by monoclonal antibodies. *J. Immunol.* **130**, 145–152.

17. Dutcher, T. F. (1985). Automated differentials: a strategy. *Blood Cells* **11**, 49–59.

18. England, J. M., Rowan, R. M., Bins, M., Bull, B. S., Coulter, W. H., Groner, W., Jones, A. R., Koepke, J. A., Lewis, S. M., Shinton, N. K., Thom, R., van Assendeft, O. W., and Verwilghen, R. L. (1988). The assignment of values to fresh blood used for calibrating automated blood cell counters. *Clin. Lab. Haematol.* **10**, 203–212.

19. Ehrlich, P., and Lazarus, A. (1900). "Histology of the Blood", Cambridge University Press, Cambridge.

20. Goyert, S. M., Ferrero, E., Rettig, W. J., Yenamandra, A. K., Obata, F., and Le Beau, M. M. (1988). The CD14 monocyte differentiation antigen maps to a region encoding growth factors and receptors. *Science* **239**, 497–500.

21. de Grooth, B. G., Terstappen, L. W. M. M., Puppels, G. J., and Greve, J. (1987). Light scattering polarization as a new parameter in flow cytometry. *Cytometry* **8**, 539–544.

22. Hallerman, L., Thom, R., and Gerhartz, H. (1964). Elektronische Differentialzahlung von Granulocyten und Lymphocyten nach Intravitaler Fluochromierung mit Acridineorange. *Verh. Dtsch. Ges. Inn. Med.* **70**, 217.

23. Hayhoe, F. G. J., and Flemans, R. J. (1987). "A Colour Atlas of Haematological Cytology," 2nd end. Wolfe Medical Publications, London.

24. Hosty, T. A., Harris, L. G., Stonacek, S. M., and Frazier, B. (1982). Evaluation of an automated continuous flow hematology instrument: The Technicon H6000. *J. Clin. Lab. Auto.* **2**, 408–415.

25. Jackson, J. R. (1961) Supravital blood studies, using acridine orange fluorescence. *Blood* **17**, 643.

26. Johnston, C. L. (1985). Leukocyte screening using the ELT-800/WS. *Blood Cells* **11**, 241–255.

27. Julius, H. M., Sweet, R. G., and Fatham, C. G. (1975). Fluorescence activated cell sorting and its applications. *In* "Mammalian Cells: Probes and Problems," ERDA Symposium Series CONF-731007 (C. R. Richmond, D. F. Peterson, P. F. Mullaney and L. A. Herzenberg, eds), p. 107. Technical Information Center, Oak Ridge, TN.

28. Koepke, J. A., ed. (1978). "Differential Leukocyte Counting." CAP, Chicago.

29. Koepke, J. A., ed. (1985). The white blood cell differential. I. *Blood Cells* **11**, 1–150.

30. Kosenow, W. (1952). Die Fluorochromierung mit Acridineorange, eine Methode zur Lebenbeobachtung gefarbter Blutzellen. *Acta-Haematol (Basel)* **7**, 217.

31. Lai, A. P., Martin, P. J., Richards, J. D. M., Goldstone, A. H. and Cawley, J. C. (1986). Automated Leukocyte Differential Counts in Acute Leukaemia: A Comparison of the Hemaolog D, H6000 and Coulter S-Plus IV. *Clin. Lab. Haemtol.* **8**, 33–41.

32. Lanier, L. L., and Warner, N. L. (1981). Paraformaldehyde fixation of hematopoietic cells for quantitative flow cytometry (FACS) analysis. *J. Immunol. Methods* **4**, 25–30.

33. Ledis, S. L., Crews, H. R., and Sena, T. (1988). Methods and reagents for differential determination of four populations of leukocytes in blood. USA patent, PN 4,751,179.

34. Loken, M. R., Shah, V. O., and Civin, C. I. (1986). Characterization of myeloid antigens on human bone marrow using multicolor immunofluorescence: *In* "Leucocyte Tying III: White Cell Differentiation Antigens" (A. J. McMichael, ed.), pp. 630–636. Oxford University Press, Oxford.

35. Loken, M. R., Shah, V. O., Dattilio, K. L., and Civin, C. I. (1987). Flow cytometric analysis of human bone marrow: II. Normal B lymphocyte development. *Blood* **70**, 1316–1324.

36. Loken, M. R., and Herzenberg, L. A. (1975). Analysis of cell populations with a fluorescence activated cell sorter. *Ann. NY Acad. Sci.* **254**, 163–171.

37. Loken, M. R., and Stall, A. M. (1982). Flow cytometry as an analytical and preparative tool in immunology. *J. Immunol. Methods* **50**, R85–R112.

38. Loken, M. R., Civin, C. I., Shah, V. O., Fachler, M. J., Segers-Nolten, I., and Terstappen, L. W. M. M. (1992) Characterization of erythroid, lymphoid, and monomyeloid lineages in normal human bone marrow. *In* "Flow Cytometry in Hematology" (R. Bjerkness and O. D. Laerum, eds), *Anal. Cytol. Ser.*, pp. 31–42. Academic Press, London.

39. Mansberg, H. P., Saunders, A. M., and Groner, W.

(1974). The Hemalog D white cell differential system. *J. Histochem. Cytochem.* **22**, 711–724.
40. Mayer, K. (1985). Presence of abnormal cells. *Blood Cells* **11**, 25–30.
41. McClure S., Bates, J. E., Harrison, R., Gilmer, P. R., and Bessman, J. D. (1988). Use of Microcomputer Analysis to Triage Blood Specimens for Microscopic Examination. *Am. J. Clin. Pathol.* **90**, 163–168.
42. Miller, M. M. (1976). Design and clinical results of Hematrek an automated differential counter. *IEEE. Trans. Biomed. Eng.* **23**(5), 400–405.
43. Miroli, A. A., and Spitz, M. (1987). Studies on the basophils IgE receptor generation and characterization of monoclonal antibodies against the receptor. *J. Immunol. Methods* **97**, 185–190.
44. Mullaney, P. F., and Dean, P. N. (1970). The small angle light scattering of biological cells. *Biophys. J.* **10**, 764–772.
45. National Committee for Clinical Laboratory Standards (1984). Tentative standard H20-T, Leukocyte differential counting. Villanova, PA, Vol. 4, No. 22.
46. Payne, B. A., Robert, M. T., Pierre, V., and Lee, W. L. (1987). Evaluation of the TOA E-5000 Automated Hematology Analyzer. *Am. J. Clin. Pathol.* **88**, 51–57.
47. Pierre, R. V. (1985). The routine differential leukocyte count vs automated differential counts. *Blood Cells* **11**, 11–23.
48. Rumke, C. L., Bezemer, P. D., and Kuik, D. J. (1975). Normal values and least significant differences for leukocyte counts. *J. Chron. Dis.* **78**, 662–668.
49. Rumke, C. L. (1985). Imprecision of ratio-derived differential leukocyte counts. *Blood Cells* **11**, 311–314.
50. Rumke, C. L. (1985). Statistical reflections on finding atypical cells. *Blood Cells* **11**, 141–144.
51. Rumke, C. L. (1985). The imprecision of the ratios of two percentages observed in differential white blood cell counts: a warning. *Blood Cells* **11**, 137–140.
52. Salzman, G. C., Mullaney, P. F., and Price, B. J. (1979). Light scattering approaches to cell characterization. *In* "Flow Cytometry and Sorting" (M. R. Melamed, P. F. Mullaney, and H. L. Mendelsohn, eds), pp. 105–124. Wiley, New York.
53. Salzman, G. C., Growell, J. M., and Martin, J. C. (1975). Cell classification by laser light scattering: Identification and separation of unstained leukocytes. *Acta Cytol.* (*Baltimore*) **19**, 374–377.
54. Savage, R. A., and Salazar, C. (1985). Lymphocyte percentage correlations for the Coulter S + IV™, *Brief Sci. Rep.* **4**(4), 497–498.
55. Shah, V. O., Civin, C. I., and Loken, M. R. (1988). Flow cytometric analysis of human bone marrow: IV. Differential quantitative expression of T-200 common leukocyte antigen during normal hematopoiesis. *J. Immunol.* **140**, 1861–1867.
56. Schiffer, L. M. (1962). Fluorescence microscopy with acridine orange: A study of hematopoietic cells in fixed preparations. *Blood* **19**, 200.
57. Shapiro, H. M., Schildkraut, E. R., Curbelo, R., Laird, C. W., Turner, B., and Hirchfield, H. (1976). Combined blood cell counting and classification with fluorochrome stains and flow instrumentation. *J. Histochem. Cytochem.* **24**, 396–411.
58. Sheenan, D. C., and Hrapchak, B. B. (1980). "Theory and Practice of Histotechnology". C. V. Mosby, London.
59. Simson, E., Ross, D. W., and Kocher, W. D. (1988). "Atlas of Automated Cytochemical Hematology." Technicon, Tarrytown, NY, LCCCN 88-50183.
60. Sloot, P. M. A., and Figdor, C. G. (1986). Elastic lightscattering from nucleated cells. Rapid numerical analysis. *Appl. Optics* **25**, 3559–3565.
61. Terstappen, L. W. M. M., Shah, V. O., Conrad, M. P., Recktenwald, D., and Loken, M. R. (1988). Discriminating between damaged and intact cells in fixed flow cytometric samples. *Cytometry* **9**, 447–484.
62. Terstappen, L. W. M. M., de Grooth, B. G., Vissher, K., van Kouterik, F. A., and Greve, K. (1988). Four-parameter white blood cell differential counting based on light scattering measurements. *Cytometry* **9**, 39–43.
63. Terstappen, L. W. M. M., and Loken, M. R. (1988). 5-Dimensional flow cytometry as a new approach for blood and bone marrow differentials. *Cytometry* **9**, 548–556.
64. Terstappen, L. W. M. M., Hollander, Z., Meiners, H. M., and Loken, M. R. (1990). Quantitative comparison of cell surface antigens on five blood cell lineages. *J. Leucocyte Biol.* **48**, 138–148.
65. Terstappen, L. W. M. M., Mickaels, R., Dost, R., and Loken, M. R. (1990) Increased light scattering resolution facilitated multidimensional flow cytometric analysis. *Cytometry* **11**, 506–512.
66. Terstappen L. W. M. M., de Grooth B. G., ten Napel C. H. H., van Berkel W, and Greve J. (1988). Discrimination of human cytotoxic lymphocytes from regulatory and B-lymphocytes by orthogonal light scattering. *J. Immunol. Methods* **95**, 211–216.
67. Terstappen, L. W. M. M., de Grooth, B. G., van Berkel, W., ten Napel, C. H. H., and Greve, J. (1988). Application of orthogonal light scattering for routine screening of lymphocyte samples. *Cytometry* **9**, 220–225.
68. Terstappen, L. W. M. M., de Grooth, B. G., van Berkel, W., ten Napel, C. H. H., and Greve, J. (1988). Abnormal distribution of CD8 subpopulation in B-chronic lymphocytic leukemia identified by flow cytometry. *Leuk. Res.* **12**, 551–557.

69. Terstappen, L. W. M. M., de Grooth, B. G., van Berkel, W., ten Napel, C. H. H., and Greve, J. (1988). Flow cytometric characterization of chronic lymphocytic leukaemias using orthogonal light scattering and quantitative immunofluorescence. *Blut* **56**, 201–208.
70. Terstappen, L. W. M. M., and Loken, M. R. (1988). Multi dimensional flow cytometric characterization of normal and leukemic cells in acute non lymphocytic leukemia. *Blood* **72**, suppl 1, 230a.
71. Undritz, E., (1973). "Atlas of Haematology," 2nd edn. Sandoz, Basle.
72. Visser, J. W., van den Engh, G. J., and van Bekkum, D. W. (1980). Light scattering properties of murine hematopoietic cells. *Blood Cells* **6**, 391–407.
73. Wielders, J. P. M., Beunis, M. H., and van Wersch, J. W. J. (1986). A Comparison of the Screening Ability of Two Automated Leukocyte Differential Counters. *J. Clin. Chem. Clin. Biochem.* **24**, 471–480.
74. Wilding, P., and LeBoy E. L. (1985). Use of Pattern Recognition Technology for Determination of the Human Differential Leukocyte Count. *Blood Cells* **11**, 187–201.
75. Zucker-Franklin, D., Greaves, M. F., Gross, C. E., and Marmont, A. M. (1981). "Atlas of Blood Cells, Function and Pathology," Vols 1 and 2. Edi Ermes, Milan.

2.3

Analysis of Phagocyte–Microorganism Interactions

ROBERT BJERKNES

Department of Pediatrics and The Gade Institute
Department of Pathology, University of Bergen
Haukeland Hospital, N-5021 Bergen, Norway

1. INTRODUCTION

Granulocytic and mononuclear phagocytes provide an effective host defence against both bacterial and fungal infections (*78,93*). Acquired and congenital abnormalities of phagocyte functions, as well as reduced numbers of competent cells, have been associated with syndromes of recurrent infections (*35,59,64,78,92,93,127*). The normal response of phagocytic cells to infection includes the recognition of microbial invasion, adhesion to vascular endothelium, leaving of the bloodstream and migration to the infected tissue (*78,93*). At the site of infection the

FLOW CYTOMETRY IN HEMATOLOGY
ISBN 0-12-432940-3

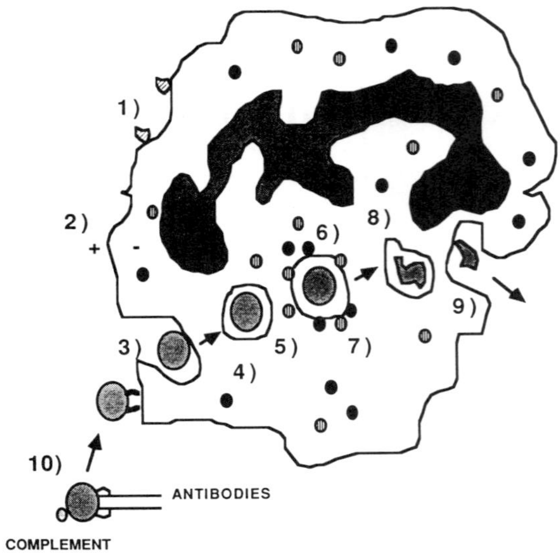

Fig. 1. The interaction between microorganisms and phagocytosing leukocytes. Cellular properties and functional processes: (1) surface receptors and regulating molecules; (2) membrane potential; (3) phagocytosis, i.e. attachment and internalization; (4) phagolysosomal pH; (5) degranulation and phagolysosomal enzyme activities; (6) oxidative burst; (7) intracellular killing; (8) intracellular degradation; (9) exocytosis; (10) microorganism opsonization by complement factors and antibodies.

pathogens are ingested, killed and degraded by the phagocytes (Fig. 1). Phagocytosis frequently requires opsonization of the microorganisms to facilitate their attachment to and ingestion by the phagocytes (Fig. 1). The direct interaction between microorganisms and phagocytes induces complex and concerted structural and metabolic alterations of the phagocytes, including altered surface receptor expression, changes of membrane potential and fluxes of intracellular ions, polymerization of microfilaments, degranulation into the phagolysosomes and the extracellular milieu, phagolysosomal acidification and the production of microbicidal oxidants (Fig. 1) (*25,78,93,122*).

In recent years, flow cytometry has become an important tool for the experimental and diagnostic examination of phagocytic cells (*25,66,109,126*). The present chapter reviews the applications of flow cytometry for the study of phagocyte interactions with microorganisms and serum opsonins.

II FUNCTIONAL PHAGOCYTE SURFACE MOLECULES

A variety of phagocyte surface molecules are involved in the process of eradicating invading microorganisms. Receptors for serum opsonins mediate attachment of microorganisms, whereas other surface structures seem to play a role in the signal transduction and architectural changes occurring during phagocyte activation and microbial ingestion and killing.

A. Methods

Together with the development of a series of monoclonal antibodies (MoAbs), flow cytometry has made possible detailed studies of phagocyte surface structures. The technology allows for single cell analyses, sorting and discrimination of leukocyte populations due to their different light scatter characteristics. Thus, the percentage of neutrophilic granulocytes or mononuclear phagocytes expressing a surface antigen can be determined, as well as the density of the surface antigens. Variations of surface antigen expression may reflect phagocyte activation, and this can be determined by flow cytometry (*20,25*). Detailed protocols for immunofluorescence staining procedures and flow cytometric analyses of immunostained cells are available, emphasizing the importance of proper controls (*66,83,89*).

B. Phagocyte Opsonin Receptors

Most pathogenic microorganisms are only effectively attached to the phagocyte surface and ingested following opsonization (*93*). The main opsonins are IgG and complement split products, but the other immunoglobulin classes and their cellular receptors may also play a role (*78*). The phagocyte surface receptors for IgG are FcγRI (CD64) present on mononuclear phagocytes (and neutrophilic granulocytes following treatment with γ-interferon) (*74*), FcγRII (CDw32) present on monocytes and granulocytes, and FcγRIII (CD16) present on granulocytes and macrophages, but not on monocytes (*65,136*). The receptors for the main complement derived opsonins are CR1 (CD35) (the receptor for the complement split product C3b) and CR3 (CD11b) (the receptor for the complement split product C3bi) (*82,142*). CR1 and CR3 are both present on mononuclear and granulocytic phagocytes (*82,142*). The relative importance of the various opsonin receptors in phagocyte activation and microbial ingestion is not completely understood but it seems that FcγRII, FcγRIII, CR1 and CR3 are the most important (*19,65,82,136,142*).

A variety of MoAbs have been described reactive with these receptors, allowing for the flow cytometric study of receptor distribution and function (*79*). Fluorescent microspheres coated with split products of complement factor C3 or Ig have provided another tool for the flow cytometric elucidation of the biology of phagocyte opsonin receptors (*34,99,100*).

a. *FcγR*

It appears that FCγRI is the FcR responsible for binding of monomeric IgG and low molecular weight immune complexes to mononuclear phagocytes (*65,107*), whereas FcγRII is important for both neutrophilic granulocyte and monocyte activation (*2,65,69,87*). Using MoAbs against FcγRII (IV.3 and mAb32), Anderson *et al.* demonstrated that monocytic cells produced superoxide when incubated with the MoAbs followed by cross linking with $F(ab)_2$ anti-murine Ig (*2*). In addition, immunecomplex-mediated superoxide production was inhibited by 80% by the same antibodies (*2*). In another study, using MoAbs against FcγRII (IV.3) and FcγRIII (CLBFcR-granl), Huizinga *et al.* demonstrated that FcγRII seems essential for both IgG-mediated phagocytosis and respiratory burst in human neutrophils, as IV.3, but not CLBFcR-granl, inhibited the uptake of IgG-opsonized bacteria and activation of the oxidative metabolism (*69*). However, the binding of some MoAbs to FcγRIII may inhibit the ingestion of Ig-opsonized bacteria, but the inhibitory effect of receptor specific MoAbs varies. In one flow cytometric study, comparing the effects of 49 anti-FcγRIII MoAbs on neutrophil granulocyte phagocytic capacity, it was demonstrated that six inhibited the uptake of serum-opsonized latex beads, two enhanced the ingestion whereas 41 did not influence phagocytosis at all (*19*). Similar diversities were demonstrated for their influence on neutrophil granulocytes respiratory burst (*19*). Recently we have shown that cross linking of both FcγRII and FcγRIII induces cytoplasmic calcium fluxes independently in human neutrophilic granulocytes (*87*). This suggests that cytoplasmic free calcium ions are involved in the signal transduction pathway of both IgG FcR in human neutrophilic granulocytes.

The influence of cytokines and chemoattractants on FcγR expression has been studied by flow cytometry. The expression of FcγRI on monocytes and granulocytes is significantly increased by γ-interferon (*65,75*), whereas the expression of FcγRII and FcγRIII is not altered by this cytokine (*75*). The expression of FcγRIII on neutrophil granulocytes is significantly reduced following stimulation of the cells with an inflammatory bacterial peptide N-formyl-met-leu-phe (FMLP) (*68*). The reduced expression has been shown to be caused by shedding of the FcγRIII following disintegration of their phosphatidyl-inositol anchors in the neutrophil granulocyte surface membrane upon cell stimulation (*68*). In contrast, TNF-α and CSF-GM do not influence neutrophil granulocyte surface FcγRIII expression (*55*).

The neutrophil granulocyte FcγRIII shows allotypic variation, directly related to the neutrophil alloantigen system, NA_1/NA_2 (*101,131*). Using flow cytometry and MoAbs against FcγRIII, it has been shown that MoAbs Leu-11 and 3G8 react with all neutrophilic granulocytes, whereas Gran-11 (anti-NA_1)

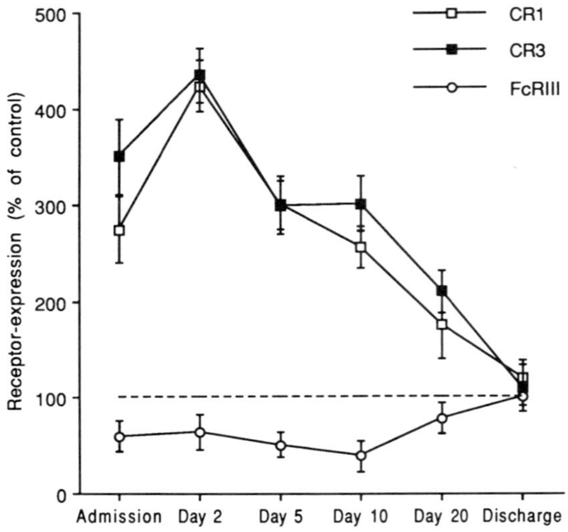

Fig. 2. The neutrophil expression of CR1, CR3 and FcRIII in patients with large burns. Neutrophils were obtained at admission, days 2, 5, 10, 20, and at discharge. The leukocytes were incubated with an anti-CR1 MoAb (Anti-CR1, Becton-Dickinson), an anti-CR3 MoAb (Mo1, Coulter) or an anti-FcRIII (Leu11c, Becton-Dickinson). The data are given as per cent of parallel controls, and as mean ± SEM; $n = 12$.

reacts with only 46% of normal neutrophils (i.e. heterozygotes NA_1/NA_2 or homozygotes NA_1/NA_1) (*101,131*). This antigenic polymorphism seems to be related to structural variations which can be accounted for by differences at the level of primary protein structure (*101*).

In recent studies we have determined the expression of FcγRIII on monocytes and neutrophilic granulocytes from patients with major burns (*31,139*). The phagocyte FcγRIII expression was significantly reduced during the first 10 days postinjury (Fig. 2), suggestive of significant phagocyte activation in the early phase postinjury (*68,139*). Another study reports normal or decreased expression of FcγRIII on neonatal neutrophilic granulocytes (*56,147*).

b. *FcαR/FcεR*

Monocyte and neutrophil granulocyte receptors for IgA have been identified and quantified by flow cytometry, using FITC-labeled IgA (*52*). In the same study, it was demonstrated that neutrophilic granulocytes obtained from the oral cavity expressed more FcαR than did blood neutrophilic granulocytes (*52*). In addition, it seemed that FcαR co-operated with FcγR in enhancing blood and oral cavity neutrophilic granulocyte phagocytosis (*52*).

The amount of cytophilic IgE bound to eosinophilic granulocytes has been determined by flow cytometry using FITC-labeled anti-human IgE antibodies (*43*). Cytophilic IgE was demonstrated both on cells expressing the high affinity FcεRI and on cell populations expressing the low affinity FcεRII (CD23) (*43,65*).

c. *CR1*

The CR1 on neutrophilic granulocytes and mononuclear phagocytes mediates or potentiates phagocytic or endocytic reactions (*24,120,142*), and inactivation of complement causes reduced ingestion of bacteria by neutrophilic granulocytes (*14*). The CR1 of unstimulated monocytes and neutrophilic granulocytes resides in clusters on the cell surface (10–20%) or in intracellular compartments (80–90%) from which it is mobilized upon phagocyte activation (*9,142*). Flow cytometry has been used to examine neutrophilic granulocyte expression of CR1, and in several studies it has been shown that the expression of CR1 increases following stimulation of cells with N-formyl-met-leu-phe (FMLP), C5a desArg, platelet activating factor (PAF) and tumor necrosis factor (TNF) (*21,53,116*). A similar response has also been demonstrated following *in vivo* activation, as cells from patients with major burns showed increased expression of CR1 during the first 3 weeks following injury (Fig. 2) (*31,139*). In contrast, neutrophilic granulocytes from patients with systemic lupus erythematosus (SLE) show significantly lower amounts of CR1 as compared with healthy controls (*143*).

d. CR3

CR3 mediates monocyte and neutrophilic granulocyte phagocytosis of C3bi-coated particles of microorganisms (*82*,*99*). In addition, this surface structure plays a role as an adhesive molecule in cellular interactions (see below). By the use of anti-CD11b MoAbs and flow cytometry, it has also been shown that binding of different MoAbs to neutrophil CR3 may have disparate functional effects (*5*,*26*,*63*,*70*,*120*). Amin Arnaout *et al.* reported inhibited neutrophil granulocyte phagocytosis of oil Red O opsonized by C3 (*5*), whereas we demonstrated no such inhibitory effect using the MoAb VIM-12 and C3 opsonized zymosan particles (*26*). Others have confirmed our results by demonstrating that VIM-12 does not inhibit the neutrophil oxidative response to serum treated zymosan, as measured by chemiluminescence (*70*). Surprisingly, both Amin Arnaout *et al.* (*5*) and we (*26*) observed inhibited neutrophil IgG-mediated phagocytosis by MoAbs to CR3. Similar inhibition occurs in monocytes (*37*), and it has been suggested that the inhibition of IgG-mediated phagocytosis by anti-CD11b MoAbs is due to the occurrence of a subpopulation of CR3 molecules which seems to be associated with FcγRs (*37*).

The expression of neutrophilic granulocyte CR3 increases after stimulation by phorbol myristate acetate (PMA), interleukin-2, cytochalasin B, TNF, FMLP, C5a and CSF-GM (*21*,*55*,*116*,*121*). The phagocyte CR3 expression is also increased in patients with burns (Fig. 2) (*31*,*139*) and hemodialysis-induced granulocytopenia (*4*), demonstrating *in vivo* activation of the phagocytes.

C. Phagocyte Adherence-related Glycoproteins

Phagocytic leukocytes possess large numbers of heterodimeric adhesion molecules, also called integrins, consisting of non-covalently associated α and β chains (*65*,*110*,*127*). The leukocyte integrins are glycoproteins and are designated LFA-1 (CD11a), CR3 (CD11b) and p150,95 (CD11c), which have α chains of 175, 165 and 150 kDa respectively (*110*,*124*). The β chain (CD18) (95 kDa) is required for these heterodimers to be expressed on the leukocyte surface (*65*,*124*). An intracellular pool of these adhesion molecules has been demonstrated by various techniques (*9*,*75*,*91*), and the surface expression of CR3 is significantly increased by a variety of stimuli inducing neutrophilic granulocyte activation (*21*,*55*,*116*,*121*,*139*).

A variety of MoAbs to these surface structures have been described and allowed for the flow cytometric study of the CD11/CD18 molecules in normal phagocytes (*36*,*79*,*108*, *123*,*133*). The CR3 mediates phagocytosis and activation of phagocytes which can be inhibited by some anti-CD11b Moabs (*26*,*71*). The CD11/CD18 molecules are also involved in the binding of unopsonized particles (*77*,*106*). In addition, they can behave as lectins in that their binding of unopsonized zymosan particles is inhibited by N-acetyl-D-glucosamin (*106*). Further, the CD11/CD18 glycoproteins can also mediate opsonin-independent attachment of *Staphylococcus epidermidis*, *Saccharomyces cerviciae*, *Histoplasma* and *Leishmania donovani* (*41*,*77*,*106*,*144*). In addition, the CD11/CD18 glycoproteins are involved in neutrophil granulocyte adherence to endothelial cells (*146*), neutrophil tissue damage (*72*) and antibody-mediated cell-mediated cytotoxicity (*138*).

A genetic disorder of leukocyte adhesion–glycoprotein deficiency (LAD) has been recognized (*3*,*47*,*59*,*127*,*132*). These patients suffer from recurrent and often life-threatening infections (*3*). Neutrophilic granulocytes from these patients show defective *in vitro* adherence, chemotaxis and C3bi-mediated phagocytosis, degranulation and oxidative burst (*3*,*127*). Monocytes show many of the same abnormalities (*127*). In patients with LAD, the synthesis of the β chain seems abnormal, resulting in impaired expression of the leukocyte integrins on the phagocytic surface (*65*).

However, a recent study by Dana *et al.* suggests that the defect may be a post-translational defect of the β chain precursor processing rather than a lack of synthesis (*48*). LAD is easily diagnosed by flow cytometry and immunofluorescence (*3,124*).

The neutrophilic granulocyte intracellular pool of CR3 molecules is deficient in congenital neutrophil-specific granule deficiency (*102*), a disease associated with recurrent life-threatening infections (*59*). In addition, diminished CR3 expression has been demonstrated in neutrophilic granulocytes from neonatal patients (*38*) and patients with the Chediak–Higashi syndrome (*42*).

D. Phagocyte Receptors for Chemoattractants

The kinetics of internalization of the bacterial derived N-formyl peptides by PMNLs has been revealed by the use of fluorescent peptides and flow cytometry (*54*). The neutrophilic granulocyte internalization of N-formyl-nle-leu-phe-nle-tyr-lys (FLPEP) was temperature dependent, and at 37°C FLPEP was detected intracellularly after 1 min (*54*). The rate of internalization was linear and proportional to the number of receptors occupied, and following 3 min of incubation 60% of the peptide receptors were internalized (*54*). In another flow cytometric study, Seligmann *et al.* demonstrated that neutrophilic granulocytes exhibiting an apparent depolarization bound more of the peptide N-formyl-met-leu-phe-lys (FMLPL) than cells being hyperpolarized (*113*). Heterogeneous binding of chemoattractant has also been demonstrated for monocytes, as only 60% of blood monocytes bind FMLPL (*84*). In addition, only two-thirds of the monocytes which bind chemoattractant do respond with directed movement (*84*).

E. Other Functional Phagocyte Surface Molecules

In recent years a variety of other phagocyte surface molecules have been characterized and demonstrated to be of functional importance (*26,44,90,96*). Using flow cytometry, we have shown that binding of monoclonal antibodies to the myeloid differentiation antigen CD15 inhibited phagocytosis of both bacteria and zymosan particles (*26*). These results have later been confirmed by others (*40*) who also have localized intracellular neutrophilic granulocyte CD15 antigens to the secondary granules (*39*). In addition, binding of anti-CD15 MoAbs to neutrophilic granulocytes increases the adherence of these cells to endothelium by an LFA-1 (CD11a/CD18)-dependent mechanism (*57*).

Other surface structures defined by MoAbs may be involved in monocyte activation (*86*), as well as neutrophilic granulocyte chemotaxis and degranulation (*44*), phagocytosis (*119*) and oxidative burst (*90,96*). Further, flow cytometric studies have shown that the expression of certain surface antigens reflects neutrophilic granulocyte heterogeneity (*114*), and that the expression of neutrophil surface structures different from adhesion molecules and opsonin receptors may reflect cellular activation (*130,137*).

III. PHAGOCYTOSIS

The phagocytosis of invading microorganisms is an essential part of host defence, and any agent or disease that depresses or alters leukocyte phagocytic capacity will increase the host's susceptibility to infections. Some microorganisms have on their surface structures which possess antiphagocytic properties (*49,93*). The capsule of *Streptococcus pneumoniae*, the M-protein containing fimbriae of *Streptococcus pyogenes* and protein A on the surface of staphylococci are all examples of factors which inhibit the attachment of microorganisms to phagocytic cells by interfering with opsonization (*49,93*). However, not all microbial inhibition of attachment to phagocytes is related to altered opsonization. *Mycoplasma* species and piliated *Neisseria gonorrheae* attach to neutrophilic and mononuclear phagocytes independently of

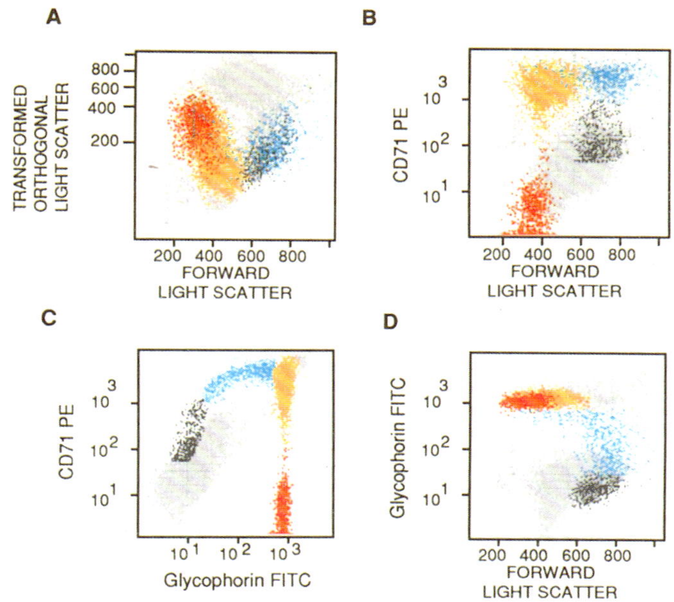

Plate 1. Four-dimensional flow cytometric analysis of low density cells from normal bone marrow stained with anti-glycophorin A FITC and CD71 PE (anti-transferin). 20 000 cells were acquired in listmode on a FACScan and displayed with the PAINT-A-GATE software. In the plate only erythroid cells are colored whereas all other cells remain gray. The correlation between forward light scattering and transformed orthogonal light scattering is shown in (A); (B) illustrates the correlation between forward light scattering and CD71 PE; (C) the correlation between glycophorin FITC and CD71 PE; and (D) the correlation between forward light scattering and glycophorin FITC. Four erythroid maturational stages were identified, each represented by a color. The most immature erythroid cells were depicted black followed by blue, gold and the most mature erythroid cells colored red.

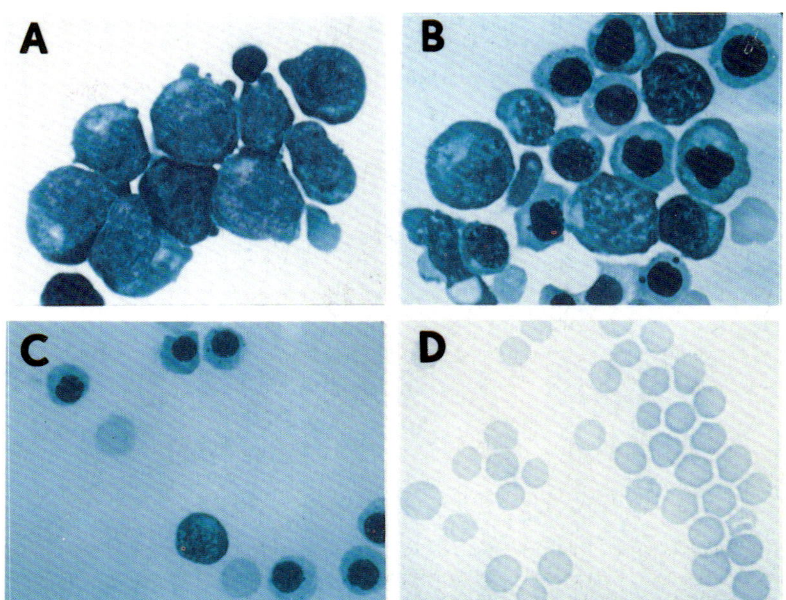

Plate 2. The cell subpopulations indicated black, blue, gold and red in Plate 1 were sorted and representative samples of Wright stained sorted cells are illustrated in (A), (B), (C), and (D) respectively. (A) Typical erythroblast; (B) early normoblast and few erythroblasts; (C) normoblast and few erythrocytes; (D) all enucleated erythrocytes.

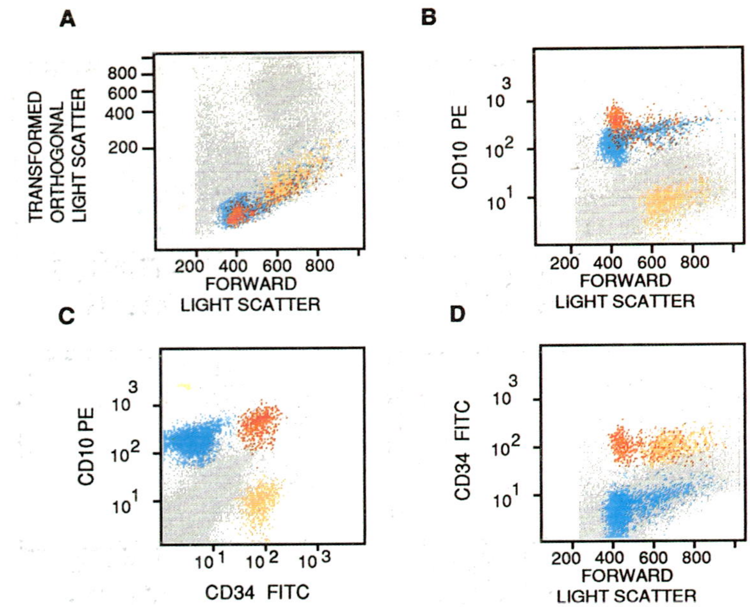

Plate 3. Four-dimensional flow cytometric analysis of a low density normal bone marrow aspirate stained with CD34 FITC (My10) and CD10 PE (Anti-CALLA). 20 000 cells were acquired in listmode on a FACScan and displayed with the PAINT-A-GATE software. The correlation between forward light scattering and transformed orthogonal light scattering is shown in (A); (B) illustrates the correlation between forward light scattering and CD10 PE; (1C) shows the correlation between CD34 FITC and CD10 PE; and (1D) the correlation between forward light scattering and CD34 FITC. The $CD34^+$, $CD10^-$ progenitors not committed to the B lymphoid cell lineage are depicted gold in the plate. The $CD34^+$, $CD10^+$ early stage BI B lymphocytes are depicted red and the $CD34^-$, $CD10^+$ stage BII and III B lymphocytes are depicted blue.

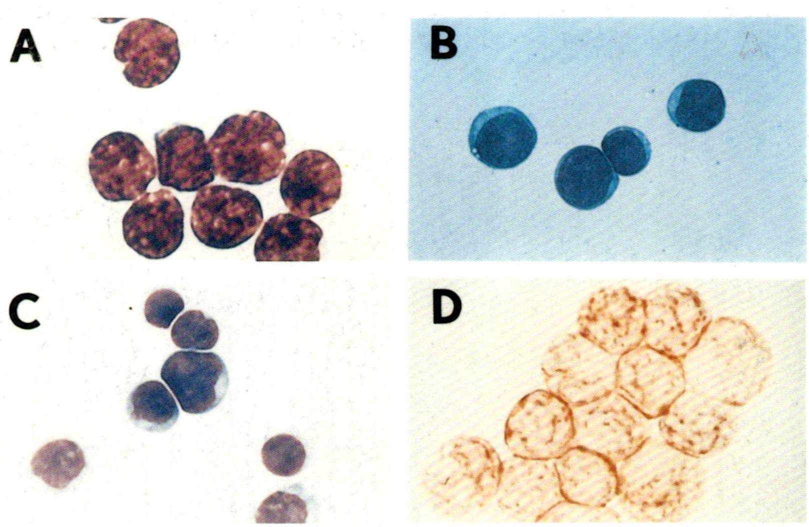

Plate 4. The cell subpopulation indicated with red (stage BI) in Plate 3 was sorted and representative samples of Wright stained cells are illustrated in (A). This population of cells was homogeneous by morphologic criteria. A slide made from the same sort was stained for the presence of nuclear TdT; all cells stained positive (D). The cell subpopulation indicated with blue (stages BII, III) in Plate 3 was sorted and representative examples of Wright stained cells are illustrated in (B). In the plate lymphoblasts are illustrated as heterogeneous in size and with distinct nucleoli. In another experiment we sorted for cells which were $CD19^+$, $CD20^+$ and $CD10^+$ (stage BIII), a representative example of which is illustrated in (C). The plate shows normal lymphocytes not clearly distinguishable from peripheral blood B lymphocytes.

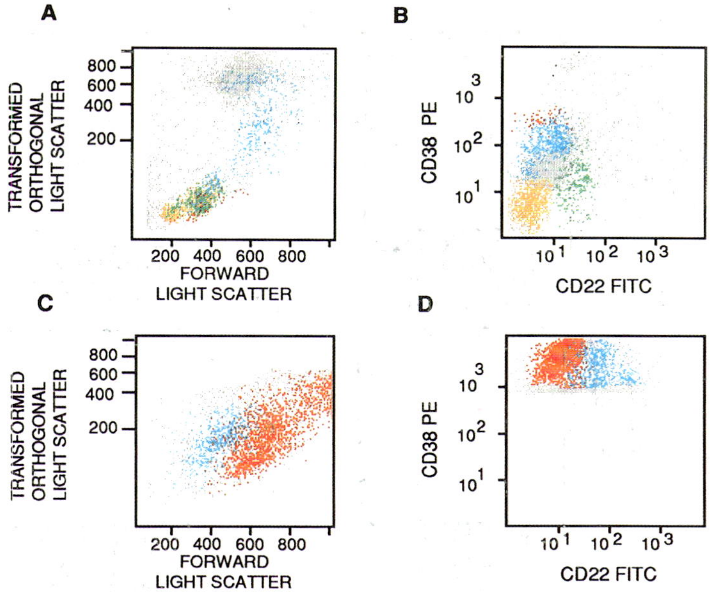

Plate 5. Four-dimensional flow cytometric analysis of an erythrocyte normal bone marrow aspirate stained with CD22 FITC (anti-Leu14) and CD38 PE (anti-Leu17). 10 000 cells were acquired in listmode on a FACScan and displayed with the PAINT-A-GATE software. The correlation between forward light scattering and transformed orthogonal light scattering is shown in (A); (B) illustrates the correlation between CD22 FITC and CD38 PE. B lymphocytes staining with CD22 and dimly or not staining with CD38 are depicted green (stage BIV). B lymphocytes brightly staining with CD38 and not staining with CD22 were depicted red (stages BI, III, III). Natural killer cells, monocytes and early monomyeloid cells are depicted blue. Lymphocytes not staining with both antibodies were depicted gold (predominantly T lymphocytes). The plasma cells are the few gray dots staining extremely brightly with CD38 in (B). To increase the number of plasma cells a threshold was used on CD38 PE in addition to a light scattering gate; two distinct populations of cells are discernable as illustrated in (C) and (D). The cells colored red were identified as mature plasma cells whereas the cells colored blue were identified as immature lymphoplasmacytoid plasma cells.

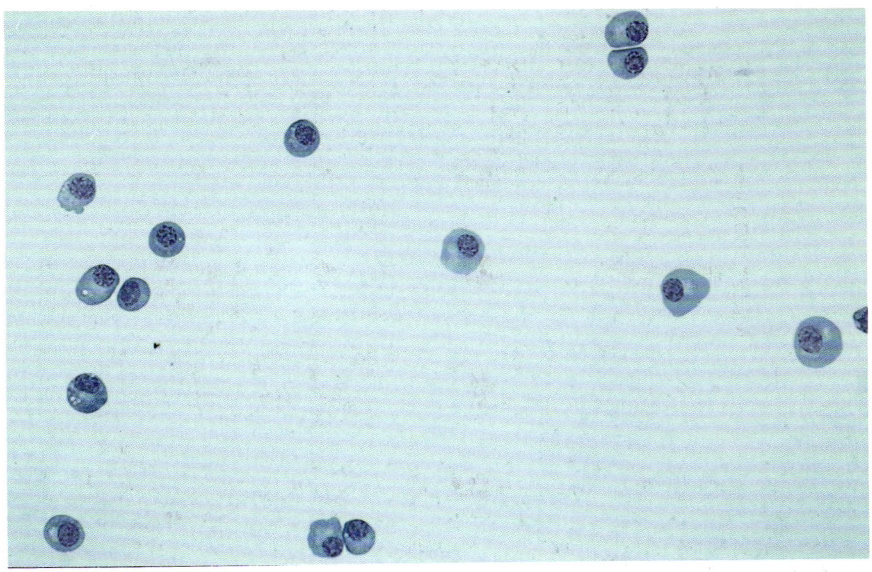

Plate 6. The cell subpopulation inferred to as plasma cells in Plate 5 and indicated with red and blue was sorted. A representative sample of Wright stained cells is illustrated in Plate 4. All cells displayed the typical morphologic features of plasma cells.

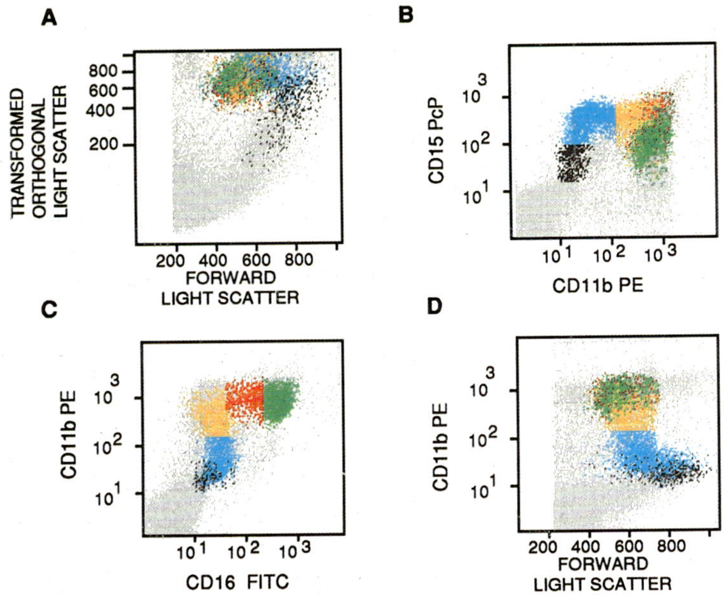

Plate 7. Five-dimensional flow cytometric analysis of an erythrocyte lysed normal bone marrow aspirate stained with CD16 FITC (My23), CD11b PE (anti-Leu15) and CD15 PerCP (L16). 30 000 cells were acquired in listmode on a FACScan and data are displayed with the PAINT-A-GATE software. The correlation between forward light scattering and transformed orthogonal light scattering is shown in (A); (B) illustrates the correlation between CD15 PerCP and CD11b PE; (1C) shows the correlation between CD16 FITC and CD11b and (1D) the correlation between forward light scattering and CD11b PE. Only the cells of neutrophilic origin were colored; all others remain gray. Five neutrophil maturational stages were identified, each represented by a color. The most immature neutrophils were depicted black followed by blue, gold, red and the most mature neutrophils were colored red.

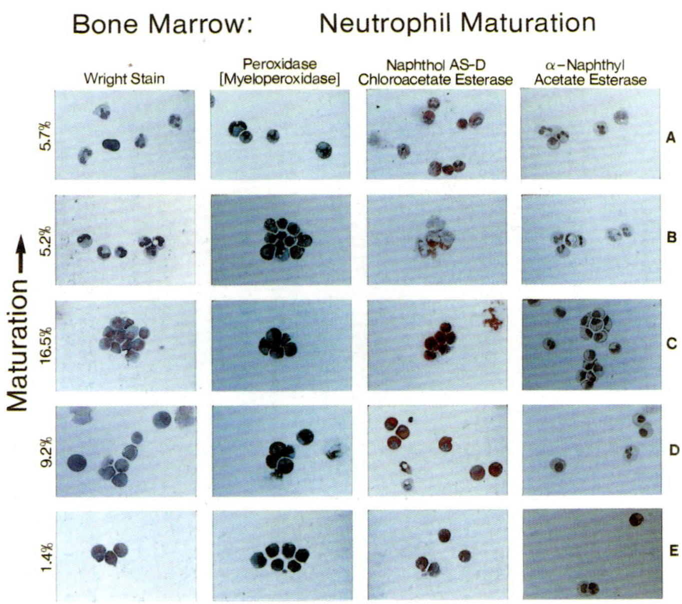

Plate 8. The cell subpopulations indicated black (E), blue (D), gold (C), red (B) and green(A) in Plate 1 were sorted and representative samples of Wright stained, myeloperoxidase stained, α-naphthol AS-D chloroacetate esterase stained and monocyte-specific α-naphthyl acetate esterase stained cells are illustrated. The different stains of each subpopulation are presented along the horizontal axis, and the different cell subpopulations are indicated along the vertical axis.

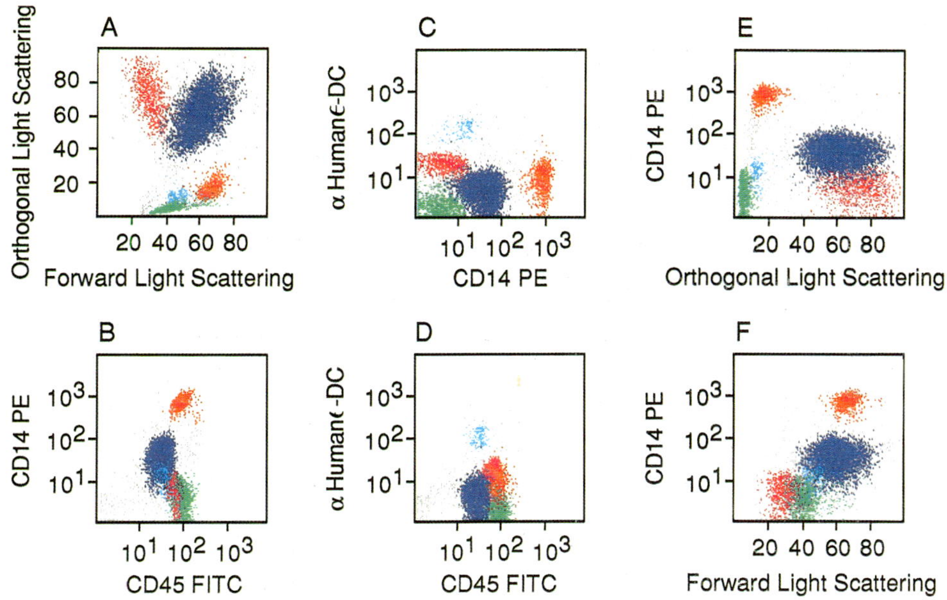

Plate 9. Erythrocyte lysed peripheral blood sample stained with α-human ϵ-biotin, Duo, CD14PE and CD45FITC. The correlation of forward and orthogonal light scattering of the cells is shown in (A), in (B) the immunofluorescence with CD14PE and CD45FITC, in (C) the immunofluorescence with α-human ϵ-biotin, Duo and CD14PE, in (D) the immunofluorescence α-human ϵ-biotin, Duo and CD45FITC, in (E) the orthogonal light scattering signals and immunofluorescence with CD14PE and in (F) the forward light scattering signals and immunofluorescence with CD14PE. The leukocyte populations were identified as unique clusters in the five-dimensional space using the PAINT-A-GATE™ computer program (Becton Dickinson, Mountain View, CA) and each cluster was assigned a color. The positions of lymphocytes, monocytes, neutrophils, eosinophils and basophils in the two-dimensional projections of the five-dimensional space are indicated with green, red, dark blue, purple and light blue respectively. Data were acquired in a FACScan and stored in list mode (Becton Dickinson, San Jose, CA).

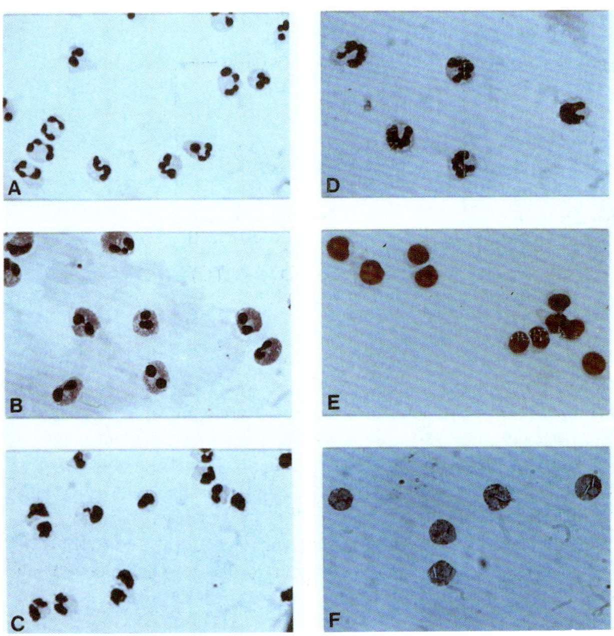

Plate 10. Wright Giemsa stain (Diff-Quik Stain Set, American Scientific Products, IL) of cells sorted by their unique characteristics of forward light scattering, orthogonal light scattering, α-human ϵ-Duochrome, CD45 fluorescein and CD14 phycoerythrine fluorescence intensity. The positions of the sorted populations are illustrated in Plate 9. (A) Neutrophils sorted on the dark blue population (Plate 9). (C) Monocytes sorted on the red population (Plate 9). (E) Eosinophils sorted on the purple population (Plate 9). (B) Lymphocytes sorted on the green population (Plate 9). (D) Basophils sorted on the light blue population (Plate 9). The granules present in the basophils did not stain with the stain set used. With different sources of Romanowsky dyes one obtains lesser or greater staining of the basophilic granules. In (F) the sorted basophils were stained with toluidine blue. Cell sorting was performed on a FACSstar-Plus (Becton Dickinson, San Jose, CA).

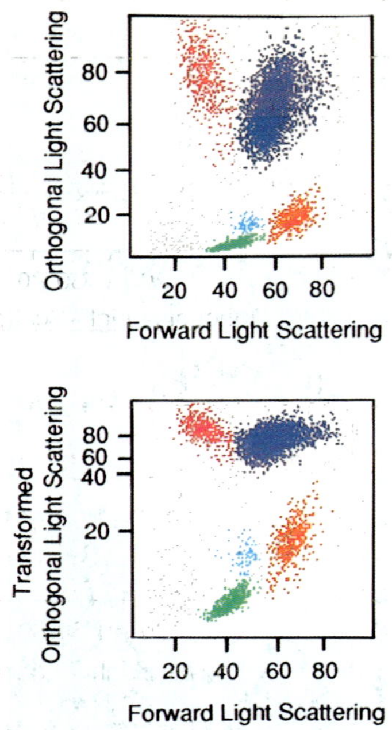

Plate 11. (A) The forward and orthogonal light scattering of the erythrocyte lysed peripheral blood cells (same preparation as is shown in Plate 9). The light scattering region typical for neutrophils is colored dark blue, for eosinophils purple, for basophils light blue, for monocytes red and for lymphocytes green. (B) The same measurement is illustrated except for a transformation of the orthogonal light scattering signals according to a polynomal function. The light scattering regions typical for the leukocytes are given the same colors as in (A). Whereas the traditional display of light scattering is monopolized by neutrophils and eosinophils (A), the lymphocyte, basophil and monocyte regions are accentuated after transformation of the orthogonal light scattering signals (B).

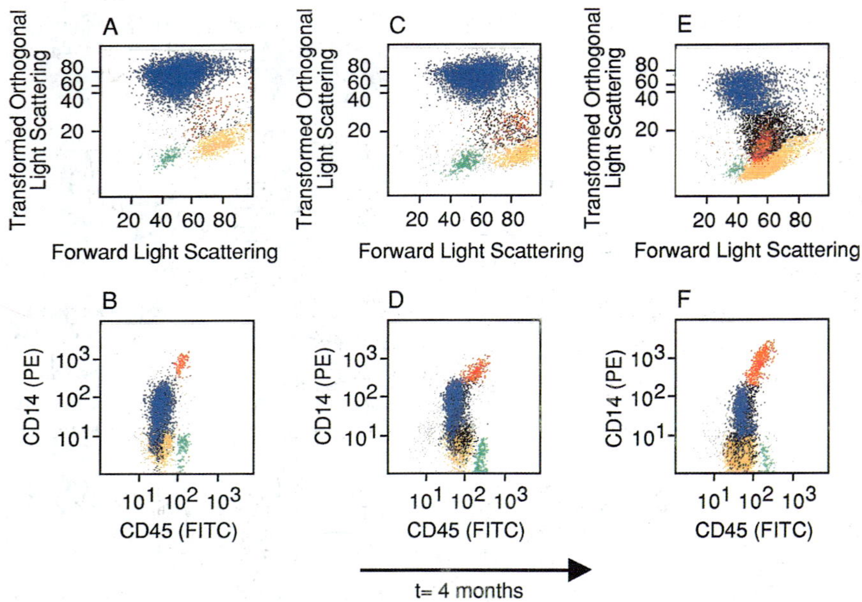

Plate 12. A patient with chronic myeloid leukemia was followed during a 4-month period during which the patient developed a blastic transformation of the leukemia. On three different time points during these 4 months peripheral blood samples were analyzed. The erythrocyte lysed samples were stained with CD45 FITC and CD14 PE. The high resolution light scattering and immunofluorescence signals of the cells are illustrated in (A) and (B) for $t = 0$, (C) and (D) for $t = 2$ months and (E) and (F) for $t = 4$ months. Five leukocyte populations were identified by their unique characteristics with forward light scattering, orthogonal light scattering and CD45 and CD14 immunofluorescence staining. Each population was assigned a color. The neutrophils are presented as dark blue, the monocytes as red, the lymphocytes as green. Two populations were identified which are not observed in peripheral blood, i.e. colored black, immature monomyeloid cells and yellow, undifferentiated cells. Note the broad distribution of the neutrophils with light scattering and CD14 immunofluorescence indicative for the presence of neutrophil precursors. Data were acquired on a FACScan (Becton Dickinson, San Jose, CA).

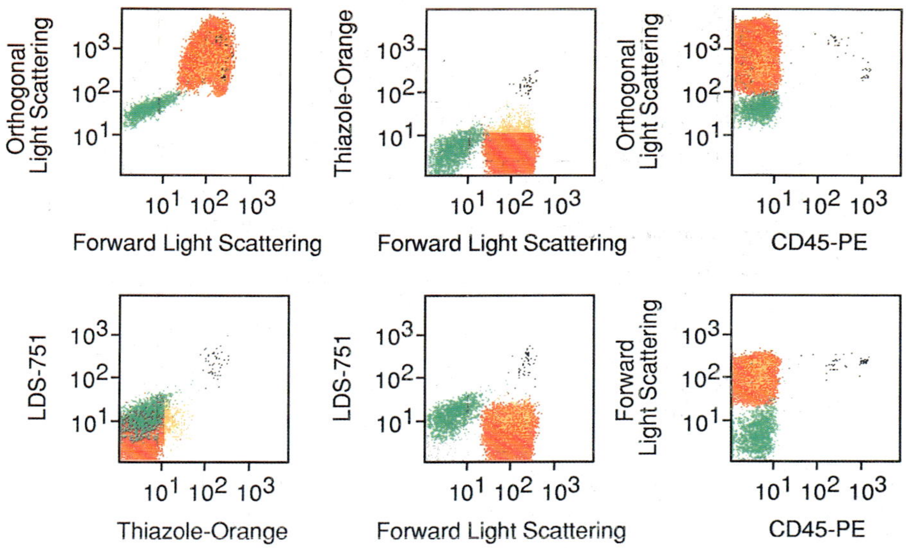

Plate 13. 10μl of a peripheral blood sample was incubated for 30 min at room temperature with 10 μl of LDS-751 (2 μg LDS-751/ml PBS), 10μl thiazole orange (1 μg thiazole orange/ml PBS) and 10 μl of a pretitered PE conjugated CD45 monoclonal antibody (Hle-1; Becton Dickinson). Forward light scattering, orthogonal light scattering, LDS-751 fluorescence, thiazole orange fluorescence and CD45 PE immunofluorescence signals were acquired into listmode on a FACScan. Six two-parameter displays of the five-dimensional data are illustrated in the plate. Four major cell clusters are identified in the five-dimensional space using the PAINT-A-GATE™ computer program (Becton Dickinson, Mountain View, CA) and each cell cluster was assigned a color. Erythrocytes are colored red (93.2%), reticulocytes are colored yellow (1.6%), nucleated cells are colored black (0.1%) and platelets are colored green (4.6%). Only a few events are not clusters into these four major cell clusters and depicted gray (0.5%). Data was acquired on a FACScan (Becton Dickinson, San Jose, CA).

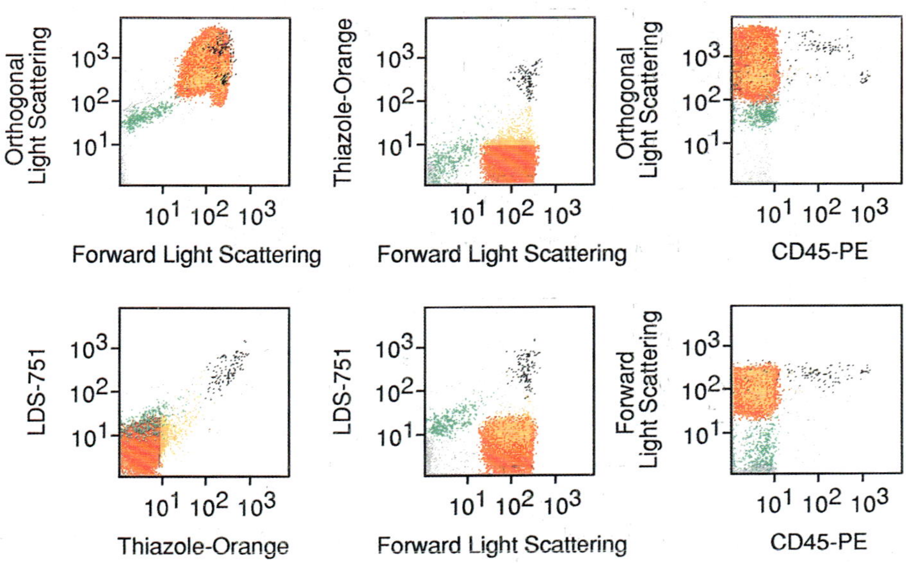

Plate 14. Bone marrow sample from the same donor as the peripheral blood sample illustrated in Plate 13. The same procedure was followed as in Plate 13. Erythrocytes are colored red (91.7%), reticulocytes are colored yellow (3.8%), nucleated cells are colored black (0.4%) and platelets are colored green (1.1%). The relative number of non-clustered events depicted as gray is larger as compared with peripheral blood (3.0%). Note that the position of the cell clusters is the same as in the peripheral blood (Plate 13). Relatively more reticulocytes are present and one can observe an increase of both thiazole orange and LDS-751 staining of the reticulocytes indicative for the presence of more immature reticulocytes. Data were acquired on a FACScan (Becton Dickinson, San Jose, CA).

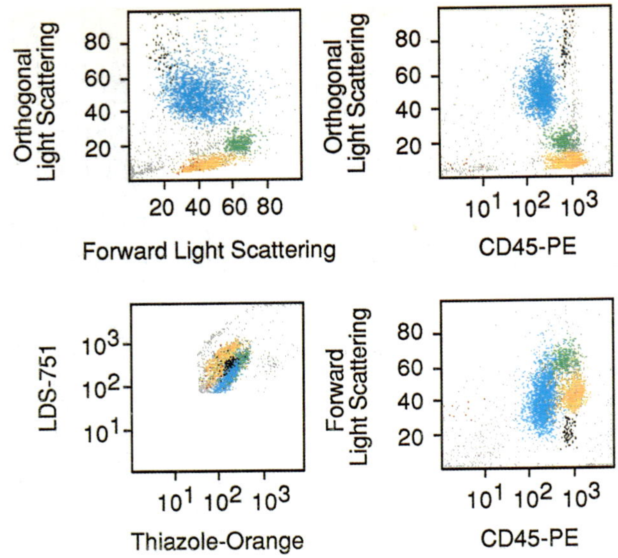

Plate 15. Detailed analysis of the nucleated cells of the peripheral blood sample presented in Plate 13. 5000 events were acquired using a gate on LDS-751 and thiazole orange fluorescence (B). In (A) the forward and orthogonal light scattering of the cells is shown; in (C) and (D) the CD45 PE fluorescence intensity with orthogonal and forward light scattering respectively. Six nucleated cell populations were identified as unique cell clusters using the PAINT-A-GATE™ computer program (Becton Dickinson, Mountain View, CA) and each cell cluster was assigned a color: neutrophils, light blue (49.9%); eosinophils, black (1.7%); monocytes, green (7.0%); lymphocytes, yellow (41.1%); nucleated red cells, red (0.2%); and immature nucleated, dark blue (0.0%). The percentages are given as a percentage of cells which are assigned a color. The events depicted as gray do not belong to the cell populations identified and likewise not clustered in one of the categories. Data were acquired on a FACScan (Becton Dickinson, San Jose, CA).

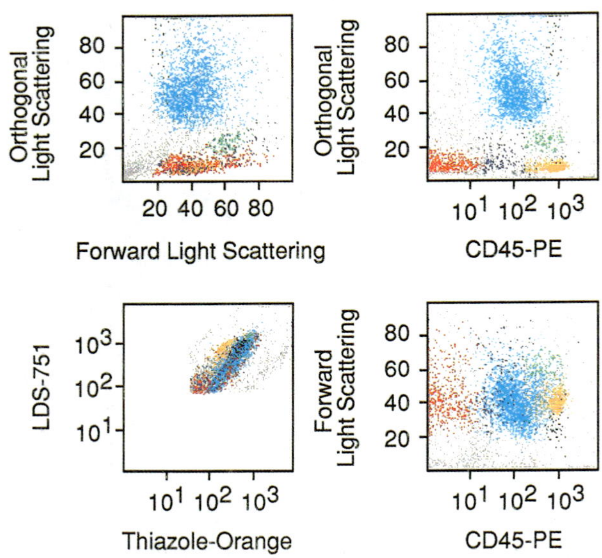

Plate 16. Detailed analysis of the nucleated cells of the bone marrow sample presented in Plate 14. The identical procedure for identification of the cell clusters was followed as in Plate 15. Neutrophils, light blue (56.3%); eosinophils, black (0.9%); monocytes, green (2.9%); lymphocytes, yellow (17.4%); nucleated red cells, red (18.5%); and immature nuleated, dark blue (4.0%). The percentages are given as a percentage of assigned cell clusters. The events depicted as gray do not belong to the cell populations identified and likewise not clustered in one of the categories. Nucleated red cells and immature nucleated cells appear in unique locations. In fact maturational pathways can be observed for all major blood cell lineages (starting with the dark blue population). Data were acquired on a FACScan (Becton Dickinson, San Jose, CA).

opsonic factors, but they resist ingestion and killing (*49*). It is therefore important when studying opsonin-dependent, as well as opsonin-independent phagocytosis, to be able to distinguish between microorganisms attached to the phagocyte surface and those actually internalized (*22,24*). In addition, since phagocytic cells in many aspects are functionally heterogeneous (*58,114*), further understanding of the pathophysiology of certain disorders of phagocyte function may require methods allowing for single cell analyses.

A. Methods

Flow cytometric assays for the measurement of phagocytosis are based on the attachment and ingestion of fluorescent particles of microorganisms by phagocytic leukocytes (*16,18,23,24,50,85,95,125,128,129,134,145*). The most widely used fluorochrome for staining of microorganisms has been FITC (*18,23*), but Texas red (*129, 134*), acridine orange (*50,145*), Lucifer yellow (*128*) and ethidium monoazide (*85*) are all examples of other applicable dyes. Following incubation of FITC-labeled bacteria or fungi with neutrophilic granulocytes or mononuclear phagocytes, free extracellular microorganisms, non-phagocytes and phagocytes are discriminated and quantified by combined measurements of FITC fluorescence and forward angle light scatter (Fig. 3) (*16,23*). The phagocytes show a distinct FITC fluorescence due to their attachment and ingestion of microorganisms, whereas the free microorganisms exhibit a less pronounced forward angle light scatter than the leukocytes, corresponding to their smaller size (Fig. 3) (*16,23*).

When determining the phagocytic capacity of a leukocyte population, it is of interest to quantify the total uptake of microorganisms, to assess the fraction of phagocytosing leukocytes and to determine whether there is significant functional heterogeneity within the phagocyte population.

Following incubation and phagocytosis, the fluorescence intensity of internalized FITC-labeled microorganisms is reduced to the phagolysosomal acidification (*15,23*). Thus, when using FITC or other pH sensitive

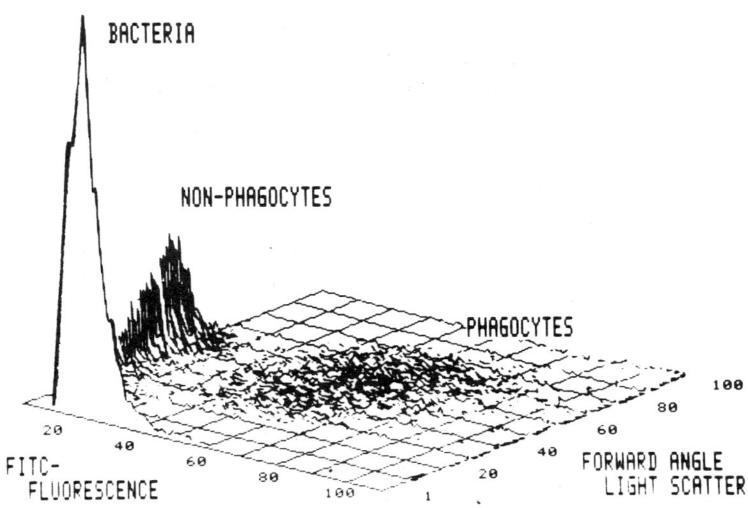

Fig. 3. Three-dimensional computer drawing showing free extracellular *Strep. pyogenes*, non-phagocyting and phagocytosing neutrophils. Preopsonized FITC-labeled bacteria (10^8/ml) were incubated with human neutrophils (5×10^6/ml) for 7.5 min before being analyzed by flow cytometry. About 16 bacteria were ingested by each neutrophil in this experiment.

fluorochromes, the uptake of microorganisms cannot be calculated from the phagocyte fluorescence. The number of bacteria or fungi ingested per cell must instead be determined from the loss of free extracellular microorganisms, i.e. by dividing the difference between the initial and final number of free extracellular microorganisms by the number of phagocytosing leukocytes (*16*,*23*). The results obtained by this assay are of the same magnitude and show good correlation with both microscopic and microbiologic methods (*16*,*17*,*23*).

If, however, the microorganisms are stained with a fluorochrome not sensitive to alterations of the environmental pH, the number of bacteria or fungi per cell can be determined from the fluorescence of each cell. Ethidium monoazide is an example of such a dye (*85*). In addition, this fluorochrome emits red fluorescence when excited at 488 nm, making it well suited for dual parameter measurements with FITC (*85*).

Since flow cytometric analyses are based on single cell measurements, the percentage of phagocytosing leukocytes (i.e. fluorescent leukocytes) and heterogeneity within the phagocyte population can be determined (*16*,*23*,*85*). In addition, the phagocytic capacity of leukocyte subpopulations can be determined by gated analyses. Monocytes, neutrophilic granulocytes and lymphocytes, identified by combined measurements of wide and forward angle light scatter, are, following ingestion of fluorescent microorganisms, analyzed for the percentage of fluorescent cells and their mean fluorescence (*23*,*85*). Laborious purification procedures, which may also interfere with phagocyte functions (*53*), can then be avoided (*23*,*85*).

However, the flow cytometric assays referred to above do not discriminate between microorganisms attached to the phagocyte surface and microorganisms actually internalized. To overcome this problem, a flow cytometric fluorescence quenching technique has been developed (*22–24*). Using FITC-labeled bacteria or fungi, the FITC fluorescence of surface bound particles was quenched with crystal violet or trypan blue, leaving the fluorescence of internalized particles unaltered (*22–24*). This flow cytometric technique allows kinetic studies of the percentage of phagocytosing leukocytes and the number of particles attached to the phagocyte surface or internalized (*22–24*).

An alternative principle for the flow cytometric discrimination between the attachment and ingestion phase of phagocytosis has been described by Sveum *et al.* (*128*). They stained bacteria with Lucifer yellow, a fluorochrome with spectral properties similar to FITC, but its fluorescence is independent of environmental pH (*128*). Following phagocytosis, mean phagocyte Lucifer yellow fluorescence was equal to the total number of phagocyte-associated bacteria. The number of surface bound bacteria was determined by using anti-Lucifer yellow antibodies conjugated with Texas red (*128*). Other investigators have used acridine orange (*145*) or combined flow cytometry with microscopic analyses of sorted phagocytes to study attachment and ingestion of microorganisms (*103*).

B. Phagocytic Cell Function

Flow cytometric phagocytosis assays have been used to study a variety of neutrophilic granulocyte and monocyte surface structures and their role in processes of attachment and ingestion. The influence of binding of MoAbs directed against specific phagocyte surface structures has been determined (*5*,*19*,*26*). When studying the possible blocking effects of a panel of MoAbs on neutrophilic granulocyte attachment and ingestion of Ig-opsonized *Staphylococcus aureus*, we demonstrated reduced attachment and ingestion by Leu-11c (FcRγIII/CD16) and IV.3 (FcγRII/CDw32), normal attachment and reduced ingestion by VIM-12 (CR3/CD11b), Mo-1 (CR3/CD11b) and VIM-D5 (CD15), whereas anti-CR1 (CR1/CD35) and VIM-2 (CDw65) did not influence the interaction between bacteria and phagocytes (Fig. 4(A)). When using C3-opsonized zymosan particles, anti-CR1

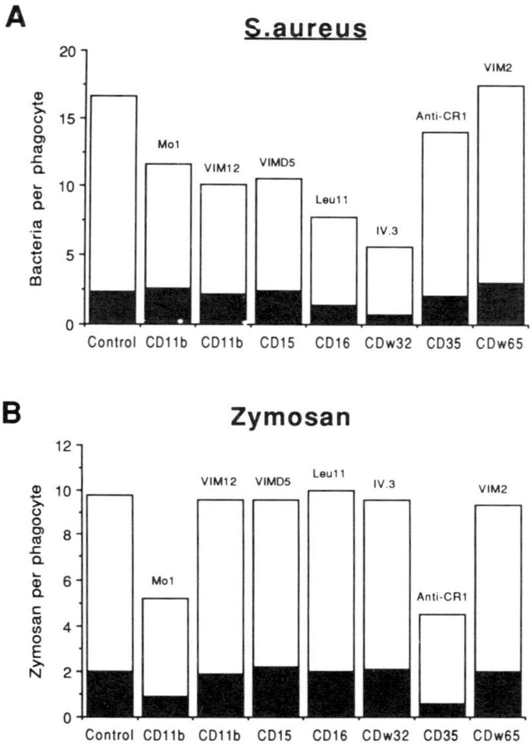

Fig. 4. The influence of antimyeloid MoAbs on human neutrophil phagocytosis of *Staph aureus* (A) and zymosan particles (B). The neutrophils were incubated with each of the MoAbs for 30 min before the cells were exposed to either preopsonized bacteria or zymosan for 15 min. The hatched area of each bar represents the mean number of bacteria or zymosan particles attached to the phagocyte surface, whereas the open area represents the mean number of bacteria or zymosan particles ingested. SE was $<10\%$ in all experiments; $n = 4$. The MoAbs used were: Mo1 (Coulter), Leu11 and Anti-CR1 (Becton Dickinson), IV.3 (MedaRex), and VIM2, VIM12 and VIMD5 (kindly provided by Dr W. Knapp, Vienna).

(CR1/CD35) and Mo-1 (CR3/CD11b) reduced attachment and ingestion, whereas VIM-D5 (CD15) reduced ingestion but not attachment (Fig. 4(B)). Neither VIM-12 (CR3/CD11b), Leu-11c (FcγRIII/CD16), IV.3 (FcγRII/CDw32) nor VIM-2 (CDw65) influenced neutrophilic granulocyte attachment or ingestion of C3-opsonized zymosan particles (Fig. 4(B)). Thus, reduced attachment and ingestion of microorganisms by neutrophilic granulocytes following binding of MoAbs to the phagocyte surface can only partly be explained by interference with opsonin receptor function alone (*26*).

It has been claimed that complement receptors (CR1 and CR3) only mediate attachment to the phagocyte surface and not internalization (*111*). Using the flow cytometric fluorescence quenching technique described above, we could show that these receptors effectively mediated internalization of C3-opsonized zymosan particles (*24*). In addition, complement-mediated attachment of new particles seemed to require internalization of particles already bound to the surface, indicating liberation or exposure of new receptors concurrently with internalization (*24*). These findings are in concordance with recent results showing that anti-CR1 MoAbs, as well as anti-CR3 MoAbs, inhibit the intracellular killing of group B streptococci by both adult and neonatal neutrophilic granulocytes (*120*).

Patients with thermal injury often die from sepsis, and impaired neutrophilic granulocyte function has been claimed to play a role in the increased susceptibility to infections (*97*). We studied the phagocytic capacity of neutrophilic granulocytes from 16 patients with large burns, and demonstrated an increased phagocytic capacity during the first days postinjury (*32,139*). Parallel measurements of neutrophilic granulocyte CR1, CR3 and FcγRIII expression demonstrated increased CR1 and CR3 density and decreased FcγRIII density (*139*), suggesting significant neutrophil activation in the early phase after injury (*53,68,139*). The increased phagocytic capacity was the result of both increased uptake of bacteria by each cell and increased percentage of phagocytosing neutrophilic granulocytes (*32*).

Using the same procedure, we have shown significant impairment of phagocytic capacity by neutrophilic granulocytes from burn patients with invasive burn wound infection and sepsis (*28*). In addition, rCSF-GM and rTNF-α increased the phagocytosis by neutrophilic granulocytes *in vitro* (unpublished results), whereas a synthetic pentapeptide and its dimer, both having hemoregulatory effects, had no effects on neutrophilic granulocyte phagocytosis (*80,81*).

C. Serum Opsonins and Phagocytosis

Phagocytes are dependent on serum opsonins for optimal function, and patients with complement or immunoglobulin deficiencies have increased susceptibility to infections (*78,93*). The application of phagocytosis assays for the study of opsonin function (*14,24*), as well as opsonic response to infection and immunization (*117,118*), can further elucidate the mechanisms involved in the interaction between microorganisms and phagocytic leukocytes (see Bjerknes, this volume for review).

IV. INTRACELLULAR KILLING

The microbicidal systems of phagocytic leukocytes are classified as being oxygen dependent (*6*) and oxygen independent (*60,122*). The oxygen-dependent mechanisms produce highly reactive oxidants from oxygen through a special metabolic pathway (*6*), whereas the oxygen-independent killing is caused by granular enzymes, cationic granule proteins and a hostile acidic phagolysosomal environment (*60,122*). The phagocyte microbicidal capacity has been estimated from the capacity of H_2O_2 production (*6,11*), but for the evaluation of the actual microbicidal capacity of a phagocyte population, it is important to directly quantify the number of microorganisms killed during phagocytosis.

A. Methods

A flow cytometric technique for the combined measurement of phagocytosis and intracellular killing of *Candida albicans* has been developed (*22*). Phagocytosis of FITC-labeled viable *C. albicans* is measured as described above. Following prescribed time intervals fungi are liberated from the phagocytes, and the fraction of fungi killed determined by flow cytometry, after staining with propidium iodide (*22*). This fluorochrome is not taken up by viable cells, whereas killed fungi are brightly stained and emit red fluorescence (*22*). Live and killed *C. albicans* are discriminated and

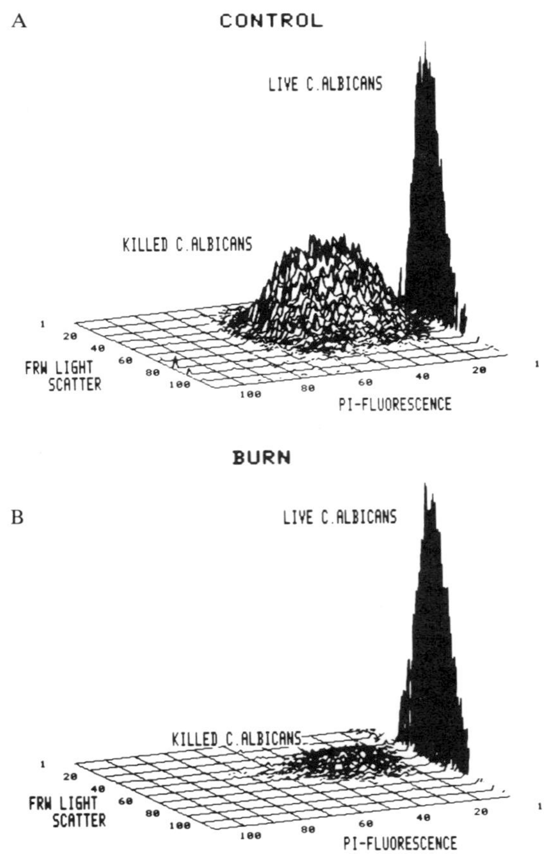

Fig. 5. Three-dimensional computer drawing showing live and phagocyte-killed *C. albicans* following incubations with neutrophils from a healthy control (A) and a patient with a 75% total body surface area burn (B). Live fungi were incubated with human neutrophils for 60 min at an initial *C. albicans* to leukocyte ratio of 5 : 1 in the presence of 10% pooled human serum. The patient leukocytes were obtained at day 15 after injury. Following 60 min of incubation the fungi were liberated from the phagocytes by hypotonic lysis and stained with propidium iodide (PI). This fluorochrome is not taken up by live fungi, whereas killed fungi are brightly stained and emit red fluorescence. About 60% of the *C. albicans* were killed in the presence of control neutrophils, whereas patient leukocytes killed about 26% of the fungi.

quantified by the combined measurement of red fluorescence and forward angle light scatter (Fig. 5) (*22*). The results obtained by this flow cytometric assay are of the same magnitude and correlate well with results obtained with a standard pour plate technique or methylene blue staining (*22*).

A flow cytometric method for the simultaneous determination of macrophage phagocytosis and intracellular killing, applying acridine orange staining, has also been reported (*67*). The authors report that, following fixation of the phagocytes with phosphate buffered 2% formalin, live intracellular *Saccharomyces cerviciae* exhibited green fluorescence, whereas live ingested yeast showed red fluorescence (*67*).

B. Experimental and Clinical Studies

The interaction between normal human neutrophilic granulocytes and *C. albicans* has been studied using flow cytometry and propidium iodide staining. Rapid phagocyte internalization and killing of *C. albicans* was demonstrated, but the capacity to kill internalized fungi had a longer time course than did ingestion (*22*). In addition, the neutrophil granulocyte capacity to kill was saturated before the capacity to ingest (*22*). A similar pattern has been observed for the neutrophilic granulocyte phagocytosis and killing of *Staphylococcus aureus* (*17*).

Infections caused by *C. albicans* have become an increasing problem in the care of burn patients (*1*). We have studied the time-related changes of neutrophilic granulocyte fungicidal capacity in patients with large burns (*32*). At admission, the intracellular killing of *C. albicans* was only slightly reduced (*32*). However, the fungicidal capacity was further compromised during the first 2 weeks after injury, with maximal impairment occurring at day 5–10 (Fig. 5) (*32*). The impairment of neutrophilic granulocyte intracellular killing was proportional to total body surface area burn (*32*), and during septic episodes, the phagocyte fungicidal capacity was further reduced (*28*).

In addition, the fungicidal capacity of granulocytes from five patients with acute myelogeneous leukemia and one patient with chronic granulomatous disease were decreased (unpublished results). In contrast, both PMNLs, monocytes and macrophages from a patient with Whipple's disease showed normal fungicidal capacity (*27,33*).

V. PHAGOLYSOSOMAL pH

Following ingestion of microorganisms, the pH of the phagolysosomal environment turns acidic (*13,15,73,112*). This acidic environment may *per se* be bactericidal to some microorganisms (e.g. pneumococci), but it is also important for the intracellular killing by the peroxidase hydrogen peroxide system (*78*). Furthermore, several digestive enzymes and microbicidal cationic proteins, dependent on a certain pH to show optimal killing capacity, are released into the phagosome during degranulation (*60,61,98,115,122,141*). Thus, the normal changes of phagolysosomal pH are important for both phagocyte intracellular killing and degradation of ingested microorganisms.

A. Methods

Several fluorochromes are sensitive to changes of pH and the combined use of these fluorochromes and flow cytometry has made measurements of the intracellular pH of individual cells possible (*94,140*). FITC is sensitive to pH changes, and the fluorescence of FITC-labeled microorganisms is reversibly reduced in an acidic environment (*13,15,23*). The reduction of FITC fluorescence of ingested microorganisms can be determined by flow cytometry, allowing for measurements of the kinetics of phagosomal acidification during phagocytosis (*13,15*). The results obtained by this assay were consistent with results obtained using microscopy and indicator stained microorganisms (*73,112*).

B. Experimental and Clinical Studies

In a study comparing the kinetics of phagolysosomal pH alterations following ingestion of seven species of bacteria it was demonstrated that the degree of phagolysosomal acidification was dependent on the type of microorganisms ingested (*13*). The phagolysosomal pH surrounding *Streptococcus pneumoniae* was measured to 4.4 after 15 min of incubation. This phagosomal pH value was significantly lower than that of the other bacteria (*13*). The phagosomal environment of *Klebsiella pneumoniae* had the highest pH, 5.7. The other bacteria studied resided in phagosomal environments of intermediate pH. For the seven microorganisms tested, maximal phagosomal acidification of normal neutrophilic granulocytes was reached within the first 15 min of incubation (*13*). In neutrophilic granulocytes from patients with chronic granulomatous disease, maximal acidification was also reached within 15 min, but the phagosomal environment was more acidic than that of the controls (*15*).

When studying the neutrophilic granulocyte microbicidal capacity in patients with thermal injury, we demonstrated significant impairment of intracellular killing (*32*). The reduced killing capacity could only partly be explained by reduced phagocyte production of reactive oxidants (*32*), and we therefore studied one aspect of the oxygen-independent microbicidal mechanisms, namely phagolysosomal changes of pH following ingestion of *Staph. aureus* (*29,32*). In control neutrophilic granulocytes, a transient phagolysosomal alkalinization with a pH of about 8.1 was documented after 5 min of incubation (Fig. 6) (*29*). This alkalinization could not be demonstrated in neutrophilic granulocytes from patients with burns, and at admission the phagolysosomal pH after 5 min of incubation was 6.0 (Fig. 6). In addition, measurements of maximal phagolysosomal acidification demonstrated a lower pH in patient phagolysosomes than in controls (*29,32*). The altered kinetics of phagolysosomal acidification persisted up to 20 days postinjury (*29*). A similar pattern of neutrophilic granulocyte dysfunction was observed when studying the pattern of phagolysosomal pH change following ingestion of *C. albicans* (*30*). An imported consequence of the normal change of phagolysosomal pH, i.e. slight alkalinization followed by neutralization and subsequent acidification, is that granular enzymes and cationic proteins will be sequentially activated and deactivated (*122*). The granules of neutrophilic granulocytes contain components of microbicidal in both alkaline (*45,122*), neutral (*61,98,141*) and acidic (*115*) environments. Hence, it is possible that the rapid fall of pH, and the more pronounced acidication of burn patient neutrophilic granulocyte phagolysosomes, alter the normal concerted and sequential activities of granular proteins, and

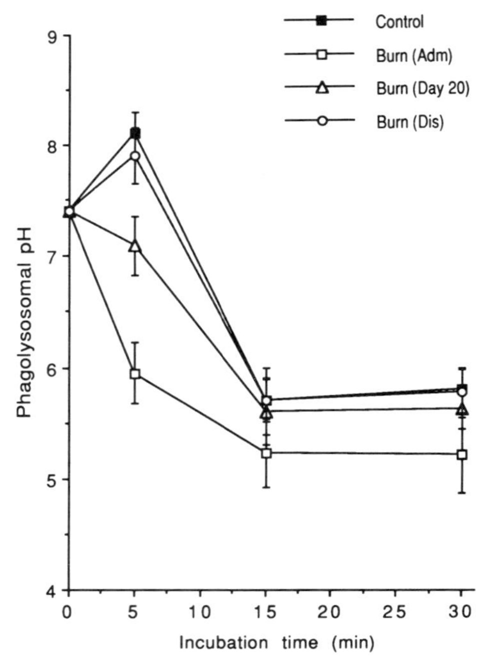

Fig. 6. The kinetics of phagolysosomal acidification after ingestion of FITC-labeled *C. albicans* in neutrophils from patients with large burns and in healthy controls. Leukocytes were obtained from the 12 patients at admission (Adm), at day 20 (Day 20) and at discharge (Dis), and incubated with fungi (10 fungi per neutrophil, and 5% pooled human serum). The data are given as means ± SEM.

thus induce reduced microbicidal activity. These results demonstrated alterations of an oxygen-independent microbicidal mechanism in phagocytic leukocytes from patients with thermal injury, which may contribute to their increased susceptibility to infection.

VI. INTRACELLULAR DEGRADATION

Following ingestion, the microorganisms are exposed to a number of hydrolytic enzymes liberated into the phagosome during degranulation (*7,78*). The killing and consecutive degradation of the microorganisms are important for effective stimulation of specific immunity for both bacteria, fungi, viruses and parasites (*51,78*)

A. Methods

Flow cytometric techniques have been developed to examine phagocytic leukocytes cytochemically, and the application of fluorogenic substrates allows time to be included for the study of enzyme kinetics (*62,76,88,105*). Recently, these principles have been applied for the measurement of phagolysosomal esterase activity (*135*). A fluorogenic substrate, decanoyl fluorescein, was coupled to microspheres, and following ingestion of the microspheres by neutrophilic granulocytes, the lysosomal esterases hydrolyzed decanoyl fluorescein (non-fluorescent) to fluorescein (fluorescent). The phagolysosomal esterase activity was then determined by direct measurement of cellular fluorescence (*135*). This technique should be applicable to other enzymes, and may be a powerful tool for the study of phagolysosomal enzyme activity, especially if combined with measurements of other phagocyte functions.

A flow cytometric technique for the study of intracellular degradation of bacterial components has also been developed (*12,13*). FITC stains cellular proteins, and the FITC fluorescence can be used to estimate the total protein content of bacteria (*8,12,46*). Similarly, ethidium bromide (EB) stains cellular DNA and RNA, and following treatment of fixed bacteria with RNAse, their EB fluorescence is proportional to their DNA content (*12,104*). For the flow cytometric study of intracellular degradation of bacteria, phagocyte-associated bacteria are liberated from the phagocytic cells by leukocyte disruption (*12,13*), and the loss of bacterial FITC and EB fluorescence estimates the phagocyte degradation of bacterial proteins and DNA, respectively (*12,13*).

B. Experimental and Clinical Studies

In a study comparing the neutrophilic granulocyte degradation of seven species of bacteria it was demonstrated that bacterial proteins were more readily digested than DNA (*13*). In addition, as has been shown by others using radiolabeling techniques (*51*), the degradation varied from one bacteria species to another (*13*). The intracellular degradation seemed to be dependent on phagolysosomal pH, as the bacteria within phagolysosomes with pH from 5.1 to 5.4 were most readily degraded (*13*). This range of pH is close to the pH optimum of RNAse, DNAse and several phagolysosomal hydrolases (*7,10*).

Monocytes and macrophages from a patient with Whipple's disease showed impaired degradation of bacterial DNA and proteins, as well as zymosan particles (*27,33*). This persistent dysfunction is consistent with the accumulation of PAS-positive bacterial degradation products seen in macrophages of affected tissue *in vivo*, suggesting a role of macrophage dysfunction in the pathogenesis of Whipple's disease (*27,33*). Defective degradation of bacteria has also been demonstrated in patients with major thermal injury (*32*), a patient with systemic lupus erythematosus (*12*), and in patients with acute myelogenous leukemia (*12*).

VII. CONCLUSION

Flow cytometric assays are now available for the study of most steps of the interaction between phagocytes, microorganisms and serum opsonins. The technology offers rapid multiparameter measurements of single cells, allowing several phagocyte functions to be analyzed simultaneously. The flow cytometric assays for the determination of phagocyte functions are applicable to both experimental and clinical work.

REFERENCES

1. Alexander, J. W. (1979). The body's response to infection. *In* "Burns: A Team Approach" (C. P. Artz, J. A. Moncrief, and B. A. Pruitt Jr, eds), pp. 107–116. W. B. Saunders, Philadelphia.
2. Anderson, C. L., Guyre, P. M., Whitin, J. C., Ryan, D. H., Looney, R. J., and Fanger, M. W. (1986). Monoclonal antibodies to Fc receptors for IgG on human mononuclear phagocytes. *J. Biol. Chem* **261**, 12856–12864.
3. Anderson, D. C., Schmalstieg, F. C., Finegold, M. J., Hughes, B. J., Rothlein, R., Miller, L. J., Kohl, S., Tosi, M. F., Jacobs, R. L., Waldorp, T. C., Goldman, A. S., Shearer, W. T., and Springer, T. A. (1985). The severe and moderate phenotypes of heritable Mac-1, LFA-1 deficiency: their quantitative definition and relation to leukocyte dysfunction and clinical features. *J. Infect. Dis.* **152**, 668–689.

Dana, N., and Colten, H. R. (1985). Increased expression of an adhesion-promoting surface glycoprotein in the granulocytopenia of hemodialysis. *N. Engl. J. Med.* **312**, 457–462.

5. Arnaout, M. A., Todd III, R. F., Dana, N., Melamed, J., Schlossman, S. F., and Colten, H. R. (1983). Inhibition of phagocytosis of complement C3- or immunoglobulin G-coated particles and of C3bi binding by monoclonal antibodies to a monocyte-granulocyte membrane glycoprotein (Mo1). *J. Clin. Invest.* **72**, 171–179.
6. Babior, B. M. (1984). Oxidants from phagocytes: Agents of defence and destruction. *Blood* **64**, 959–966.
7. Baggioloni, M. (1973). The enzymes of the granules of polymorphonuclear leukocytes and their functions. *Enzyme* **13**, 132–160.
8. Baily, J. E., Fazel-Madjlessi, J., McQuity, D. N., Lee, L. Y., Allred, J. C., and Oro, J. A. (1977). Characterization of bacterial growth by means of flow microfluorometry. *Science* **198**, 1175–1176.
9. Bainton, D. F., Miller, L. J., Kishimoto, T. K., and Springer, T. A. (1987). Leukocyte adhesion receptors are stored in peroxidase-negative granules of human neutrophils. *J. Exp. Med.* **166**, 1641–1653.
10. Barrett, A. J. (1969). Properties of lysosomal enzymes. *In* "Lysosomes in Biology and Pathology" (J. T. Dingle, and H. B. Fell, eds), pp. 244–312. North-Holland, Amsterdam.
11. Bass, D. A., Parce, J. W., DeChatelat, L. R., Szejda, P., Seeds, M. C., and Thomas, M. Flow cytometric studies of oxidative product formation by neutrophils: A graded response to membrane stimulation. *J. Immunol.* **130**, 1910–1917.
12. Bassøe, C.-F. (1984). Processing of *Staphylococcus aureus* and zymosan particles by human leukocytes measured by flow cytometry. *Cytometry* **5**, 86–91.
13. Bassøe, C.-F., and Bjerknes, R. (1985). Phagocytosis by human leukocytes, phagosomal pH, and degradation of seven species of bacteria measured by flow cytometry. *J. Med. Microbiol.* **19**, 115–125.
14. Bassøe, C.-F., and Bjerknes, R. (1984). The effect of serum opsonins on the phagocytosis of *Staphylococcus aureus* and zymosan particles measured by flow cytometry. *Acta Pathol. Microbiol. Immunol. Scand. Sect. C* **92**, 51–58.
15. Bassøe, C.-F., Laerum, O. D., Glette, J., Hopen, G., Haneberg, B., and Solberg, C. O (1983). Simultaneous measurement of phagocytosis and phagosomal pH by flow cytometry: role of polymorphonuclear neutrophilic leukocyte granules in phagosome acidification. *Cytometry* **41**, 254–262.
16. Bassøe, C.-F., Laerum, O. D., Solberg, C. O., and Haneberg, B. (1983). Phagocytosis of bacteria by leukocytes measured by flow cytometry *Proc. Soc. Exp. Biol. Med.* **174**, 182–186.
17. Bassøe, C.-F., and Solberg, C. O. (1984). Phagocytosis of Staphylococcus aureus by human leukocytes: Quantitation by a flow cytometric and a microbiological method. *Acta Pathol. Microbiol. Immunol. Scand. Sect. C.* **92**, 43–50.
18. Bassøe, C.-F., Solsvik, J., and Laerum, O. D. (1980). Quantitation of single cell phagocytic capacity by flow cytometry. *In*: 'Flow Cytometry IV' (O. D. Laerum, T. Lindmo, and E. Thorud, eds), pp. 170–174. Universitetsforlaget, Bergen.
19. Bender, J. G., Sebring, P., and Crowell, R. (1989). Granulocyte reactive monoclonal antibodies: Effects on superoxide anion production and phagocytosis. *Tissue Antigens* **33**, 210 (abstract).
20. Berger, M., O'Shea, J., Cross, A. S., Folks, T. M., Chused, T. M., Brown, E. J., and Frank, M. M. (1984). Human neutrophils increase expression of C3bi as well as C3b receptors upon activation. *J. Clin. Invest.* **74**, 1566–1671.
21. Berger, M., Wetzler, E. M., and Wallis, R. S. (1988). Tumor necrosis factor is the major monocyte product that increases complement receptor expression on mature human neutrophils. *Blood* **71**, 151–158.
22. Bjerknes, R. (1984). Flow cytometric assay for combined measurement of phagocytosis and intracellular

killing of *Candida albicans. J. Immunol. Methods* **72**, 229–241.

23. Bjerknes, R., and Bassøe, C.-F. (1983). Human leukocyte phagocytosis of zymosan particles measured by flow cytometry. *Acta Pathol. Microbiol. Immunol. Scand. Sect. C* **91**, 341–348.
24. Bjerknes, R., and Bassøe, C.-F. (1984). Phagocyte C3-mediated attachment and internalization: Flow cytometric studies using a fluorescence quenching technique. *Blut* **49**, 315–323.
25. Bjerknes, R., Bassøe, C.-F., Sjursen, H., Laerum, O. D., and Solberg, C. O. (1989). Low cytometry for the study of phagocyte functions. *Rev. Infect. Dis.* **11**, 16–33.
26. Bjerknes, R., Laerum, O. D., and Knapp, W. Inhibition of phagocytosis by monoclonal antibodies to human myeloid differentiation antigens. *Exp. Hematol.* **12**, 856–862.
27. Bjerknes, R., Laerum, O. D., and Ødegaard, S. (1985). Impaired bacterial degradation by monocytes and macrophages from a patient with treated Whipple's disease. *Gastroenterology* **89**, 1139–1146.
28. Bjerknes, R., and Vindenes, H. (1991). Impairment of human nuetrophils in patients with burns and sepsis. Submitted for publication.
29. Bjerknes, R., and Vindenes, H. (1989). Neutrophil dysfunction after thermal injury. Alteration of phagolysosomal acidification in patients with large burns. *Burns* **15**, 77–81.
30. Bjerknes, R., Vindenes, H., and Laerum, O. D. (1990). Altered neutrophil functions in patients with large burns. *Blood Cells* **16**, 127–143.
31. Bjerknes, R., Vindenes, H., Pitkänen, J., Laerum, O. D., and Åbyholm, F. (1988). Multiparameter determination of monocyte functions in patients with large burns. *Cytometry* Suppl. 2, 74 (abstract).
32. Bjerknes, R., Vindenes, H., Pitkänen, J., Ninneman, J., Laerum, O. D., and Åbyholm, F. (1989). Alterations of polymorphonuclear neutrophilic granulocyte functions in patients with large burns. *J. Trauma* **29**, 847–855.
33. Bjerknes, R., Ødegaard, S., Bjerkvig, R., Børkje, B., and Laerum, O. D. (1988). Whipple's disease: Demonstration of a persisting monocyte and macrophage dysfunction. *Scand. J. Gastroenterol.* **23**, 611–619.
34. Bonnefoy, J. Y., Banchereau, J., Aubry, J. P., and Wijdenes, J. (1986). A flow cytometric micromethod for the detection of Fcε receptors and IgE binding factors using fluorescent microspheres. *J. Immunol. Methods* **88**, 25–32.
35. Boxer, L. A., and Smolen, J. E. (1988). Neutrophil granule constituents and their role in health and disease. *Hemat. Oncol. Clin. North Am.* **2**, 101–134.
36. Breard, J., Reinhertz, E. L., Kung, P. C., Goldstein, G., and Schlossman, S. F. (1980). A monoclonal antibody reactive with human peripheral blood monocytes. *J. Immunol.* **124**, 1943–1948.
37. Brown, E. J., Bohnsack, J. F., and Gresham, H. D. (1988). Mechanism of inhibition of immuoglobulin G-mediated phagocytosis by monoclonal antibodies that recognize the Mac-1 antigen. *J. Clin. Invest.* **81**, 365–375.
38. Bruce, M. C., Baley, J. E., Medvik, K. A., and Berger, M. (1987). Impaired surface membrane expression of C3bi but not C3b receptors on neonatal neutrophils. *Pediatric Res.* **21**, 306–311.
39. Buescher, E. S., Livesey, S. A., Skubitz, K. M., and McIlheran, S. M. (1989). Biochemical, physical and ultrastructural localization of human granulocyte CD-15 antigens to the secondary granule. *Tissue Antigens* **33**, 213 (abstract).
40. Buescher, E. S., and McIlheran, S. M. (1989). Studies of the effects of putative CD-15 murine monoclonal antibodies on human polymorphonuclear leukocyte phagocytosis. *Tissue Antigens* **33**, 213 (abstract).
41. Bullock, W. E., and Wright, S. D. (1987). Role of the adherence-promoting receptors, CR3, LFA-1, and p150,95, in the binding of Histoplasma capsulatum by human macrophages. *J. Exp. Med.* **165**, 195–210.
42. Cairo, M. S., Vandeven, C., Toy, C., Tischler, D., and Sender, L. (1988). Fluorescent cytometric analysis of polymorphonuclear leukocytes in Chediak-Higashi syndrome: Diminished C3bi receptor expression (OKM1) with normal granular cell density. *Pediatric Res.* **241**, 673–676.
43. Capron, M., Kusnierz, J. P., Prin, L., Spiegelberg, H. L., Ovlaque, G., Gosset, P., Tonnel, A.-B., and Capron, A. (1985). Cytophilic IgE on human blood and tissue eosinophils: Detection by flow microfluorometry. *J. Immunol.* **134**, 3013–3018.
44. Cotter, T. G., Spears, P., and Henson, P. M. (1981). A monoclonal antibody inhibiting human neutrophil chemotaxis and degranulation. *J. Immunol.* **127**, 1355–1360.
45. Craven, N., Williams, M. R., Field, T. R., Bunch, K. J., Mayer, S. J., and Bourne, F. J. (1986). The influence of extracellular and phagolysosomal pH changes on the bactericidal activity of bovine neutrophils against *Staphylococcus aureus. Vet. Immunol. Immunopharmacol.* **13**, 97–110.
46. Crissman, H. A., Darzynkiewicz, Z., Toby, R. A., and Steinkamp, J. A. (1985). Correlated measurements of DNA, RNA, and protein in individual cells by flow cytometry. *Science* **228**, 1321–1324.
47. Crowley, C. A., Curnette, J. T., Rosin, R. E., André-Schwartz, J., Gallin, J. I., Klempner, M., Snyderman, R., Southwick, F. S., Stossel, T. P., and Babior, B. M. (1980). An inherited abnormality

of neutrophil adhesion. *N. Engl. J. Med.* **302**, 1163–1168.

48. Dana, N., Clayton, L. K., Tennen, D. G., Pierce, M. W., Lachman, P. J., Law, S. J., and Arnaout, M. A. (1987). Leukocytes from four patients with complete or partial leu-CAM deficiency contain the common beta-subunit precursor and beta-subunit precursor messenger RNA. *J. Clin. Invest.* **79**, 1010–1015.
49. Denson, P., and Mandell, G. L. (1980). Phagocyte strategy vs microbial tactics. *Rev. Infect. Dis.* **2**, 817–838.
50. Derer, M., Walker, C., Kristensen, F., and Reinhardt, M. C. (1983). A simple and rapid flow cytometric method for routine assessment of Baker's yeast by human polymorphonuclear leukocytes. *J. Immunol. Methods* **61**, 359–365.
51. Elsbach, P. (1980). Degradation of microorganisms by phagocytic cells. *Rev. Infect. Dis.* **2**, 106–128.
52. Fanger, M. W., Goldstine, S. N., and Shen, L. (1983). Cytofluorographic analysis of receptors for IgA on human polymorphonuclear cells and monocytes and the correlation of receptor expression with phagocytosis *Mol. Immunol.* **20**, 1019–1027.
53. Fearon, D. T., and Collins, L. A. (1983) Increased expression of C3Bb receptors on polymorphonuclear leukocytes induced by chemotactic factors and by purification procedures. *J. Immunol.* **130**, 370–375.
54. Finney, D. A., and Sklar, L. A. (1983). Ligand/receptor internalization: A kinetic, flow cytometric analysis of the internalization of N-formyl peptides by human neutrophils. *Cytometry* **4**, 54–60.
55. Fischer, G. G., Majdic, O., and Knapp, W. (1987). Altered expression of surface antigens after stimulation of human granulocytes. *In* "Leukocyte Typing III" (A. L. Mc.Michael, ed.), pp. 698–702. Oxford University Press, Oxford.
56. Fleit, H. B. (1989). Fc and complement receptor (CR1 and CR3) expression on neonatal human polymorphonuclear leukocytes. *Biol. Neonate.* **55**, 156–163.
57. Forsyth, K. D., Simpson, A. C., and Levinsky, R. J. (1989). CD15 antibodies increase neutrophil adhesion to endothelium by an LFA-1-dependent mechanism. *Eur. J. Immunol.* **19**, 1331–1334.
58. Gallin, J. I. (1980). Human neutrophil heterogeneity exists, but is it meaningful? *Blood* **63**, 977–983.
59. Gallin, J. I. (1985) Leukocyte adherence-related glycoproteins LFA-1, Mo-1, and p150,95: A new group of monoclonal antibodies, a new disease and a possible opportunity to understand the molecular basis of leukocyte adherence. *J. Infect. Dis.* **152**, 661–664.
60. Ganz, T., Selsted, M. E., and Lehrer, R. I. (1990) Defensins. *Eur. J. Haematol.* **44**, 1–8.
61. Ganz, T., Selsted, M. E., Szklarek, D., Harwig, S. S. L., Daher, K., Bainton, D. F., and Lehrer, R. I. (1985). Defensins: natural peptide antibiotics of human neutrophils. *J. Clin. Invest.* **76**, 1427–1435.
62. Haskill, S., and Becker, S. (1982). Flow cytometric analysis of macrophage heterogeneity and differentiation: utilization of electronic cell volume and fluorescent substrates corresponding to common macrophage markers. *J. Reticuloendothel. Soc.* **32**, 273–285.
63. Hickstein, D. D., Locksley, R. M., Beatty, P. G., Smith, A., Stone, D. M., and Root, R. K. (1986). Monoclonal antibodies binding to the human neutrophil C3bi receptor have disparate functional effects. *Blood* **67**, 1054–1062.
64. Hill, H. R. (1987). Biochemical, structural, and functional abnormalities of polymorphonuclear leukocytes in the neonate. *Pediatr. Res.* **22**, 375–382.
65. Hogg, N. (1988). The structure and function of Fc receptors *Immunol. Today* **9**, 185–187.
66. Hudson, J. L., Duque, R. E., and Lovett III, E. J. (1985). Applications of flow cytometry in immunotoxicology. *In* "Immunotoxicology and Immunopharmacology" (J. Dean *et al.*, eds), pp. 159–177. Raven Press, New York.
67. Hudson, J. L., Schol, H. M., Stennis, R. V., Steele, R. V., Anson, J. A., Miller, G. A., Hinson, W. G., and Pipkin Jr, J. L. (1981). Multiparameter flow-cytometric analysis of macrophage phagocytosis and killing. *Cytometry* **2**, 2 (abstract).
68. Huizinga, T. W. J., van der Schoot, C. E., Jost, C., Klaassen, R., Kleijer, M., von dem Borne, A. E. G. K., Roos, D., and Tetteroo, P. A. T. (1988). The PI-linked receptor FcRIII is released on stimulation of neutrophils. *Nature* **333**, 667–669.
69. Huizinga, T. W. J., van Kemenade, F., Koenderman, L., Dolman, K. M., von dem Borne, A. E. G. K., Tetteroo, P. A. T., and Roos, D. (1989). The 40-kDa Fcγ receptor (FcRII) on human neutrophils is essential for the IgG-induced respiratory burst and the IgG-induced phagocytosis. *J. Immunol.* **142**, 2365–2369.
70. Huu, T. P., Chollet-Martin, S., Amit, N., Debre, P., and Hakim, J. (1987). Monoclonal antibodies to granulocyte determinants: Effects on granulocyte respiratory burst. *In* "Leukocyte Typing III" (A. L. Mc.Michael, ed.), pp. 695–697. Oxford University Press, Oxford.
71. Huu, T. P., Chollet-Martin, S., Perianin, A., Marquetty, C., Sourbier, P., Babin-Chevaye, C., Olive, D., Gougerot-Pocidalo, M.-A., Debre, P., and Hakim, J. (1987). Comparison of blocking effects of monoclonal antibodies anti-MO1-α and anti-LFA1-α on human neutrophil functions. *Immunology* **62**, 61–67.
72. Ismail, G., Moranroth, M. L., Todd III, R. F., and Boxer, L. A. (1987). Prevention of pulmonary

injury in isolated perfused rat lungs by activated human neutro-phils preincubated with anti-Mo1 monoclonal antibody. Blood *69*: 1167–1174.

73. Jacques, Y. V., and Bainton, D. F. (1978). Changes in pH within the phagocytic vacuoles of human neutrophils and monocytes. *Lab. Invest.* **39**, 179–185.

74. Jayram, U., Buckle, A. J., and Hogg, N. (1989). The Fc receptor, FcRI, and other activation molecules on human mononuclear phagocytes after treatment with interferon-gamma. *Clin. Exp. Immunol.* **75**, 414–420.

75. Jones, D. H., Anderson, D. A., Burr, B. L., Rudloff, H. E., Smith, C. W., Krater, S. S., and Schmalstieg, F. C. (1988). Quantitation of intracellular Mac-1 (CD11b/CD18) pools in human neutrophils. *J. Leukocyte Biol.* **44**, 535–544.

76. Kaplow, L. S. (1977). The application of cytochemistry to automation. *J. Histochem. Cytochem.* **25** 990–1000.

77. Klebanoff, S. J., Beatty, P. G., Schreiber, R. D., Ochs, H. D., and Waltersdorph, A. M. (1985). Effect of antibodies directed against complement receptors on phagocytosi by polymorphonuclear leukocytes: Use of iodination as a convenient measure of phagocytosis. *J. Immunol.* **134**. 1153–1159.

78. Klebanoff, S. J., and Clark, R. A. (1978). "The Neutrophil: Function and Clinical Disorders." Elsevier/North-Holland, Amsterdam.

79. Knapp, W. (1989). "Leukocyte Typing IV". Oxford University Press, Oxford.

80. Laerum, O. D., Sletvold, O., Bjerknes, R., Eriksen, J. A., Johansen, J. H., Schanche, J. S., Tveterås, T., and Paukovits, W. R. (1988). The dimer of hemoregulatory peptide (HP5B) stimulates mouse and human myelopoiesis *in vitro*. *Exp. Hematol* **16**, 274–280.

81. Laerum, O. D., Sletvold, O., Bjerknes, R., and Paukovits, W. R. (1987). A synthetic hemoregulatory peptide (HP5B) inhibits human myelopoietic colony formation (CFU-GM) but not leukocyte phagocytosis *in vitro*. *Eur. J. Haematol.* **89**, 86–93.

82. Lambris, J. D. (1988). The multifunctional role of C3, the third component of complement. *Immunol. Today* **9**, 387–393.

83. Landay, A. L., and Muirhead, K. A. (1989). Procedural guidelines for performing immunophenotyping by flow cytometry. *Clin. Immunol. Immunopathol.* **52**, 48–60.

84. Leonard, E. J., Noer, K., and Skeel, A. (1985). Analysis of human monocyte chemoattractant binding by flow cytometry. *J. Leukocyte Biol.* **38**, 403–413.

85. Lund-Johansen, F., Bjerknes, R, and Laerum, O. D. (1990). Flow cytometric assay for the measurement of human bone marrow phenotype, function and cell cycle. *Cytometry* **11**, 610–616.

86. Lund-Johansen, F., Olweus, J., Aarli, Å, and Bjerknes, R. (1990) Signal transduction in human monocytes and granulocytes through the PI-linked antigen CD14. *FEBS Lett.* **273**, 610–616.

87. Lund-Johansen, F., Olweus, J., Aarli, Å. and Bjerknes, R. (1991) The IgG FcRII and the PI-linked IgG FcRIII trigger cytoplasmic calcium fluxes independently in human granulocytes. *Scand. J. Immunol.* **33**, 261–266.

88. Malin-Berdel, J., and Valet, G. (1980). Flow cytometric determination of esterase and phosphatase activities and kinetics in hematopoietic cells with fluorogenic substrates. *Cytometry* **1**, 222–228.

89. Melamed, M. R., Lindmo, T. and Mendelsohn, M. L. (1990) "Flow Cytometry and Sorting", 2nd edn. Wiley-Liss, New York.

90. Melnick, D. A., Nasueef, W. M., Markowitz, S. D., Gardner, J. P., and Malech, H. L. (1985). Biochemical analysis and subcellular localization of a neutrophil-specific antigen, PMN-7, involved in the respiratory burst. *J. Immunol.* **134**, 3346–3355.

91. Miller, L. J., Bainton, D. F., Borregaard, N., and Springer, T. A. (1987). Stimulated mobilization of monocyte Mac-1 and p150,95 adhesion proteins from an intracellular vesicular compartment to the cell surface *J. Clin. Invest.* **80**, 535–544.

92. Mills, E. L., and Quie, P. G. (1980). Congenital disorders of the functions of polymorphonuclear neutrophils. *Rev. Infect. Dis.* **2**, 505–517.

93. Mims, C. A. (1982). "The Pathogenesis of Infectious Disease". Academic Press, New York.

94. Musgrove, E., Rugg, C., and Hedley, D. (1986). Flow cytometric measurement of cytoplasmic pH: A critical evaluation of available fluorochromes. *Cytometry* **7**, 347–355.

95. Nagel, J. E., Han, K., Coon, P. J., Adler, W. H., and Bender, B. S. (1986). Age differences in phagocytosis by polymorphonuclear leukocytes measured by flow cytometry. *J. Leukocyte Biol.* **39**, 399–407.

96. Nauseef, W. M., Root, R. K., Newman, S. L., and Malech, H. L. (1983). Inhibition of zymosan activation of human neutrophil oxidative metabolism by a mouse monoclonal antibody. *Blood* **62**, 635–644.

97. Ninnemann, J. L. (1981). "The Immune Consequences of Thermal Injury." Williams and Wilkins, Baltimore.

98. Odeberg, H., and Olsson, I. (1975). Antibacterial activity of cationic proteins from human granulocytes. *J. Clin. Invest.* **56**, 1118–1124.

99. Ogle, J. D., Noel, J. G., Sramkoski, R. M., Ogle, C. K., and Alexander, J. W. (1988). Phagocytosis of opsonized fluorescent microspheres by human neutrophils. A two-color flow cytometric method for the determination of attachment and ingestion. *J. Immunol. Methods* **115**, 17–29.

100. Ogle, J. D., Ogle, C. K., Noel, J. G., Hurtubise, P., and Alexander, J. W. (1985). Studies on the binding of C3b-coated microspheres to human neutrophils. *J. Immunol. Methods* **76**, 47–62.

101. Ory, P. A., Goldstein, I. M., Kwoh, E. E., and Clarkson, S. B. (1989). Characterization of polymorphic forms of Fc receptor III on human neutrophils. *J. Clin. Invest* **83**, 1676–1681.

102. O'Shea, J. J., Brown, E. J., Seligmann, B. E., Metcalf, J. A., Frank, M. M., and Gallin, J. I. (1985). Evidence for distinct intracellular pools of receptors for C3b and C3bi in human neutrophils. *J. Immunol.* **134**, 2580–2587.

103. Parod, R. J., and Brain, J. D. (1986). Immune opsonin-independent phagocytosis by pulmonary macrophages. *J. Immunol.* **136**, 2041–2047.

104. Paau, A. S., Lee, D., and Cowles J. R. (1977). Comparison of nucleic acid content in populations of free-living and symbiotic *Rhizobium meliloti* by flow microfluorometry. *J. Bacteriol.* **129**, 1156–1158.

105. Ross, D. W. (1986). Measurement of myeloid maturation by flow cytochemistry in HL-60 leukemia: esterase is inducible, myeloperoxidase is not. *Cytometry* **7**, 263–267.

106. Ross, G. D., Cain, J. A., and Lachmann, P. J. (1985). Membrane complement receptor type 3 (CR3) has lectin-like properties analogous to bovine conglutinin and functions as a receptor for zymosan and rabbit erythrocytes as well as a receptor for C3b. *J. Immunol.* **134**, 3307–3315.

107. Rossman, M. D., Chen, E., Chien, P., and Schreibner, A. D. (1989). Modulation of Fcγ receptors on the human macrophage cell line U-937. *Cell. Immunol.* **119**, 174–187.

108. Russ, G. R., Haddad, A. P., Tait, B. D., and d'Apice, A. J. F. (1985). Polymorphism of the complement receptor for C3bi. *J. Clin. Invest.* **76**, 1965–1970.

109. Ryan, D. H., Fallon, M. A., and Horan, P. K. (1988). Flow cytometry in the clinical laboratory. *Clin. Chim. Acta* **171**, 125–174.

110. Sanchez-Madrid, F., Nagy, J. A., Robbins, E., Simon, P., and Springer, T. A. (1983). A human leukocyte differentiation antigen family with distinct a-subunits and a common β-subunit. *J. Exp. Med.* **158**, 1785–1803.

111. Scribner, D. J., and Fahrney, D. (1976). Neutrophil receptors for IgG and complement: Their roles in the attachment and ingestion phases of phagocytosis. *J. Immunol.* **116**, 892–897.

112. Segal, A. W., Geisow, M., Garcia, R., Harper, A., and Miller, R. (1981). The respiratory burst of phagocytic cells is associated with a rise in vacuolar pH. *Nature* **290**, 406–409.

113. Seligmann, B. E., Chused, T. M., and Gallin, J. I. (1984). Differential binding of chemoattractant peptide to subpopulations of neutrophils. *J. Immunol.* **133**, 2641–2646.

114. Seligmann, B., Malech, H. L., Melnick, D. A., and Gallin, J. I. (1985). An antibody binding to human neutrophils demonstrates antigenic heterogeneity detected early in myeloid maturation which correlates with functional heterogeneity of mature neutrophils. *J. Immunol.* **135**, 2647–2653.

115. Shafer, W. H., Martin, L. E., and Spitznagel, J. K. (1986). Late intraphagosomal hydrogen ion concentration favors the *in vitro* antimicrobial capacity of a 37-kilodalton cationic granule protein of human neutrophil granulocytes. *Infect. Immun.* **53**, 651–655.

116. Shalit, M., von Allmen, C., Atkins, P. C., and Zweiman, B. (1988). Platelet activating factor increases expression of complement receptors on human neutrophils. *J. Leukocyte Biol.* **44**, 212–217.

117. Sjursen, H., Bjerknes, R., Haltensen, A., Naess, A., Frøholm, O., Rosenqvist, E., and Solberg, C. O. (1987). Serum opsonins to group B meningococci. *Acta Pathol. Microbiol. Immunol. Scand. Sect. C* **95**, 283–289.

118. Sjursen, H., Bjerknes, R., Halstensen, A., Naess, A., Soernes, S., and Solberg, C. O. (1989). Flow cytometric assay for the measurement of serum opsonins to *Neisseria meningitidis* serogroup B, serotype 15. *J. Immunol. Methods* **116**, 235–243.

119. Skubitz, K. M., Weisdorf, D. J., and Peterson, D. K. (1985). Monoclonal antibody AHN-1 inhibits phagocytosis by human neutrophils. *Blood* **65**, 333–339.

120. Smith, C. L., Baker, C. J., Anderson, D. C. and Edwards, M. S. (1990). Role of complement receptors in opsonophagocytosis of group B streptococci by adult and neonatal neutrophils. *J. Infect. Dis.* **162**, 489–495.

121. Socinski, M. A., Cannistra, S. A., Sullivan, R., Elias, A., Antman, K., Schnipper, L., and Griffin, J. D. (1988). Granulocyte-macrophage colony-stimulating factor induced the expression of the CD11b surface adhesion molecule on human granulocytes *in vivo*. *Blood* **72**, 691–697.

122. Spitznagel, J. K. (1984). Nonoxidative antimicrobial reactions of leukocytes. *In* "Regulation of Leukocyte Function. Contemporary Topics in Immunobiology" (R. Snyderman, ed.), pp. 283–343. Plenum Press, New York.

123. Springer, T. A., Galfre, G., Secher, D. S., and Milstein, C. (1979). Mac-1: a macrophage differentiation antigen identified by a monoclonal antibody. *Eur. J. Immunol.* **9**, 301–306.

124. Springer, T. A., Thompson, W. S., Miller, L. J., Schmalstieg, F. C., and Anderson, D. C. (1984). Inherited deficiency of the Mac-1, LFA-1, p150,95 glycoprotein family and its molecular basis. *J. Exp. Med.* **160**, 1901–1918.

125. Steinkamp, J. A., Wilson, J. S., Saunders, G. C., and Stewart, C. C. (1982). Phagocytosis: flow cytometric quantitation with fluorescent microspheres. *Science* **215**, 64–66.

126. Steltzer, G. T., and Robinson, J. P. (1988). Flow cytometric evaluation of leukocyte function. *Diagn. Clin. Immunol.* **5**, 223–231.

127. Styrt, B. (1989). History and implications of the neutrophil glycoprotein deficiencies. *Am. J. Hematol.* **31**, 288–297.

128. Sveum, R. J., Chused, T. M., Frank, M. M., and Brown, E. J. (1986). A quantitative fluorescent method for measurement of bacterial adherence and phagocytosis. *J. Immunol. Methods* **90**, 257–264.

129. Szejda, P., Parce, J. W., Seeds, M. S., and Bass, D. A. (1984). Flow cytometric quantitation of oxidative product formation by polymorphonuclear leukocytes during phagocytosis. *J. Immunol.* **133**, 3303–3307.

130. Tetteroo, P. A. T., Bos, M. J. E., Visser, F. J., and von dem Borne, A. E. G. K. (1986). Neutrophil activation detected by monoclonal antibodies. *J. Immunol.* **136**, 3427–3432.

131. Tetteroo, P. A. T., van der Schoot, C. E., Visser, F. J., Bos, M. J. E., and von dem Borne, A. E. G. (1987). Three different types of Fcγ receptors on human leucocytes defined by Workshop antibodies: FcγR$_{low}$ of neutrophils, FcγR$_{low}$ of K/NK lymphocytes, and FcγRII. *In* "Leukocyte Typing III" (A. L. Mc.Michael, ed.), pp. 702–706. Oxford University Press, Oxford.

132. Thompson, R. A., Candy, D. C. A., and McNeish, A. S. (1984). Familial defect of polymorph neturophil phagocytosis associated with absence of a surface glycoprotein antigen (OKM1). *Clin. Exp. Immunol.* **58**, 229–236.

133. Todd III, R. F., Nadler, L. M., and Schlossman, S. F. (1981). Antigens on human monocytes identified by monoclonal antibodies. *J. Immunol.* **126**, 1435–1442.

134. Trinkle, L. A., Wellhausen, S. R., and McLeish, K. R. (1987). A simultaneous flow cytometric measurement of neutrophil phagocytosis and oxidative burst in whole blood. *Diagn. Clin. Immunol.* **5**, 62–68.

135. Uchida, T., Hosaka, S., and Miura, K. (1985). Direct measurement of phagolysosomal esterase activity. *Biochem. Biophys. Res. Commun.* **127**, 584–589.

136. Unkeless, J. C. (1989). Function and heterogeneity of human Fc receptors for immunoglobulin G. *J. Clin. Invest.* **83**, 355–361.

137. van de Schoot, C. E., Huzinga, T., Daams, M., and von dem Borne, A. (1989). A granulocyte specific-PI-linked activation antigen, recognized by 3 different workshop antibodies. *Tissue Antigens* **33**, 221 (abstract).

138. van Kessel, K. P., van Strijp, J. A. G., van der Tol, M. E., van Kats-Renaud, H. J., Fluit, A. C., and Verhoef, J. (1989). Quantitation of conjugate formation between human polymorphonuclear leukocytes and antibody-coated target cells by flow cytometry. *J. Leukocyte Biol.* **46**, 467–475.

139. Vindenes, H., and Bjerknes, R. (1991). Activation of neutrophils in patients with thermal injury: altered expression of surface complement and Fc receptors. Submitted for publication.

140. Visser, J. W. M., Jongelind, A. A. M. and Tanke, H. J. (1979). Intracellular pH-determination by fluorescence measurements. *J. Histochem. Cytochem.* **27**, 32–35.

141. Weiss, J., Kao, L., Victor, M., and Elsbach, P. (1985). Oxygen-independent intracellular and oxygen-dependent extracellular killing of *Escherichia coli* S15. *J. Clin. Invest.* **76**, 206–212.

142. Wilson, J. G., Andriopoulos, N. A., and Fearon, D. T. (1987). CR1 and the cell membrane proteins that bind C3 and C4. *Immunol. Res.* **6**, 192–209.

143. Wilson, J. G., Ratnoff, W. D., Schur, P. H., and Fearon, D. T. (1986). Decreased expression of the C3b/C4b receptor (CR1) and the c3d receptor (CR2) on B lymphocytes and of CR1 on neutrophils of patients with systemic lupus erythematosus. *Arthritis Rheum.* **29**, 239–747.

144. Wilson, M. E., and Pearson, R. D. (1988). Roles of CR3 and mannose receptors in the attachment and ingestion of *Leishmania donovani* by human mononuclear phagocytes. *Infect. Immunol.* **56**, 363–369.

145. Wilson, R. M., Galvin, A. M., Robins, R. A., and Reeves, W. G. (1985). A flow cytometric method for the measurement of phagocytosis by polymorphonuclear leucocytes. *J. Immunol. Methods* **76**, 247–253.

146. Zimmerman, G. A., and Mcintyre, T. M. (1988). Neutrophil adherence to human endothelium *in vitro* occurs by CDw18 (Mo1, MAC-1/LFA-1/GP150,95) glycoprotein-dependent and -independent mechanisms. *J. Clin. Invest.* **81**, 531–537.

147. Carr, R., and Davies, J. M. (1990). Abnormal FcRIII expression by neutrophils from very premature neonates. *Blood* **76**, 607–611.

2.4

Measurements of Granulocyte Oxidative Metabolism

MICHAEL C. SEEDS

Department of Medicine/Section on Pulmonary and Critical Care
Bowman Gray School of Medicine
Winston-Salem, North Carolina, USA

DAVID A. BASS

Department of Medicine/Section on Pulmonary and Critical Care
Bowman Gray School of Medicine
Winston-Salem, North Carolina, USA

I INTRODUCTION

The biochemical characterization of granulocyte activation has been aided by increased application of flow cytometric studies over the last decade (*102*). This includes the identification and quantitation of surface receptors (see Chap. 2.3), changes in membrane potential, calcium fluxes, degranulation, cell shape change, actin filament assembly, and oxidative metabolism. With a variety of spectrally unique dyes available for many cellular processes, several events have been examined simultaneously at the single cell level. Such applications allow examination of the complex interaction between different signal transduction mechanisms in the activation of granulocytes.

This review focuses on oxidative metabolism

FLOW CYTOMETRY IN HEMATOLOGY
ISBN 0-12-432940-3

of activated granulocytes from the perspective of flow cytometry. Where data are available, correlations to other cellular processes are also reviewed in the context of signal transduction in neutrophils. While oxidative metabolism stimulated by phagocytosis will be discussed below, all other aspects of phagocytosis have been discussed in the previous chapter. Rapid, diagnostic assays of granulocyte function continue to be of clinical interest. Such applications are also discussed where available.

II. GRANULOCYTE OXIDATIVE METABOLISM

Granulocytes perform as professional phagocytes by virtue of two bactericidal strategies. These are artificially divided into oxygen-dependent and -independent processes, although the two often act synergistically (*25,45,50,84*). Oxygen-dependent processes involve transfer of electrons through an NADPH oxidase resulting in the one electron reduction of oxygen to form O_2^- which dismutates rapidly to H_2O_2.

$$NADPH + O_2 \xrightarrow{\text{NADPH oxidase}} NADP^+ + O_2^- \rightarrow H_2O_2$$

NADPH is replenished by parallel activation of the hexose monophosphate shunt. These combined processes comprise the cell's oxidative burst (*2,96*).

Oxygen-independent bactericidal mechanisms involve release of granule contents including lytic proteases (*25,45*), defensins (*21,79*) bactericidal permeability increasing proteins (*65*), and others (*1*).

A. Measuring Oxidative Metabolism with Dichlorofluorescin Diacetate

Many fluorophores can be oxidized by reduced oxygen products of the granulocyte's oxidative burst, generating a change in fluorescence. The most widely adapted dye is dichlorofluorescin diacetate, DCFH-DA (*4,9,37*). As a diacetate, DCFH-DA easily diffuses into cells where the acetates are hydrolyzed, trapping non-fluorescent, polar dichlorofluorescin (DCFH). Upon activation of the cell, increased production of H_2O_2 results in the oxidation of DCFH to highly green fluorescent dichlorofluorescin (DCF). DCFH and DCF have been identified in neutrophil extracts using GC mass spectrometry (*4*). Unstimulated cells have relatively low fluorescence, while maximal stimulation of neutrophils with phorbol myristate acetate (PMA) will increase DCF production up to 20-fold, making quantitation of oxidative responses possible over a wide range of agonist concentrations.

a. *Method of the DCFH Oxidation Assay*

Details of the methods have been published elsewhere (*4*), but the assay is not critically dependent on type of buffer, etc. so the following brief summary provides the important aspects. Isolated cells ($1-5 \times 10^6$ cells/ml) are preincubated with 5 μM DCFH-DA for 15 min at 37°C. An initial fluorescence value is obtained for an aliquot of cells with 488 nm excitation and 510–550 nm emission detection. Stimuli are added and subsequent fluorescence readings made at desired times. Extracellular Ca^{2+} and Mg^{2+} are not required for dye oxidation by neutrophils, and unless required for specific stimuli, are deleted from buffers to decrease aggregation of stimulated cells, avoiding clogging of flow cells. The assay can be used for isolated granulocytes (*4*), buffy coat cells (*95*), or whole blood (*115*). Higher concentrations of DCFH-DA may be needed for the latter to ensure linear results (*10,115*).

b. *Possible Mechanisms of Intracellular Oxidation of DCFH*

Stimulation of normal polymorphonuclear leukocytes (PMN) caused oxidation of DCF which paralleled other aspects of the oxidative

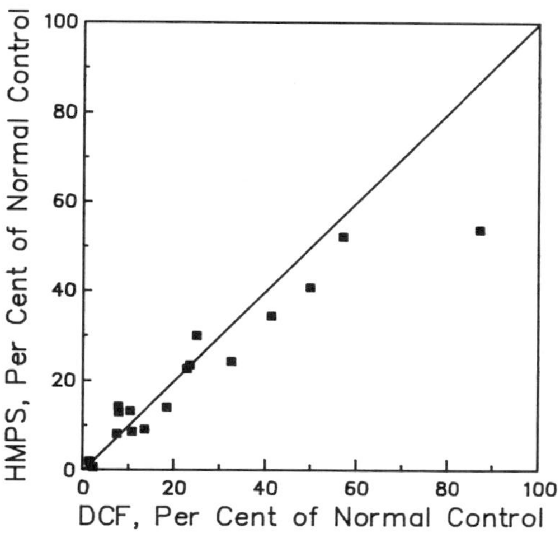

Fig. 1. Cells from patients whose PMNs have oxidative metabolic defects of varying severity were examined in two parallel assays of the oxidative burst. Hexose monophosphate shunt and DCFH oxidation was determined for parallel aliquots of PMNs stimulated by PMA under identical conditions. The regression line (slope of 1) is shown for a theoretical fit of the data. All patients have chronic granulomatous disease or are X-linked carriers of the trait (D. A. Bass, M. S. Cohen, L. R. DeChatelet, P. Szejda, and M. C. Seeds, unpublished).

burst (e.g. the hexose shunt) after varying doses of stimulant (*4*). Additionally, DCF production closely paralleled hexose shunt activity in cells from patients with partial metabolic defects of varying severity (Fig. 1). These data suggested the DCFH oxidation was quantitatively related to the oxidative burst of stimulated normal neutrophils. To attempt definition of intracellular oxidants causing DCF production, oxidants were added extracellularly with monitoring intracellular DCF production. Responses in normal neutrophils were linear with: (1) exogenous reagent H_2O_2 concentrations of 1–60 μM; (2) H_2O_2 production from exogenous xanthine oxidase plus acetaldehyde or glucose oxidase plus glucose (*4*). DCFH oxidation in these systems was inhibitable by catalase (which destroys H_2O_2) but not by superoxide dismutase (which destroys O_2^-). These experiments suggested DCF production was probably mediated by H_2O_2. Similar results have been obtained by other laboratories (*24,115*).

In recent cell-free assays we have observed oxidation of DCFH by hypochlorous acid (HOCl). HOCl is produced by stimulated cells as a myeloperoxidase-generated byproduct of H_2O_2. Although myeloperoxidase also rapidly catalyzes H_2O_2-dependent DCFH oxidation in cell-free assays, in normal healthy PMN, we speculated that MPO would not be a significant factor in this assay due to its compartmentalization within granules. The concept assumed that (1) the DCF was localized to cytosol and could not cross membranes (*4*); (2) H_2O_2 (and hence HOCl) would be generated outside the cell or inside vacuoles and (3) little highly reactive HOCl could diffuse across membranes to react with cytosolic DCF. Recent data suggest this model was not totally valid: a patient with total absence of myeloperoxidase but normal generation of O_2^- and H_2O_2 had about half of normal activity in the DCF assay. Thus, a portion of the DCFH oxidation is myeloperoxidase (and hence HOCl) independent, yet a significant fraction of the reaction involves MPO-dependent events. In cells from uninfected individuals, relative contribution by MPO appears to be a constant fraction. As noted below, possible MPO-mediated events become more complex when assaying PMN from patients with severe infection.

Hydroperoxides can also serve as an electron donor for DCFH in the presence of appropriate enzymatic catalysts (*15*), but are unlikely to contribute to whole cell DCF production.

Since with some agonists pinocytosis of extracellular dye occurs, gentle washing of the cells by centrifugation may be desired. Washing affects the quantitative estimates of DCF production, but does not affect the qualitative results.

B. DCFH Oxidation in Normal Cells

Neutrophil activation by phorbol myristate acetate (PMA) (*4*), chemotactic factors such as N-formal-methionyl-leucyl-phenylalanine (fMLP) (*95*), or phagocytic stimuli including opsonized *Staphylococcus aureus* (*34,111,115*), *Mycobacterium leprae* and BCG (*36*), and IgG-opsonized latex (*35,94*) have been examined. Typical histograms of PMA-stimulated neutrophils are shown in Fig. 2. PMA is a popular agonist, because it is a soluble substance which activates protein kinase C-dependent pathways and causes abrupt and (at concentrations over 100 ng/ml) maximal stimulation of the PMN oxidative burst. PMA was used to show that PMNs respond in a graded, rather than all-or-none increase in DCF production (i.e. 1 ng/ml PMA caused a mean response of 34% maximal with a single population of responding PMNs rather than a maximal response in only 34% of cells). Similarly, increasing DCF production was seen when neutrophils phagocytized increasing numbers of serum-opsonized *Staph. aureus* or IgG-opsonized latex beads (Fig. 3) (*94,111*). Thus, we have concluded that the neurophil's oxidative burst is a graded response, closely linked to receptor occupancy.

Patel *et al.* have used DCFH oxidation to suggest that PMNs have different lag times and different stimulation maxima for both fMLP and PMA (*76*). Using maximal PMA concentrations, they found >95% of cells responded unimodally. However, using minimal stimulatory concentration of 1 ng/ml, they found that only a proportion of cells responded and that the proportion increased with time. Thus, they concluded that cells had distinct, individual lag times (59 ± 17 s for PMA) and distinct, individual thresholds for stimulus concentrations necessary to activate the respiratory burst (42% of cells responded by 30 min). These conclusions were based on fluorescent microscopic

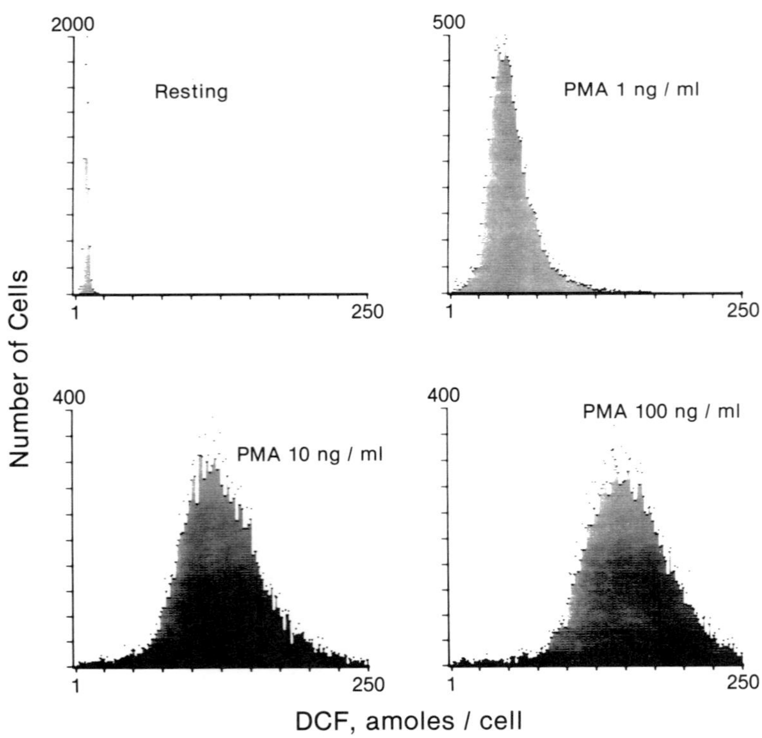

Fig. 2. The unimodal fluorescence distribution of DCF in PMN stimulated for 15 min with increasing concentrations of PMA demonstrated a graded response (*4*).

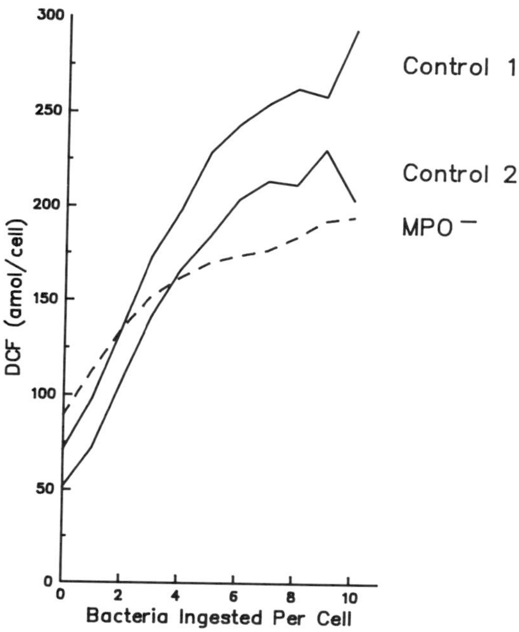

Fig. 3. DCF fluorescence was measured in PMN which had phagocytized serum-opsonized, Texas Red-stained *Staph. aureus*. Analysis of DCF (amol/cell) was gated from the red fluorescence of the ingested bacteria using a dual laser approach (488 nm excitation of DCF from an argon laser, 568 nm excitation of Texas Red from a krypton laser). Increased phagocytosis (1–10 bacteria ingested per cell) resulted in graded increases in oxidative metabolism. The response of PMNs from an MPO-deficient patient (MPO^-) demonstrated that for phagocytosis, DCFH oxidation was largely MPO independent (*111*).

observations of adherent PMNs while population distributions were corroborated by flow cytometry.

a. *Application of DCFH Oxidation to Other Cell Types*

There are few published reports of DCFH oxidation in other cells other than in abstract form. However, these preliminary reports offer reasonable promise that end-differentiated cells which arise from common stem cell progenitors of the neutrophil will have significant oxidative metabolic responses at some point in their ontogeny. Examples are given in Table I.

C. DCFH Oxidation in Patient Studies

a. *Chronic Granulomatous Disease*

As a family of genetic disorders, chronic granulomatous disease (CGD) is hallmarked by the failure of CGD neutrophils to produce an oxidative burst (*19,57,96,112*). Chronic infections and early mortality of affected individuals supports the significance of oxygen metabolism in bactericidal functions of granulocytes. Defects in CGD neutrophils can occur in any of three identified proteins of the NADPH oxidase electron transport chain. X-linked CGD neutrophils are predominantly deficient for an unusual b cytochrome found in the plasma membrane (*54,97*). Most autosomal recessive CGD neutrophils lack functional cytosolic proteins, either a 47 kDa phosphoprotein (*14,72,73*) or a 65 kDa protein (*68,69*). The function of either protein is unknown, although during activation these proteins translocate to plasma membrane containing the b cytochrome (*18,88*). All three proteins have been recently cloned (*51,89,117*), although the preliminary analysis of the cytosolic proteins' primary sequences did not suggest functional homology with any recorded peptides. Other defects may as yet be unidentified, as chromatography of PMN cytosols suggests a third cytosolic oxidase component exists (*69,96*) (Seeds, Harless, Lindheimer and Bass, unpublished).

Identification of CGD initially requires demonstration of defective PMN oxidative burst. Preincubation of CGD neutrophils with DCFH-DA results in normal trapping of DCFH (*4*). Resting cells are indistinguishable from healthy control cells. Upon activation with PMA, DCF formation in CGD cells will increase only minimally, in stark contrast to the 15–20-fold increase seen in normal cells (Fig. 4). Thus, CGD neutrophils are quantifiably unreactive. Similarly, carriers of X-linked CGD display the expected genetic mosaic of two subpopulations of cells. One subpopulation is unreactive, the other normal. Complete resolution of subpopulations is aided

Table I

APPLICATION OF DCFH OXIDATION ASSAY TO OTHER CELLS

Cell type	Observed DCFH oxidation	References
Human		
Neutrophils	Parallels oxidative burst with PMA, opsonized *Staph. aureus*, IgG latex	(*4,12,28,76,115*)
	Defective in CGD neutrophils	(*4,64*)
U937 cell line	Enabled in induced cells	(*33*)
Monocytes	PMA > > heat aggregated IgG: gamma interferon lowered PMA response, GM-CSF enhanced Ag-IgG response	(*48*)
	Only 30% of DCFH-DA hydrolyzed in cells vs 100% in PMN	(*82*)
Non-human		
Rat PMNs	Responsive to PMA, PMN > > macrophage	(*44*)
	PMN from 150 μl whole blood samples from carotid catheters; 3.2 ng/ml ED_{50} PMA	(*10*)
	Rats on essential fatty acid-free diets showed low DCFH oxidation, chemotaxis, and aggregation to fMLP or A23187; PMA and LTB_4 responses normal. Percentage of responsive cells unchanged	(*31*)
Rat alveolar macrophages	Three-fold homogeneous increase to PMA; DCFH response greater than hydroethidine	(*44*)
Rabbit PMNs	Responsive in whole blood; catalase from rbc lysis blocks low response	(R. Taylor, personal comm)
Rabbit alveolar macrophages	Less responsive than chondrocytes	(*80*)
Rabbit chondrocytes	Dose and time-dependent responses to PMA and concanvallin A	(*80*)
Dog PMNs	Sublethal irradiation of animal yielded no significant change (40–90% increase) in DCF to 18 days	(*20*)
Sheep PMNs	Infusion of animals with *N. naja* cobra venom factor left cells refractory to C5a and PMA *in vitro* for 2 h	(*38*)

by the addition of catalase to the cell suspensions. Catalase prevents H_2O_2 released by a stimulated cell from perfusing into and oxidizing dye in an unreactive cell: innocent bystander oxidation (*4,71*).

Mizuno *et al.* studied X-linked CGD children and their carrier mothers, and found that the number of cytochrome b_{558} positive cells (binding of monoclonal 7D5 raised against cytochrome b_{558}) paralleled the number of DCF positive cells stimulated by PMA (*64*). One cytochrome b positive female studied, presumably autosomal recessive, was completely defective for DCFH oxidation. Surprisingly the mother of two partially defective, cytochrome b negative males was completely normal for both cytochrome and DCFH oxidation. The combination of cytochrome b surface expression and DCFH oxidation was useful for diagnostic determination of PMN defects and determining some factors of the mode of genetic transmission of the disease, factors significant for genetic counselling.

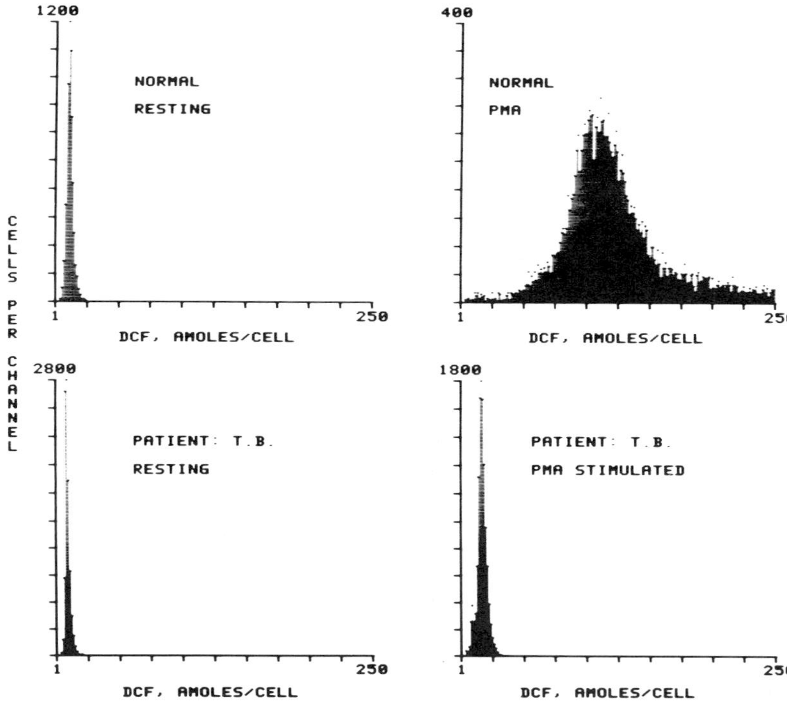

Fig. 4. Fluorescence distribution of resting and stimulated PMNs from a normal donor and from a patient with CGD. Cells were examined after 15 min stimulation (*4*).

b. The Oxidative Burst of Neutrophils from Patients with Sepsis

Though less dramatic than in CGD, altered granulocyte function might contribute to host defense against infection (by enhancing or "priming" bactericidal mechanisms) or conversely worsen host defense (by reduced PMN responsiveness) or even directly damage host tissues (by over-production of O_2 derivatives by primed PMNs). Older data on bulk cell populations suggested that priming of PMNs could occur during acute infection (*59*). Given appropriate metabolic assays, flow cytometry provides the possibility of greater precision by detection of subpopulations of primed or deactivated PMNs in blood of such patients.

Initially, we applied the DCFH oxidation assay to cells from patients with sepsis. Isolated peripheral blood neutrophils were obtained and compared with normal cells using PMA and IgG latex as routine stimuli. The results of studying PMNs from patients (with unavoidable variability among the patient population) demonstrated a spectrum of DCFH oxidation responses. Stimulation of neutrophils with PMA resulted in an increase in mean DCFH oxidation; 180 ± 4.5 amol/cell for controls, 312 ± 14 amol/cell for patients. Parallel increases were seen in these patients' hexose shunt activity (*3*).

Examination of the green fluorescence histograms (Fig. 5) demonstrated that the mean DCF values underestimated the oxidative metabolic responses in patient samples. Many patients developed two populations of responding neutrophils; one normal subpopulation and one "primed" subpopulation with up to 4.6 times control DCF production. Resting cells in

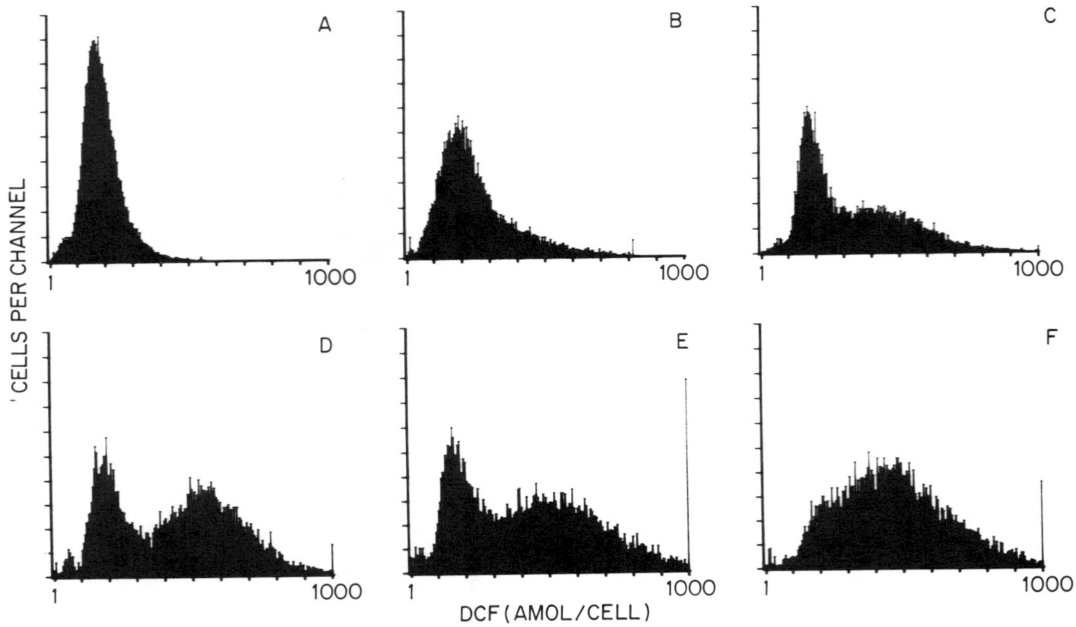

Fig. 5. DCF fluorescence distributions in PMNs from patients with acute bacteremia often demonstrated bimodal responses which varied between 20% and 80% of cells elevated (panels B–F). The "normal" distribution in panel A represents the summation of six healthy control PMN samples. All histograms were obtained after 15 min of stimulation by PMA (adapted from reference *3*).

all patients were indistinguishable from controls. On average, 40% of patients studied had elevated DCFH oxidation, with 0–80% of their cells appearing as a distinct, elevated subpopulation. Appearance of primed subpopulations has not yet correlated with any class of infecting organism.

c. *Peroxidase-mediated Dye Oxidation in PMNs of Patients with Sepsis*

Initially we interpreted these data as demonstrating primed oxidative metabolism in septic patients' cells. Precautions were taken to examine myeloperoxidase function in the patients' PMNs. Total MPO was only minimally increased in patients' PMNs. Also, when H_2O_2 was added exogenously to either control or patients' PMNs, single identical populations of DCF production were seen. The unimodal result made it unlikely that significant differences existed in competing intracellular antioxidant pathways such as the glutathione system or catalase. However, we later reexamined patients' PMN oxidative metabolism by oxygen consumption and extracellular H_2O_2 release (by electrode methods) and found that these responses did not always correlate with DCF production. We then examined PMA-stimulated patients' cells by fluorescence microscopy and observed spheroidal vacuoles containing green DCF in addition to the diffusely stained cytoplasm (*93*). Histochemistry at the electron microscopic (EM) level demonstrated vacuoles which stained positive for peroxidase. The percentage of cells with peroxidase-positive vacuoles was the same for both DCF fluorescence and EM MPO histochemistry, suggesting the DCF-positive vacuoles resulted from MPO-dependent oxidation of the dye. Although vacuoles also appeared in normal PMN stimulated by PMA, as previously reported (*81*), over 95% of them were peroxidase negative by both DCF fluorescence and EM MPO histochemistry.

The incidence of MPO-mediated DCFH oxidation is unknown for many of the diseased states to which this assay has been applied. The utility of DCF as a strict measurement of stimulated H_2O_2 production is limited by the possibly variable contribution by MPO-dependent events.

There have been differences in the population profiles of PMA-stimulated DCF production reported by various laboratories. Thus, one study reported that only 45–55% of PMNs were responsive to PMA when examined after 30 min (*11*). In septic patients, the proportion of responsive PMNs increased to 60–85%. We have always found > 95% of normal, isolated PMNs to be unimodally responsive to PMA examined after 15 min (with similar results over a time course of 1–20 min post-stimulation) (*4,95*). We also found no change in the subpopulation distributions of eight patient–control pairs when results with buffy coat cells were compared to isolated PMNs. Furthermore, Duque and Ward found unimodal responses in normal buffy coat PMNs stimulated with PMA (*24*). These differences between laboratories may have resulted from technical details regarding cell isolation (whole blood vs buffy coats vs isolated cells), the ability of serum albumin to interfere with PMA-stimulated DCFH oxidation (albumin may bind phorbols), or the length of the assay (15 vs 30 min). In the latter instance, we have observed destruction of DCF fluorescence by MPO in a cell-free system, but no data are available on the fate of intracellular DCF.

We speculated that prolonged neutropenia (as after marrow suppression due to leukemia chemotherapy) might induce maximal levels of cytokines which could prime PMNs, thus, with recovery of myelopoiesis, maximally functional (primed) PMNs would appear in the blood. In 22 patients, studied at the time that PMNs reappeared in the blood after chemotherapy-induced neutropenia, most patients had primed PMN subpopulations, and in 10 of these, priming occurred without detectable infection. All patients with significantly primed PMNs survived that hospitalization. In contrast, four patients had a significant (18–22% of PMNs) population with reduced responsiveness. Three of those patients, with over 70% subnormal cells, died during their induction attempts. Blast cells did not account for the unresponsive populations (see Fig. 1 in (*77*)). Observing the responsiveness of > 95% of normal PMNs to PMA stimulation allowed detection of PMNs which were truly unresponsive. The clinical significance of these phenomena is yet to be determined. The number of patients studied is small, however the potential prognostic value of the DCFH oxidation assay warrants further investigation.

Intensive care patients have been examined for granulocyte function in a preliminary attempt to predict organ failure. Rothe *et al.* examined six different flow cytometric criteria of granulocyte function in 265 samples from 64 patients (reported in (*85*)). During sepsis, granulocytes had an alkaline pH, normal phagocytosis of *Escherichia coli*, and oxidative metabolism was depressed in phagocytizing cells but normal in phorbol ester-stimulated cells. Cells from post-traumatic patients showed increased phagocytosis, increased $[Ca^{2+}]$, decreased peroxidase activity, lower pH and increased death of granulocytes during phagocytosis. Automated analysis of the multifunctional pattern of cellular responses predicted 92% of organ failures up to 3 days in advance.

Cells from septic patients have also been examined by Trinkle *et al.* using simultaneous monitoring of phagocytosis of Texas red-labeled *Staph. aureus* and resultant DCFH oxidation (*115*). Those assays were performed on 1 ml samples of whole blood, using higher DCFH-DA concentrations (50 μM) due to dye sequestration by other cells. Red cells were lysed immediately prior to reading each sample on the flow cytometer. PMNs were discerned using combined forward and 90° scattered light. Both phagocytosis and DCF formation increased in PMNs from patients with documented infections. In patients with sepsis, we have also observed increased oxidation of DCFH but normal phagocytosis of IgG-opsonized red latex beads with DCFH-DA loaded neutrophils, both isolated and buffy coat (*94*). The apparent priming of these cells probably does not reflect increased immunoglobulin receptor (FcRIII) expression as we found actually decreased binding of anti-receptor monoclonal antibody (3G8) to PMNs from septic patients (unpublished).

d. Other Patient Studies

DCFH oxidation has been applied to examine altered PMN function for diseases other than sepsis. For example, Shalit *et al.* have examined neutrophil activation in human inflammatory skin reactions with the DCFH activation assay (*101*). Oxidation of DCFH by cells from pollen antigen sites was significantly greater than PMNs from buffer sites. Both PMN groups were significantly elevated relative to resting autologous peripheral blood neutrophils, reflecting increased oxidative metabolism *in vivo* in the blisters. Upon stimulation with PMA, the same maximal responses were found for all PMNs. The use of flow cytometry allowed study of skin chambers with low cell numbers ($< 1 \times 10^6$ cells/site).

The effects of HIV infection on PMN function has also been addressed by flow cytometry (*90*). Parallel samples from nine HIV seropositive homosexual males with HIV-associated periodontitis and intraoral candidiasis were compared to HIV seronegative, uninfected age-matched heterosexual controls. Oxidative metabolism (to PMA), percentage of cell phagocytizing latex beads (but not number of beads/cell), and actin polymerization (see method below) were all elevated in the HIV seropositive patients relative to controls, suggesting that some *in vivo* priming of these cells had occurred.

Patients with chronic renal failure often suffer infections by *Staph aureus*, *Pseudomonas aeruginoso*, *Proteus mirabillis*, and *Candida albicans*. Since neutrophils offer the first line of defense against such organisms, PMN function was evaluated in patients by flow cytometric assays of phagocytosis of fluorescent microspheres and DCFH oxidation in response to stimulation with PMA. Hirabayashi *et al.* found that in patients undergoing chronic hemodialysis, DCFH oxidation was 60% of controls just prior to dialysis but returned to normal following the procedure (*35*). In patients who had never been on dialysis, PMN oxidation of DCFH was also statistically, though minimally, below controls. No correlations were found with laboratory measurements of renal function (serum creatinine, blood urea nitrogen, or uric acid) in chronic hemodialysis patients. However, in first-time dialysis patients, an inverse correlation was found between DCFH oxidation and both the BUN and serum creatinine levels, suggesting depressed PMN function correlated with depressed renal function.

The oxidative responses of monocytes and PMNs to stimulation by *Mycobacterium bovis bacillus Calmette–Guerin* (BCG) and *M. leprae* have been examined using DCFH oxidation to follow single cell responses. After 30 min of phagocytosis, *M. leprae* stimulated almost no DCFH oxidation in PMN, while BCG were almost as stimulatory as *Staph. aureus*, all at a ratio of 100 : 1. Microscopic analysis revealed that while *Staph. aureus* were taken up rapidly, both BCG and *M. leprae* were mostly adherent but not internalized. BCG were internalized slowly and greater than *M. leprae*. Adherence of BCG stimulated the oxidative burst of PMNs while *M. leprae* adhered but essentially did not stimulate the PMNs. Thus, both mycobacteria, which are co-components of proposed leprosy vaccines, can to differing degrees directly stimulate phagocytic cell oxidative responses (*36*).

Granulocyte function analyzed by flow cytometric analysis on very small whole blood samples further demonstrates the potential clinical utility of the system. Phagocytic cell deficits may be related to periodontal disease. Robinson *et al.* have examined the function of PMNs isolated from washings of gingiva crevicular pockets (reported in (*83*)). They found that PMNs from diseased pockets had elevated basal levels of DCFH oxidation relative to autologous peripheral blood PMNs. Upon stimulation, the diseased socket sites were less responsive. However, PMNs from healthy sites were primed relative to PMNs from diseased sites, suggesting possible virulence in host defence against periodontal disease. In another study on limited cell number, 100 μl samples of whole blood from healthy adults was examined for PMN function using DCFH oxidation and phagocytosis of fluorescenated *Staph. aureus* (*34*). DCF production increased in parallel with phagocytosis and the assay was insensitive to absolute PMN number.

Priming of PMN function has been modeled *in vitro* by studying the effects of cytokines and bacterial products on the oxidative burst. Livingston *et al.* demonstrated tumor necrosis factor primed phagocytosis of *Staph. aureus* and resultant oxidative metabolism (DCFH oxidation), and increased C3b receptor expression (*53*). TNF alone also gave dose-dependent PMN activation. Interferon gamma did not activate cells directly, nor did it prime to phagocytosis. However, IFN gamma did prime TNF stimulation of PMNs, demonstrating significant cytokine interaction. Trinkle *et al.* observed LPS priming of phagocytosis (89%) and DCFH oxidation (179%) *in vitro* (*115*). Since priming occurred in whole blood samples incubated for 1 h with LPS, the mediator(s) were unknown.

In all of these studies, the consistent observation of increased activation of the cells parallels dye oxidation in dose- and time-dependent manners. The relative contributions of peroxidase catalysis of dye oxidation and kinetics of H_2O_2 remains to be established. Conservative application of the assay suggests it has great utility as a screen for neutrophil disfunction.

D. Other Assays of Oxidative Metabolism

a. *Autofluorescence*

Autofluorescence has been utilized as a measurement of pyridine nucleotide metabolism. There are inherent differences in the resting autofluorescence of granulocytes. Eosinophils have greater autofluorescence than neutrophils (excited by 350 nm and 450 nm), a factor that led to early attempts to sort eosinophils from neutrophils for further biochemical characterization (*120*). The granule contents responsible for this phenomenon have never been identified (*108*).

Autofluorescence in neutrophils has been applied to follow the initial depletion of reduced pyridine nucleotides (NADH, NADPH) upon activation of the cells. Hafeman *et al.* found an all-or-none activation of neutrophils over a narrow dose range of PMA (*32*). Using UV-excitable ($>$ 425 nm emission) autofluorescence as an estimate of reduced pyridine nucleotide, they found activation of the cells produced a single shift in blue fluorescence. This possibly reflects only the initial shift in steady state levels of NADP/NADPH prior to activation of the hexose shunt.

Bender and Van Epps examined blue autofluorescence in parallel to direct assays of hexose monophosphate shunt (*5*). Using 6-aminonicotinamide as an inhibitor of the hexose shunt, they demonstrated that HMPS activity (measured with ^{14}C glucose), O_2^- production, and the cyclic changes in autofluorescence were all inhibited. These observations were consistent with a parallel activation of the HMPS which supplies NADPH fueling the oxidase. PMA-stimulated decrease in blue autofluorescence was unimodal while fMLP stimulated a more skewed population response with some cells reacting to a lesser degree, but all cells responded. The resolution of the assay may not be sufficient for accurate statistical analysis.

Finally, autofluorescence has also been applied to human mononuclear cells by Bender *et al.* (*6*). Stimulation of monocytes with PMA led to dose-dependent selection of responsive cells, consistent with differential threshold for responsiveness. Again, once a cell responds, its blue autofluorescence drops to a steady final level for a 15 min assay. In monocytes, the same reduced pyridine nucleotide metabolism appears to fuel the cells oxidative burst in a manner completely analogous to the neutrophil. Since lymphocytes did not respond in those assays, it was also suggested that autofluorescence would allow sorting of those two cells.

b. *Nitroblue Tetrazolium*

The reduction of nitroblue tetrazolium (NBT) dyes has provided a routine screening of neutrophil function. As a histochemical assay, it requires small blood samples with little or no cell purification, is inexpensive, and is rapid. Histochemical NBT reduction monitors the spontaneous ingestion of NBT-heparin complexes, with subsequent reduction of the NBT to formazan if superoxide is present (*58*). Ingestion becomes its own stimulus, although further stimulation by PMA is possible. The effect of the formazan on light scatter provides a flow cytometric measurement of the cells ability to produce superoxide. This has been useful in maturation studies of differentiating cell lines such as HL-60 cells (*7,8*). Unfortunately, NBT assays are not highly quantitative, giving positive or negative results for oxidative responses in cell populations. Nor does NBT monitor maximal or primed oxidative responses. The advantage of DCFH oxidation over NBT reduction is that it allows quantitative measurement over a large range of stimulus concentrations.

Others have obtained more quantitative results by using NBT reduction to quench the fluorescence of PMN nuclear DNA prelabeled with dyes such as Hoechst 33258 (HO). In one recent abstract, O'Connell *et al.* found that PMA and fMLP caused a dose-dependent decrease in DNA-associated HO blue fluorescence (*70*). The NBT quenching of HO was superoxide dismutase inhibitable and correlated with parallel assays of cytochrome c reduction. While flow cytometry offers subpopulation discrimination of differentiating cells, the indirect nature of quenching assays, the need for extra reagents, cell manipulations, and the unknown biochemistry of these measurements limits interpretation of these data beyond semi-quantitative reporting.

In a similar manner, hydroethidine (HE) has been used to label the DNA of superoxide-producing PMNs (*87*). Hydroethidine is a cell permeant, reduced form of ethidium bromide (EB). Once inside the cell, oxygen radicals produced upon stimulation of the cell will oxidize HE to EB providing a diffusable label for DNA. It has been suggested that superoxide, not hydrogen peroxide, mediates the dye oxidation; however, even mitochondrial oxidation in resting cells contributes to dye oxidation. Thus, while potentially a useful technique, much of the biochemistry is unknown.

Rothe, Oser and Valet have also used dihydrorhodamine 123 to follow oxidative metabolism in granulocytes (*86*). Dihydrorhodamine 123 is no-fluorescent and uncharged; oxidation produces highly fluorescent rhodamine 123 which is charged and thus trapped intracellularly. Stimulation of prelabeled cells by fMLP produced an 11-fold increase in R123 fluorescence vs a 3-fold increase in DCFH oxidation in parallel samples. Both dyes required peroxidase catalysis in cell-free experiments with H_2O_2. No further details of dihydrorhodamine biochemistry were presented. Transmembrane potential sensitivity and possible mitochondrial trapping of R123 have not been addressed.

In summary, a variety of cell-permeant dyes have been tested as substrates for oxidation by oxygen radicals produced by stimulated cells. Of these, DCFH-DA has seen the widest use. However, every dye thus reported has limitations. These include undefined biochemistry (HE), non-linearity with oxygen metabolism (NBT), poor response or population resolution at extremely low levels of stimulation (DCFH-DA), dependence on (or interference by) peroxidases (DCFH-DA, DHR, others unknown), requirements to co-labeling with other dyes (NBT quenching), toxicity to the user (most DNA stains), and possible interactions with mitochondria (R123). Yet it is possible to determine dose dependency for most agonists, determine subpopulations of responsive cells, and quantitate relative oxidative metabolism, if not H_2O_2 or O_2^- precisely. Most promising is the potential use of these dyes for identifying altered cell function during disease. With the increased distribution of flow cytometers in hospital clinical laboratories, assays of oxidative metabolism may be useful diagnostic tools in documenting impaired granulocyte function during compromised host defense. As empirically defined clinical tests, precise definition of biochemical mechanisms, though ideal, may not be necessary.

III. OTHER ASSAYS OF GRANULOCYTE ACTIVATION

A. Membrane Potential

Granulocytes maintain a differential membrane potential across their plasma membrane. Cationic dyes will partition across vital cell membranes in a manner largely dependent upon this transmembrane potential. Stimulation of cells produces changes in potential and results in a re-equilibration of the dyes between the aqueous media and the cells, measured as a change in cell-associated fluorescence. Several families of fluorescent, cationic dyes have been used to follow membrane potential changes in granulocytes. The carbocyanines, developed by Alan Waggoner, were the first such applicable dyes offering the advantage of different excitation/emission maxima, dependent upon the number of methene groups in the molecules (*119*). Early studies used a thiocarbocyanine, 3,3′-dipentylthiocarbocyanine (di-S-C_5(3)), however, measurements depended upon fluorescence quenching of high concentrations of the dye, and the dye was destroyed by melyoperoxidase and H_2O_2 in stimulated cells (*121*). These complications prompted a shift to oxacarbocyanines and indocarbocyanines, di-O-C_5(3) and di-I-C_5(3), which are green and red fluorescent, respectively (*99*). These dyes are highly fluorescent below 10 n*M*, avoiding toxic effects on the cells under study. However, at low concentrations, mitochondrial staining becomes a larger proportion of the dye signal (*46*), complicating data interpretations and calculation of membrane potential (*17*).

Negatively charged oxonol dyes have also been utilized (*17*). Like the carbocyanines, they are symmetric membrane-permeant fluorochromes with a delocalized charge. In contrast to cationic carbocyanines, the opposite cell association occurs with anionic oxonol dyes: resting cells have very little dye uptake while depolarized cells take up more dye. While unsuited for cuvette studies, these dyes are well suited for flow cytometry where fluctuation of extracellular dye is minimal, and most flow cells and nozzles physically minimize background media from analysis so signal-to-noise ratios are optimized. Also, the oxonal's negative charge keeps them strongly repulsed from mitochondria. However, their low cell association requires the sensitivity of flow cytometer optics. Chused *et al.* recommended the use of bis-(1,3-dibutylbarbiturate) trimethine at 100–200 n*M* for green fluorescence and bis-(1,3-dibutylbarbiturate pentamethine oxonol for red fluorescence (*17*). Both dyes gave linear fluorescence changes over a range of 0 to -100 mV. Calibration, not possible with valinomycin due to dye–drug–K^+ complex interaction, can be achieved with T lymphocytes which have a calcium-activated K^+ channel.

Many studies of membrane potential have been performed on neutrophils. Neutrophils maintain a negative inside transmembrane potential owing largely to the leakage of K^+ out of the cell (*17*). When stimulated by PMA, all cells undergo a rapid depolarization which does not recover. Stimulation by fMLP produces a depolarization of a subpopulation of cells which varies from 35% to 65% of the cells, depending upon the isolation conditions, cation content of the buffer, or use of cytochalasin b (*95,99*). The other cells were unresponsive or slightly hyperpolarized. This provided the strongest argument for suggesting heterogeneity amongst mature PMN (*27*).

In the early 1980s it was observed that cells

from CGD patients failed to depolarize when stimulated (*98*). Since depolarization preceded oxygen reduction in normal cells (*47,95*), the failed depolarization was thought to be one significant step closer to the real lesion in those cells. Since that time, the defects in CGD have been defined as defects in the membrane cytochrome b, or the cytosolic components of the NADPH oxidase. Why CGD PMNs do not depolarize remains undefined. Membrane depolarization had been dissociated from the oxidative burst in a number of more recent studies (*56,95,113*), discussed further in Section IV). Thus, depolarization of membrane potential appears to be neither necessary nor sufficient for activation of the respiratory burst. It remains a useful marker of early cell response to stimulation and may be linked to homeostasis or other cell functions.

B. Calcium

Using probes developed by Roger Tsien, including Quin 2 (*116*), Indo 1, and Fura 2 (*29*), an abrupt, transient increase in cytosolic free calcium has been documented as one of the earliest measurable changes upon receptor-mediated stimulation of PMNs (*52,67,100,103, 104,118*). The initial calcium increase comes from release of intracellular stores, possibly "calciosomes", although details of the mechanisms remain undefined (*16,109*). One pathway involves release from membrane of polyphosphoinositols, which are potential endogenous calcium ionophores (*62*). Calcium-dependent systems of neutrophils include phospholipases, protein kinase C, Ca^{2+}-dependent protein kinases, calcium channels, and actin assembly/disassembly (*78*). Many of these signal mediators in turn regulate cell shape change, chemotaxis, degranulation and oxygen metabolism. Although increased $[Ca^{2+}]_i$ is by itself not sufficient to activate the NADPH oxidase, cytosolic $[Ca^{2+}]$ is necessary for the response (at least in unprimed PMNs) and the relative role of $[Ca^{2+}]_i$ in normal receptor-mediated events is a topic of intense current research.

The probes are calcium chelators which report $[Ca^{2+}]_i$ as fluorescence magnitude. The initial kinetics of calcium fluxes occur in less than 10 s and must be obtained with standard spectrofluorometers. Fluorometer studies demonstrated that stimulation of neutrophils with fMLP, LTB_4, IgG, or PAF resulted in rapid increase in $[Ca^{2+}]_i$ which decayed over several minutes (*49,52,55,67,118*, and Bass, unpublished). Microscopically, Ca^{2+} fluxes were greatest in the periphagosomal region (*66*) and near the leading edge of cells in a chemotactic gradient (*92*).

While Quin-2 does not have absorbance maxima suitable to most available lasers, Indo-1 offers suitable spectral qualities for flow cytometry (*17*). Indo-1 is 30-fold brighter than Quin-2, and offers the unique advantage of an emission wavelength shortening upon increased calcium binding. Dye loaded cells are analyzed by taking the ratio of UV excited Indo 1 emission at 404 : 485 nm. Flow cytometric analysis of calcium fluxes using Indo-1 have been reported for neutrophils (*49,55*).

Recently, Lopez *et al.* studied the calcium spike following stimulation of neutrophils with chemoattractants fMLP and LTB4 (*55*). They found a rapid, unimodal increase in $[Ca^{2+}]_i$ as measured by Indo-1 when either stimulus was used. Free intracellular calcium peaked within 30 s of fMLP introduction, then fell to a plateau at 40 s which was maintained for the next 6 min. While 98% of the PMNs responded to fMLP with increased calcium, the population distribution of blue fluorescent ratios was broad and heterogeneous although subpopulations were not resolved. LTB_4 stimulation produced a similar peak response, however less than 51% of the cells remained above the baseline after 130 s, confirming LTB_4 short-lived effect on calcium (*74*). Similar results were obtained whether PMNs were isolated or gated from mixed leukocyte preparations.

As an additional experiment, Lopez *et al.* also looked for cytokine activity in a cell line supernatant by studying PMN calcium fluxes (*55*). Normally, such studies are difficult by spectrofluorometry due to contributions of the proteins of interest and the culture media to fluorescence signals. With flow cytometry, media interference with fluorescent signals is greatly reduced. That study indicated 5% conditioned media from three different sources of a human fibrous histiocytoma cell line (GCT) gave a calicium spike similar to LTB_4. No other cell line media produced these responses, nor did recombinant G-CSF, GM-CSF, IL-1alpha, nor any conditioned media containing these factors. The identity of the factor is unknown, but the study demonstrates the usefulness of flow cytometric measurement of PMN calcium fluxes as a screening bioassay for cytokines relevant to granulocyte activation.

Very recently, Kao, Tsien and co-workers have announced a new group of fluorescent calcium indicators which combine a tetracarboxylate chelating site with a xanthene chromophore to give compounds with either fluorescein- or rhodamine-like spectral qualities (*43,63*). These dyes are excitable with visible light and are more convenient for flow cytometry and fluorescence microscopy. These dyes do not undergo a wavelength shift, so ratios cannot be used to eliminate concerns regarding dye concentration. However, the dye increase in fluorescence is up to 40-fold upon binding calcium, and the dissociation constant in the range of 0.37–2.3 μM makes them better suited for high $[Ca^{2+}]_i$ than either Quin-2, Fura-2, or Indo-1.

C. Shape Change and Actin Polymerization

Neutrophils exposed to chemoattractants initially undergo a shape change detectable as an elongation as they attempt to reorient in the chemotactic gradient. Cells will develop ruffled membranes within 15 s and will elongate within 5–30 min in suspension. These morphologic events are detectable as changes in light scatter measured by flow cytometry (*60,61,106,122*). Sklar *et al.* have reported that cells undergo a biphasic right angle scatter response when stimulated with chemotactic peptides (*106*). Maximal at 15 s, this early response required ligation of less than 100 receptors/cell and resulted in a 20% decrease in right angle scatter with a parallel doubling of relative f-actin content. A prolonged second phase of decreased right angle scatter was maximal after 1 min and required continual occupation of up to 5000 receptors/cell/min (10% of total). Interruption of ligand binding resulted in rapid decay of all responses. Similar results have been reported by McNeil *et al.* (*60*). The initial phase of right angle scatter decline correlated with membrane ruffling. Elongation of cells correlated with saturation of peptide receptors between 3 and 15 min and was recorded as a 15% increase in 90° scatter with a 5% decrease in forward scatter. Scatter changes are measurable simultaneously with other fluorescently tagged cell functions with the advantage of good temporal resolution on large numbers of cells.

Actin polymerization has been studied using NBD-phallacidin binding to f-actin in formalin fixed cells (*39–41*). Actin polymerization was seen to double in fMLP stimulated cells within 45 s and decayed to an intermediate value between 5 and 10 min. The maximum polymerization, the rate of depolymerization, but not the initial rate of polymerization were related to fMLP concentration. F-actin was redistributed with cellular shape change. The assay is simple and applicable to flow cytometry with standard FITC optics. Parallel studies with methanol extraction of bound NBD-phallacidin allows quantitative corroboration of flow cytometric data (*40*).

D. Degranulation and Other Assays of Cell Activation

Degranulation has been studied by flow cytometry as a functional response of cell stimulation. The most simple approach relies on careful measurements of right angle scatter of the incident laser light. Decreased 90° scatter may be correlated to degranulation in neutrophils stimulated by fMLP with cytochalsin B present (*26,105,107*). This method is simple as it requires no stains, can be performed on freshly isolated cells, and is available on all modern instruments without special adaptations. The separation of shape change and degranulation, both measured by changes in 90° scatter, is confusing since they move in opposite directions. They may be separable by different requirements for receptor occupancy. Clearly both assays are empirical, though overlapping.

Fluid pinocytosis has been identified in fMLP-stimulated PMNs. Fluoresceinated dextran (MW 70 000) uptake was found to increase in a dose-dependent manner within 15–30 s after the addition of fMLP. This stimulated pinocytosis was independent of extracellular cations, was temperature dependent, and was increased in cells exposed to cytochalasin and also in adherent cells (*22*).

Enzymes studied by flow cytometry include the following. Glutathione forms a cyclic compound with UV excitable o-phthaldialdehyde (*114*). Myeloperoxidase has been examined fluorometrically and microscopically with homovanillic acid (*30,75*), and esterases have been studied using trapping of fluorescein diacetate or 4-methylumbelliferone acetate (*23*) Lactoferrin was studied using monoclonal antibodies to neutrophil lactoferrin in formalin-fixed, saponin-permeabilized PMNs which were counterstained with FITC-GAM-IgG2b (*13*). Other intracellular antigens should be amenable to flow cytometric analysis as antibodies become available to cellular components of interest (*42*). Antibodies against recombinant components of the NADPH oxidase will undoubtedly be used and reported in the near future to study such problems as granulocyte differentiation, alteration of cell function in disease, and further study of activation of oxidative metabolism in normal cells.

IV. APPLICATION OF TWO OR MORE ASSAYS OF GRANULOCYTE ACTIVATION

Phagocytosis and resultant oxidative metabolism can be simultaneously measured using fluorescently tagged particles and the DCFH oxidation assay. With flow cytometry, it has been possible to quantify DCFH oxidation as a function of the number of fluorescently tagged particles phagocytized by each cell. Population averages demonstrate that DCF production increases with increasing phagocytosis (*111*). This increases not only as the proportion of cells in a population phagocytizing one or more particles, but also with the number of particles phagocytized by each responding cell. When examined by fluorescence microscopy, phagocytosis demonstrated a localized activation of DCF formation in and around the phagosome (*91*), confirming that neutrophil responses are graded with receptor occupation. Technical requirements for simultaneous monitoring of these events require a red label on the particle. Texas red-labeled *Staph. aureus* (*53,111,115*), and red fluorescent latex beads (*35,90,94*) have been used. We have coupled a red carbocyanine, di-I-C_{18}(3), to latex to obtain red fluorescent beads. These beads have extremely low signal overlap with DCF, can be prepared in bulk, and are stable for over 12 months. While they do excite weakly with 488 nm light, better excitation is obtained from a 568 nm krypton laser, thus dual beam methods were required. Trinkle *et al.* also investigated phagocytosis and DCFH oxidation in parallel (*115*).

Other simultaneous monitoring of stimulated neutrophil functions include membrane potential and DCFH oxidation. Using cells preincubated with both DCFH-DA (green) and di-I-C_5(3), we found that membrane potential depolarizations which normally occur in stimulated PMNs were neither necessary nor sufficient for the activation of oxidative metabolism (*95*). Low (0.03–0.1 n*M*) PMA depolarized cells completely without any increase in DCFH oxidation. PMA at 0.17 n*M* was required to stimulate oxidative metabolism (Fig. 6). With fMLP as stimulant, a biphasic population distribution was found for membrane potential, however simultaneous oxidation of DCFH was always unimodal. Additionally, fMLP stimulation of buffy coat PMN produced DCFH oxidation without depolarization. The dis-

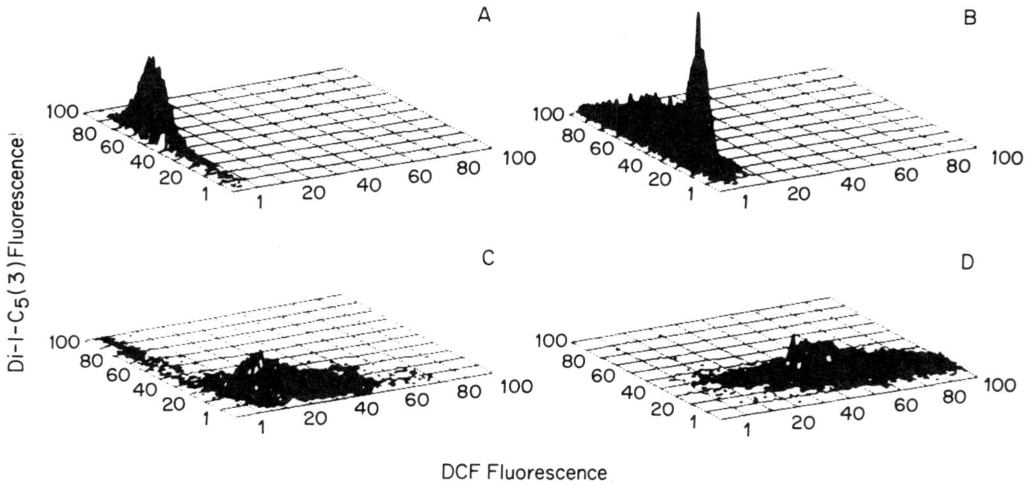

Fig. 6. PMNs were double labeled with di-I-$C_5(3)$ (membrane potential) and DCFH-DA (H_2O_2 production) and stimulated with varying doses of PMA for 15 min. Depolarization was seen as a decrease in di-I-$C_5(3)$ red fluorescence while an increase in DCF green fluorescence represented an increased oxidative burst. The concentrations of PMA used were 0, 0.1, 0.17, and 1.7 n*M* in panels A–D, respectively (*95*).

sociation of these events has been confirmed elsewhere (*56,113*), and membrane potential, although possibly linked to other cell functions, plays no required signal transduction role in activating PMN oxidative metabolism.

In contrast, membrane depolarizations do accompany phagocytosis in a dose-dependent manner. When cells are gated according to the number of particles ingested, increased phagocytosis correlates with depolarization (Seeds and Bass, unpublished). Also, Sullivan *et al.* have shown independent stimulation of membrane-bound calcium release, depolarization, and chemotaxis (*110*). The functional relation between phagocytosis, depolarization, and oxidative metabolism is implied, not established. Yet such applications are suitable for bioassays of normal cell function in patient studies, where one or more events may be altered.

Many other studies have examined different functions in parallel samples using the assays for actin assembly, oxidative metabolism, and FITC-conjugated anti-receptor antibody studies. Since all of these probes utilize green fluorescence, parallel samples are required. Development of a red fluorochrome for oxidative metabolism would enhance the variety of dual parameter studies available.

ACKNOWLEDGEMENT

This work was supported by National Institute of Health grants AI14929, AI10732 and CA41532.

REFERENCES

1. Alam, M., Ranadive, N. S., and Pruzanski, W. (1987). Influence of neutrophil cationic proteins on generation of superoxide by human polymorphonuclear cells during phagocytosis. *Inflammation* **11**, 131–142.

2. Babior, B. M. (1984). The respiratory burst of phagocytes. *J. Clin. Invest.* **73**, 599–601.

3. Bass, D. A., Olbrantz, P., Szejda, P., Seeds, M. C., and McCall, C. E. (1986). Subpopulations of neutrophils with increased oxidative product formation in blood of patients with infection. *J. Immunol.* **136**, 860–866.

4. Bass, D. A., Parce, J. W., DeChatelet, L. R., Szejda, P., Seeds, M. C., and Thomas, M. (1983). Flow cytometric studies of oxidative product formation by neutrophils: a graded response to membrane stimulation. *J. Immunol.* **130**, 1910–1917.

5. Bender, J. G., and Van Epps, D. E. (1985). Inhibition of human neutrophil function by 6-aminonicotinamide: The role of the hexose monophosphate shunt in cell activation. *Immunopharmacology* **10**, 191–199.

6. Bender, J. G., Van Epps, D. E., and Steinkamp, J. A. (1985). Characterization of human mononuclear cells using reduced pyridine nucleotide fluorescence and flow cytometry. *J. Leuk. Biol.* **38**, 603–611.

7. Blair, O. C., Carbone, R., and Sartorelli, A. C. (1985). Differentiation of HL-60 promyelocytic leukemia cells monitored by flow cytometric measurement of nitro blue tetrazolium (NBT) reduction. *Cytometry* **6**, 54–61.

8. Blair, O. C., Carbone, R., and Sartorelli, A. C. (1986). Differentiation of HL-60 promyelocytic leukemia cells: Simultaneous determination of phagocytic activity and cell cycle distribution by flow cytometry. *Cytometry* **7**, 171–177.

9. Brandt, R., and Keston, A. S. (1965). Synthesis of diacetyldichlorofluorescin: A stable reagent for fluorometric analysis. *Anal. Biochem.* **11**, 6–9.

10. Bruner, L. H., Robinson, J. P., Ward, P. A., and Hudson, J. L. (1987). Characterization of H_2O_2 production from rat blood neutrophils using dichlorofluorescin oxidation. *Cytometry* Suppl. 1, 76A.

11. Burrow, S., and Valet, G. (1987). Flow-cytometric characterization of stimulation, free radical formation, peroxide activity and phagocytosis of human granulocytes with 2,7-dichlorofluorescein (DCF). *Eur. J. Cell Biol.* **43**, 128–133.

12. Bussolino, F., Aglietta, M., Sanavio, F., Stacchini, A., Lauri, D., and Camussi, G. (1985). Alkyl-ether phosphoglycerides influence calcium fluxes into human endothelial cells. *J. Immunol.* **135**, 2748–2753.

13. Butler, T. W., Heck, L. W., Huster, W. J., Grossi, C. E., and Barton, J. C. (1988). Assessment of total immunoreactive lactoferrin in hematopoietic cells using flow cytometry. *J. Immunol. Methods* **108**, 159–170.

14. Caldwell, S. E., McCall, C. E., Hendricks, C. L., Leone, P. A., Bass, D. A., and McPhail, L. C. (1988). Coregulation of NADPH oxidase activation and phosphorylation of a 48-kD protein(s) by a cytosolic factor defective in autosomal recessive chronic granulomatous disease. *J. Clin. Invest.* **81**, 1485–1496.

15. Cathcart, R., Schwiers, E., and Ames, B. N. (1983). Detection of picomole levels of hydroperoxides using a fluorescent dichlorofluorescein assay. *Anal. Biochem.* **134**, 111–116.

16. Chandler, D. E., and Kazilek, C. J. (1987). Calcium signals in neutrophils can be divided into three distinct phases. *Biochim. Biophys. Acta* **931**, 175–179.

17. Chused, H. M., Wilson, H. A., Seligmann, B. E., and Tsien, R. (1986). Probes for use in the study of leukocyte physiology by flow cytometry. *In* "Applications of Fluorescence in the Biomedical Sciences," pp. 531–544. Alan R. Liss, New York.

18. Clark, R. A., Volpp, B. D., Leidal, K. G., and Nauseef, W. M. (1990). Translocation of cytosolic components of neutrophil NADPH oxidase. *J. Clin. Invest.* **85**, 714–721.

19. Curnutte, J. T. and Babior, B. M. (1986). Chronic granulomatous disease. *Adv. Human Genet.* **16**, 229–297.

20. D'Alesandro, M. M., Gruber, D. F., and MacVittie, T. J. (1987). Canine neutrophil dysfunction following sublethal irradiation. *Cytometry* Suppl. 1, 76A.

21. Daher, K. A., Lehrer, R. I., Ganz, T., and Kronenberg, M. (1988). Isolation and characterization of human defensin cDNA clones. *Proc. Natl. Acad. Sci. USA* **85**, 7327–7331.

22. Davis, B. H., McCabe, E., and Langweiler, M. (1986). Characterization of f-met-leu-phe-stimulated fluid pinocytosis in human polymorphonuclear leukocytes by flow cytometry. *Cytometry* **7**, 251–262.

23. Dive, C., Workman, P., and Watson, J. V. (1987). Improved methodology for intracellular enzyme reaction and inhibition kinetics by flow cytometry. *Cytometry* **8**, 552–561.

24. Duque, R. E., and Ward, P. A. (1987). Quantitative-assessment of neutrophil function by flow cytometry. *Anal. Quan. Cytol. Histol.* **9**, 42–48.

25. Elsbach, P., and Weiss, J. (1989). Oxygen-dependent and oxygen-independent mechanisms of microbicidal activity of neutrophils. *Immunol. Lett.* **11**, 159.

26. Fletcher, M. P., and Seligmann, B. E. (1985). Monitoring human neutrophil granule secretion by flow cytometry: Secretion and membrane potential changes assessed by light scatter and fluorescent probe of membrane potential. *J. Leuk. Biol.* **37**, 431–447.

27. Gallin, J. I. (1984). Human neutrophil heterogeneity exists, but is it meaningful? *Blood* **63**, 977–983.

28. Gomez-Cambronero, J., Inarrea, P., Alonso, F., and Sanchez Crespo, M. (1984). The role of calcium ions in the process of acetyltransferase activation during the formation of platelet-activating factor (PAF-acether). *Biochem. J.* **219**, 419–424.

29. Grynkiewicz, G., Poenie, M., and Tsien, R. Y.

(1985). A new generation of Ca^{2+} indicators with greatly improved fluorescence properties. *J. Biol. Chem.* **260**, 3440–3450.

30. Guilbault, G. G., Brignac, P., Jr, and Zimmer, M. (1967). Homovanillic acid as a fluorometric substrate for oxidative enzymes. Analytical applications of the peroxidase, glucose oxidase, and xanthine oxidase systems. *Anal. Chem.* **40**, 190–196.
31. Gyllenhammar, H., Palmblad, J., Ringertz, B., Hafstrom, I., and Borgeat, P. (1988). Rat neutrophil function, and leukotriene generation in essential fatty acid deficiency. *Lipids* **23**, 89–95.
32. Hafeman, D. G., and McConnell, H. M. (1982). Neutrophil activation monitored by flow cytometry: stimulation by phorbol diester is an all-or-none event. *Science* **215**, 673–675.
33. Hassan, N. F., Campbell, D. E., and Douglas, D. S. (1988). Phorbol myristate acetate induced oxidation of 2′,7′-dichlorofluorescin by neutrophils from patients with chronic granulomatous disease. *J. Leuk, Biol.* **43**, 317–322.
34. Hasui, M., Hirabayashi, Y., and Kobayashi, Y. (1989). Simultaneous measurement by flow cytometry of phagocytosis and hydrogen peroxide production of neutrophils in whole blood. *J. Immunol. Methods* **117**, 53–58.
35. Hirabayashi, Y., Kobayashi, T., Nishikawa, A., Okazaki, H., Aoki, T., Takaya, J., and Kobayashi, Y. (1988). Oxidative metabolism and phagocytosis of polymorphonuclear leukocytes in patients with chronic renal failure. *Nephron* **49**, 305–312.
36. Holzer, T. J., Kizlaitis, L., Vachula, M., Weaver, C. W., and Andersen, B. R. (1988). Human phagocytic cell responses to Mycobacterium leprae and Mycobacterium bovis bacillus calmette-guerin. An *in vitro* comparison of leprosy vaccine components. *J. Immunol.* **141**, 1701–1708.
37. Homan-Muller, J. W. T., Weening, R. S., and Roos, D. (1975). Production of hydrogen peroxide by phagocytizing human granulocytes. *J. Lab. Clin. Med.* **85**, 198–207.
38. Horn, J. K., Lim, J. R., Goldstein, I. M., and Waldman, F. M. (1987). Flow cytometric analysis of oxidation in circulating neutrophils following intravascular complement-C5 activation in sheep. *Cytometry* Suppl. 1, 76. (abstract)
39. Howard, T. H., and Oresajo, C. O. (1985). The kinetics of chemotactic peptide-induced change in F-actin content, F-actin distribution, and the shape of neutrophils. *J. Cell Biol.* **101**, 1078–1085.
40. Howard, T. H., and Oresajo, C. O. (1985). A method of quantifying F-actin in chemotactic peptide activated neutrophils: study of the effect of tBOC peptide. *Cell Motility* **5**, 545–557.
41. Howard, T. H., and Wang, D. (1987). Calcium ionophore, phorbol ester, and chemotactic peptide-induced cytoskeleton reorganization in human neutrophils. *J. Clin. Invest.* **79**, 1359–1364.
42. Jacobberger, J. W., Fogleman, D., and Lehman, J. M. (1986). Analysis of intracellular antigens by flow cytometry. *Cytometry* **7**, 356–364.
43. Kao, J. P. Y., Harootunian, A. T., and Tsien, R. Y. (1989). Photochemically generated cytosolic calcium pulses and their detection by Fluo-3. *J. Biol. Chem.* **264**, 8179–8184.
44. Kindt, R., and Goecke, W. (1988). Suitabilty of dichlorofluorescin diacetate (DCF) and hydroethidine (HE) for monitoring the respiratory burst of rat alveolar macrophages (AM). *Cytometry* Suppl. 2, 65A.
45. Klebanoff, S. J., and Clark, R. A. (1978). *In* "The Neutrophil: Function and Clinical Disorders." Elsevier/North-Holland, Amsterdam.
46. Korchak, H. M., Rich, A. M., Wilkenfeld, C., Rutherfford, L. E., and Weissmann, G. (1982). A carbocyanine dye, $DiOC_6(3)$, acts as a mitochondrial probe in human neutrophils. *Biochem. Biophys. Res. Commun.* **108**, 1495–1501.
47. Korchak, H. M., and Weissman, G. (1978). Changes in membrane potential of human granulocytes antecede the metabolic responses to surface stimulation. *Proc. Natl. Acad. Sci. USA* **75**, 3818–3822.
48. Landay, A., Rothberg, L., and Spear, G. (1988). Evaluation of oxidative metabolism of monocytes utilizing a flow cytometric assay. *Cytometry* Suppl. 2, 21A.
49. Lazzari, K. G., Proto, P. J., and Simons, E. R. (1986). Simultaneous measurement of stimulus-induced changes in cytoplasmic Ca^{2+} and in membrane potential of human neutrophils. *J. Biol. Chem.* **261**, 9710–9713.
50. Lehrer, R. I., Ganz, T., Selsted, M. E., Babior, B. M. and Curnutte, J. T. (1988). Neutrophils and host defense. *Ann. Intern. Med.* **109**, 127–142.
51. Leto, T. L., Lomax, K. J., Nunoi, H., Sechler, J. M. G., Nauseef, W. M., Clark, R. A., Gallin, J. I., and Malech, H. L. (1990). Cloning of a 67-kD neutrophil oxidase factor with similarity to a non-catalytic region of $p60^{C\text{-}Src}$. *Science* **248**, 727–730.
52. Lew, P. D., Dayer, M. J., Wolheim, C. B., and Pozzan, T. (1984). Effect of leukotriene B4, prostaglandin E2 and arachidonic acid on cytosolic-free calcium in human neutrophils. *FEBS Lett.* **166**, 44–48.
53. Livingston, D. H., Appel, S. H., Sonnenfeld, G., and Malangoni, M. A. (1989). The effect of tumor necrosis factor-alpha and interferon-gamma on neutrophil function. *J. Surg. Res.* **46**, 322–326.
54. Lomax, K. J., Burch-Whitman, C., Tiffany, H. L., Gallin, J. I., and Malech, H. L. (1988). Analysis of chronic granulomatous disease kindreds reveals

distinct genetic lesions affecting the same gene product. *Clin. Res.* **36**, 413A.

55. Lopez, M., Olive, D., and Mannoni, P. (1989). Analysis of cytosolic ionized calcium variation in polymorphonuclear leukocytes using flow cytometry and indo-1 AM. *Cytometry* **10**, 165–173.
56. Lyman, C. A., Simons, E. R., Meinick, D. A., and Diamond, R. D. (1987). Unopsonized *Candida albicans* hyphae stimulate a neutrophil respiratory burst and a cytosolic calcium flux without membrane depolarization. *J. Infect. Dis.* **156**, 770–776.
57. Malech, H. L., and Gallin, J. I. (1987). Neutrophils in human diseases. *N. Engl. J. Med.* **317**, 687–694.
58. McCall, C. E., DeChatelet, L. R., Butler, R., and Brown, D. (1974). Enhanced phagocytic capacity. The biologic basis for the elevated histochemical nitroblue tetrazolium reaction. *J. Clin. Invest.* **54**, 1227–1234.
59. McCall, C. E., DeChatelet, L. R., Cooper, M. R. and Shannon, C. (1973). Human toxic neutrophils. III. Metabolic characteristics. *J. Infect. Dis.* **127**, 26–33.
60. McNeil, P. L., Kennedy, A. L., Waggoner, A. S., Taylor, D. L., and Murphy, R. F., (1985). Light-scattering changes during chemotactic stimulation of human neutrophils: Kinetics followed by flow cytometry. *Cytometry.* **6**, 7–12.
61. Meshulam, T., Proto, P., Diamond, R. D., and Melnick, D. A. (1986). Calcium modulation and chemotactic response: Divergent stimulation of neutrophil chemotaxis and cytosolic calcium response by the chemotactic peptide receptor. *J. Immunol.* **137**, 1954–1960.
62. Michell, B. (1986). A second messenger function for inositol tetrakisphosphate. *Nature* **324**, 613.
63. Minta A., Kao, J. P. Y., and Tsien, R. Y. (1989). Fluorescent indicators for cytosolic calcium based on rhodamine and fluorescein chromophores. *J. Biol. Chem.* **264**, 8171–8178.
64. Mizuno, Y., Hara, T., Nakamura, M., Ueda, K., Minakami, S., and Take, H. (1988). Classification of chronic granulomatous disease on the basis of monoclonal antibody-defined surface cytochrome b deficiency. *J. Pediatr.* **113**, 458–462.
65. Moore, M. A. S., and Warren, D. J. (1987). Synergy of interleukin 1 and granulocyte colony-stimulating factor: *In vivo* stimulation of stem-cell recovery and hematopoietic regeneration following 5-fluorouracil treatment of mice. *Proc. Natl. Acad. Sci. USA* **84**, 7134–7138.
66. Murata, T., Sullivan, J. A., Sawyer, D. W., and Mandell, G. L. (1987). Influence of type and opsonization of ingested particle on intracellular free calcium distribution and superoxide production by human neutrophils. *Int. Immun.* **55**, 1784–1791.
67. Naccache, P. H., Molski, M. M., Volpi, M., Shefcyk, J., Molski, T. F. P., Loew, L., Becker, E. L., and Sha'afi, R. I. (1986). Biochemical events associated with the stimulation of rabbit neutrophils by platelet-activating factor. *J. Leuk. Biol.* **40**, 533–548.
68. Nunoi, H., Gallin, J. I., and Malech, H. L. (1988). Different defects in cytosolic factors may be responsible for autosomal forms of chronic granulomatous disease (CGD). *Clin. Res.* **36**, 415A.
69. Nunoi, H., Rotrosen, D., Gallin, J. I., and Malech, H. L. (1988). Two forms of autosomal chronic granulomatous disease lack distinct neutrophil cytosol factors. *Science* **242**, 1298–1301.
70. O'Connell, S. M., Laskin, J. D., Beavis, A. J., Robertson, F. M., and Laskin, D. L. (1988). Measurement of superoxide anion (O_2) production by flow cytometry. *Cytometry* Suppl. 2, 58A.
71. Ohno, Y., and Gallin, J. I. (1985). Diffusion of extracellular hydrogen peroxide into intracellular compartments of human neutrophils. Studies utilizing the inactivation of myeloperoxidase by hydrogen peroxide and azide. *J. Biol. Chem.* **260**, 8438–8446.
72. Okamura, N., Curnutte, J. T., Roberts, R. L., and Babior, B. M. (1988). Relationship of protein phosphorylation to the activation of the respiratory burst in human neutrophils. Defects in the phosphorylation of a group of closely related 48-kDa proteins in two forms of chronic granulomatous disease. *J. Biol. Chem.* **263**, 6777–6782.
73. Okamura, N., Malawista, S. E., Roberts, R. L., Rosen, H., Ochs, H. D., Babior, B. M., and Curnutte, J. T. (1988). Phosphorylation of the oxidase-related 48K phosphoprotein family in the unusual autosomal cytochrome-negative and X-linked cytochrome-positive types of chronic granulomatous disease. *Blood* **72**, 811–816.
74. Omann, G. M., Traynor, A. E., Harris, A. L, and Sklar, L. A. (1987). LTB_4-induced activation signals and responses in neutrophils are short-lived compared to formylpeptide. *J. Immunol.* **138**, 2626–2632.
75. Papadimitriou, J. M., van Duijn, P., Brederoo, P., and Streefkerk, J. G. (1976). A new method of the cytochemical demonstration of peroxidase for light, fluorescence and electron microscopy. *J. Histochem. Cytochem.* **24**, 82–90.
76. Patel, A. K., Hallett, M. B., and Campbell, A. K. (1987). Threshold responses in production of reactive oxygen metabolites in individual neutrophils detected by flow cytometry and microfluorimetry. *Biochem. J.* **248**, 173–180.
77. Powell, B. L., Olbrantz, P., Bicket, D., and Bass, D. A. (1986). Altered oxidative product formation in neutrophils of patients recovering from therapy for acute leukemia. *Blood* **67**, 1624–1630.
78. Rasmussen, H. (1986). The calcium messenger system. *N. Engl. J. Med.* **314**, 1094–1170.
79. Rice, W. G., Ganz, T., Kinkade, Jr., J. M., Selsted,

M. E., Lehrer, R. I., and Parmley, R. T. (1987). Defensin-rich dense granules of human neutrophils. *Blood* **70**, 757–765.

80. Robertson, F. M., Liesch, J. B., Tiku, K., and Tiku, M. L. (1988). Production of reactive oxygen intermediates by rabbit articular chondrocytes. *Cytometry* Suppl. 2, 30.

81. Robinson, J. M., Badwey, J. A., Karnovsky, M. L., and Karnovsky, M. J. (1985). Release of superoxide and change in morphology by neutrophils in response to phorbol esters: Antagonism by inhibitors of calcium-binding proteins. *J. Cell Biol.* **101**, 1052–1058.

82. Robinson, J. P., Bruner, L. H., Bassoe, Cl-F., Hudson, J. L., Ward, P. A., and Phan, S. H. (1988). Measurement of intracellular fluorescence of human monocytes relative to oxidative metabolism. *J. Leuk. Biol.* **43**, 304–310.

83. Robinson, J. P., Loesche, W., and Hudson, J. R. (1988). Isolation and characterization of neutrophils collected from gingival crevicular spaces: Application of flow cytometry in periodontal disease. *Cytometry* Suppl. 2, 71A.

84. Root, R. K., and Cohen, M. S. (1981). The microbicidal mechanisms of human neutrophils and eosinophils. *Rev. Inf. Dis.* **3**, 565–597.

85. Rothe, G., Kellermann, W., and Valet, G. (1988). Early prediction of organ failure in intensive care patients by multifunctional flow-cytometry of granulocytes. *Cytometry* Suppl. 2, 78A.

86. Rothe, G., Oser, A., and Valet, G. (1988). Dihydrorhodamine 123: A new low-cytometric indicator for respiratory burst activity in neutrophils. *Cytometry* Suppl. 2, 60A.

87. Rothe, G., and Valet, G. (1987). Use of hydroethidine and 2,7-dichlorofluorescin for the flow cytometric measurement of NADPH oxidase and mitochondrial oxygen radical formation in phagocytes. *Cytometry* Suppl. 1, 77A.

88. Rotrosen, D., Nunoi, H., Gallin, J. I., and Malech, H. L. (1989). Evidence for interaction of neutrophil 47 kD phosphoprotein with cytochrome b558 during activation of the respiratory burst. *Clin. Res.* **37(2)**, 438A.

89. Royer-Pokora, B., Kunkel, L. M., Monaco, A. P., Goff, S. C., Newburger, P. E., Baehner, R. L., Cole, F. S., Curnutte, J. T., and Orkin, S. H. (1986). Cloning the gene for an inherited human disorder — chronic granulomatous disease — on the basis of its chromosomal location. *Nature* **322**, 32–38.

90. Ryder, M. I., Winkler, J. R., and Weinreb, R. N. (1988). Elevated phagocytosis, oxidative burst, and F-actin formation in PMNs from individuals with intraoral manifestations of HIV infection. *J. Acq. Immun. Defic. Synd.* **1**, 346–353.

91. Salata, R. A., Sullivan, J. A., and Mandell, L. (1983). Visualization of hydrogen peroxide in living polymorphonuclear neutrophils utilizing leucodiacetyl 2′,7′-dichlorofluorescin: Photomicrographic and microphotometrometric studies. *Trans. Assoc. Am. Phys.* **96**, 375–383.

92. Sawyer, D. W., Sullivan, J. A., and Mandell, G. L. (1986). Regional intracellular calcium changes in neutrophils (PMN) related to morphology and F-actin formation. *Clin. Res.* **34**, 723A.

93. Seeds, M. C., and Bass, D. A. (1987). Phorbol myristate acetate induced azurophil degranulation in neutrophils from patients with acute bacterial infections. *Clin. Res.* **35**, 490A.

94. Seeds, M. C., Olbrantz, P. J., Thomas, M., McCall, C. E., and Bass, D. A. (1985). Phagocytosis-mediated stimulation of oxidative metabolism is elevated in neutrophils from patients with bacteremia. *Clin. Res.* **33**, 419A.

95. Seeds, M. C., Parce, J. W., Szejda, P., and Bass, D. A. (1985). Independent stimulation of membrane potential changes and the oxidative metabolic burst in polymorphonuclear leukocytes. *Blood* **65**, 233–240.

96. Segal, A. W. (1989). The electron transport chain of the microbicidal oxidase of phagocytic cells and its involvement in the molecular pathology of chronic granulomatous disease. *J. Clin. Invest.* **83**, 1785–1793.

97. Segal, A. W., Cross, A. R., Garcia, R. C., Borregaard, N., Valerius, N. H., Soothill, J. F., and Jones, O. T. G. (1983). Absence of cytochrome b-245 in chronic granulomatous disease. A multicenter European evaluation of its incidence and relevance. *N. Engl. J. Med.* **308**, 245–251.

98. Seligmann, B. E., and Gallin, J. I. (1980). Use of lipophilic probes of membrane potential to assess human neutrophil activation. Abnormality in chronic granulomatous disease. *J. Clin. Invest.* **66**, 493–503.

99. Seligmann, B. E., and Gallin, J. I. (1983). Comparison of indirect probes of membrane potential utilized in studies of human neutrophils. *J. Cell. Physiol.* **115**, 105–115.

100. Sha'afi, R. I., Shefcyk, J., Yassin, R., Molski, T. F. P., Volpi, M., Naccache, P. H., White, J. R., Feinstein, M. B., and Becker, E. L. (1986). Is a rise in intracellular concentration of free calcium necessary or sufficient for stimulated cytoskeletal-associated actin? *J. Cell Biol.* **102**, 1459–1463.

101. Shalit, M., Campbell, D. E., von Allmen, C., Atkins, P. C., Douglas, S. D. and Zweiman, B. (1987). Neutrophil activation in human inflammatory skin reactions. *J. Allergy Clin. Immunol.* **80**, 87–93.

102. Shapiro, H. M. (1985). "Practical Flow Cytometry." Alan R. Liss, New York.

103. Sklar, L. A., Hyslop, P. A., Oades, Z. G., Omann, G. M., Jesaitis, A. J., Painter, R. G., and Cochrane, C. G. (1985). Signal transduction and

ligand–receptor dynamics in the human neutrophil. Transient responses and occupancy–response relations at the formyl peptide receptor. *J. Biol. Chem.* **260**, 11461–11467.

104. Sklar, L. A., and Oades, Z. G. (1985). Signal transduction and ligand-receptor dynamics in the neutrophil. Ca^{2+} modulation and restoration. *J. Biol. Chem.* **260**, 11468–11475.

105. Sklar, L. A., Oades, Z. G., and Finney, D. A. (1984). Neutrophil degranulation detected by right angle light scattering: Spectroscopic methods suitable for simultaneous analyses of degranulation of shape change, elastase release, and cell aggregation. *J. Immunol.* **133**, 1483–1487.

106. Sklar, L. A., Omann, G. M., and Painter, R. G. (1985). Relationship of actin polymerization and depolymerization to light scattering in human neutrophils: dependence on receptor occupany and intracellular Ca^{++}. *J. Cell Biol.* **101**, 1161–1166.

107. Smolen, J. E., Stoehr, S. J., Traynor, A. E., and Sklar, L. A. (1987). The kinetics of secretion from permeabilized human neutrophils: Release of elastase and correlations with other granule constituents and right angle light scatter. *J. Leuk. Biol.* **41**, 8–13.

108. Spry, C. J. F. (1988). "Eosinophils. A Comprehensive Review and Guide to the Scientific and Medical Literature." Oxford University Press, Oxford.

109. Stendahl, O., Krause, K. H., Volpe, P., Gil, J., Pittet, D., Meldolesi, J., and Lew, P. D. (1988). Calciosomes, the inositol 1,4,5-trisphosphate sensitive Ca^{++} store, are redistributed in human neutrophils during phagocytosis and degranulation. *Clin. Res.* **36**, 472A.

110. Sullivan, G. W., Donowitz, R. G., Sullivan, J. A., and Mandell, G. L. (1984). Interrelationships of polymorphonuclear neutrophil membrane-bound calcium, membrane potential, and chemiluminescence: Studies in single living cells. *Blood* **64**, 1184–1192.

111. Szejda, P., Parce, J. W., Seeds, M. S., and Bass, D. A. (1984). Flow cytometric quantitation of oxidative product formation by polymorphonuclear leukocytes during phagocytosis. *J. Immunol.* **133**, 3303–3307.

112. Tauber, A. I., Borregaard, N., Simons, E., and Wright, J. (1983). Chronic granulomatous disease: A syndrome of phagocyte oxidase deficiencies. *Medicine* **62**, 286–309.

113. Tauber, A. I., and Simons, E. R. (1983). Dissociation of human neutrophil membrane depolarization, respiratory burst stimulation and phospholipid metabolism by quinacrine. *FEBS Lett.* **156**, 161–164.

114. Treumer, J., and Valet, G. (1986). Flow-cytometric determination of glutathione alterations in vital cells by o-phthaldialdehyde (OPT) staining. *Exp. Cell Res.* **163**, 518–524.

115. Trinkle, L. S., Wellhausen, S. R., and McLeish, K. R. (1987). A simultaneous flow cytometric measurement of neutrophil phagocytosis and oxidative burst in whole blood. *Diag. Clin. Immunol.* **5**, 62–68.

116. Tsien, R. Y. (1980). New calcium indicators and buffers with high selectivity against magnesium and protons: design synthesis, and properties of prototype structures. *Biochemistry* **19**, 2396–2404.

117. Volpp, B. D., Nauseef, W. M., Donelson, J. E., Moser, D. R., and Clark R. A. (1989). Cloning of the cDNA and functional expression of the 47-kilodalton cytosolic component of human neutrophil respiratory burst oxidase. *Proc. Natl. Acad. Sci. USA* **86**, 7195–7199, correction p. 9563.

118. von Tscharner, V., Deranleau, D. A., and Baggiolini, M. (1986). Calcium fluxes and calcium buffering in human neutrophils. *J. Biol. Chem.* **261**, 10163–10168.

119. Waggoner, A. S. (1979). Dye indicators of membrane potential. *Ann. Rev. Biophys. Bioeng.* **8**, 47.

120. Well, G. J., and Chused, T. M. (1981). Eosinophil autofluorescence and its use in isolation and analysis of human eosinophils using flow microfluorometry. *Blood* **57**, 1099–1100.

121. Whitin, J. C., Clark, R. A., Simons, E. R., and Cohen, H. J. (1981). Effects of the myeloperoxidase system on fluorescent probes of granulocyte membrane potential *J. Biol. Chem.* **156**, 8904–8906.

122. Zimmerli, W., Seligmann, B. E., and Gallin, J. I. (1987). Neutrophils are hyperpolarized after exudation and show an increased depolarization response to formyl-peptide but not to phorbol myristate acetate. *Eur. J. Clin. Invest.* **17**, 435–441.

2.5

Analysis of Platelets

KENNETH A. AULT, JANE G. MITCHELL, H. M. RINDER, C. RINDER AND ROBERT S. HILLMAN

Maine Cytometry Research Institute
Maine Medical Center
Portland, ME 04012, USA

I. INTRODUCTION

This review will address the subject of the use of flow cytometers for the analysis of platelets. We will emphasize the measurement of platelet activation, although the general principles may be applied to the measurement of many different platelet properties. For those who are interested in the biology of platelets we believe that flow cytometry offers a new technique which may make a significant contribution to the study of platelets for a number of reasons that will become obvious. For those who may be primarily interested in flow cytometry, we believe that platelets represent a challenging and clinically important application of this technology. In the course of attempting to study platelets we and others have discovered that there are other small particles in blood which may also be clinically important. We will therefore include a discussion of these various forms of "debris".

The rationale for the flow cytometric approach to the study of platelets is that platelets are, by their nature, difficult to study, and the flow cytometer allows us to study them in ways which require minimal disturbance of these very reactive cells. In addition, platelets frequently become important clinically in situations in which they are deficient, i.e. thrombocytopenia. The flow cytometer allows us to obtain useful information about platelets even when they are so deficient that it is difficult to obtain enough of them to study by other methods.

It is in the nature of platelets to react to abnormal surfaces which they encounter in the

FLOW CYTOMETRY IN HEMATOLOGY
ISBN 0-12-432940-3

circulation. For this reason, their physiology is altered whenever they encounter a foreign surface. It is well known that inducing blood stasis with a tourniquet, puncturing a vein, and drawing platelets through needles or tubing into a container all induce physiological changes in platelets. This realization has plagued and clouded attempts to study platelets by any means. The use of flow cytometry certainly does not overcome all of these difficulties, but it adds an important new dimension, and we believe it will take its place amongst the use of aggregometers and the measurement of the products of platelet secretion as a standard method for platelet study both in the clinic and in research.

Aside from the technical issues of labeling and analyzing them, probably the most important aspect of platelet study by any method is the preparation of the platelets and thus we will address this issue first.

II. PREPARATION OF PLATELETS FOR ANALYSIS

It is axiomatic that platelets should be disturbed as little as possible in the course of preparing them for analysis. In the extreme case they are capable of becoming aggregated into macroscopic clumps which renders analysis impossible. Although they respond to almost any foreign surface they seem to be least responsive when they remain in their normal setting of whole blood. Thus our approach has been to reduce the steps required to "purify" the platelets to the minimum required for the specific measurement. Fortunately, one of the great strengths of the flow cytometer is that it is capable of recognizing and accurately measuring the properties of platelets even when they are heavily contaminated by other cells. The only situation in which we have found it necessary to prepare washed platelets is in the measurement of platelet associated Ig (*2*). Because the label in this case is an anti-Ig antibody the plasma must be washed away. Even in this case we have made no attempt to "purify" the platelets by careful differential centrifugation. A washed platelet preparation which is contaminated by erythrocytes and leukocytes can be quickly prepared and results in minimal handling of the platelets.

Another principle which we have applied repeatedly in our study of platelets is that of fixation. Platelets which have been fixed in paraformaldehyde are rendered stable and will not undergo further changes. Thus we have, in every case, fixed the platelets at an early stage in their preparation, usually before labeling. This strategy has the obvious advantage that it stabilizes the platelets, but it has the disadvantage that it may induce artefacts due to changes induced in the platelet by the fixative solution, or in that some epitopes of platelet antigens may be altered or destroyed (denatured) by fixation. Thus, although we have found early fixation to be a practical and reliable method for the studies we have carried out, it must be emphasized that every time a new measurement is to be made on platelets the issue of fixation must be addressed anew. Specifically, any monoclonal antibody which is used in the study of fixed platelets must be carefully evaluated to determine if the specific epitope recognized by the antibody is stable to fixation.

III. WHOLE BLOOD TECHNIQUES

An important element of our work with platelets has been our adoption of methods of studying platelets in whole blood. This approach was first employed by Sanford Shattil (*24*) and has proven to be extremely useful. The method consists of labeling the platelets in whole blood using a platelet-specific monoclonal antibody such as anti-GPIb or anti-GPIIb/IIIa conjugated to a fluorochrome, usually fluorescein. The platelets can also be labeled at the same time with another monoclonal, a nucleic acid-specific dye, etc., which fluoresces at a different wavelength. The labeled whole blood sample is then analyzed in the flow cytometer using the technique of "triggering" on the platelet-specific marker. The result is that only particles having a level of the platelet-specific marker greater than some threshold set by the operator are analyzed by the instrument. Despite the fact that the flow cytometer is "blinded" to the presence of particles beneath the threshold, the operator must not forget that the other particles are present. For example, in a normal blood sample which has 2.5×10^5 platelets/mm^3 and 5×10^6 erythrocytes/mm^3 there will be 20 red cells passing through the machine for every platelet. This ratio can become much more extreme if the platelet count is low. Depending on the volume flow rate and the observation volume of the flow cytometer, some of these red cells may be in the observation volume at the same time as the platelet and thus they may profoundly affect the measurement made on the platelet. This can be easily demonstrated when performing the whole blood technique by varying the volume flow rate. At low flow rates the light scatter properties of the platelets will be indistinguishable from those of purified platelets, but at some critical flow rate the light scatter pattern will become markedly altered by the presence of red cells.

This limitation affects primarily the rate at which it is possible to obtain data using the whole blood technique. We have been impressed that the quality of the data obtained in this way is excellent. In the absence of any interference by substances in plasma one generally gets excellent labeling with very little of the non-specific interactions sometimes seen with separated cells.

IV. THE PROBLEM OF "DEBRIS"

Regardless of whether one chooses to study purified platelets or platelets in whole blood, they have a characteristic appearance in the flow cytometer. When identified either by light scatter or fluorescence the platelets appear as a typical sloping cluster in which there is a strong positive correlation between particle size and light scatter or fluorescence (Fig. 1). This is due of course to the fact that larger platelets scatter more light, have a greater surface area, and thus express more of most any antigen than small platelets. Compared to other blood cells the size heterogeneity of platelets is very large. For example, normal lymphocytes range in size from perhaps 10 to 20 μm. Thus there is a $2\times$ variation in diameter and a $4\times$ variation in surface area between a small lymphocyte and a large lymphocyte. However, normal platelets vary from 1 to 4 or 5 μm in diameter. This results in a 16- to 25-fold variation in surface area and a corresponding variation in fluorescence. In pathological states, or when platelets become activated, there can be a much larger

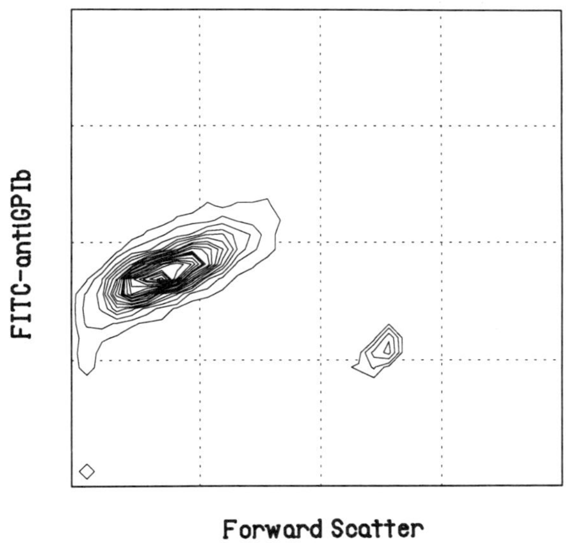

Fig. 1. The typical "platelet cluster" obtained in whole blood. This figure shows flow cytometric data obtained from fixed whole blood labeled with FITC–anti-GPIb. The instrument was triggered on FITC fluorescence so that GPIb-negative particles are not visualized. The typical platelet cluster shows a positive correlation between size (forward scatter) and fluorescence. To the left of the platelets are GPIb-positive microparticles which are sub-micron in size. To the right are some leukocytes which fall above the threshold by virtue of their relatively high levels of autofluorescence. This data was obtained using a FACScan flow cytometer (Becton Dickinson Immunocytometry Systems, Mountain View, CA). Both scales are logarithmic (four decades).

variation as platelets form clumps ranging in size up to about 20 microns.

For these reasons, the standard flow cytometric technique of "gating" on the platelets must be undertaken with caution. Small platelets, at about 1 micron diameter are not well resolved on some flow cytometers. If the small platelets cannot be clearly distinguished from instrument noise then it is very easy to contaminate one's data with noise. Even when they can be clearly resolved we have learned that the typical platelet cluster can contain particles that are not platelets but which cannot be distinguished from platelets on the basis of size or light scatter properties. The presence of this "debris" contaminating platelet data has become a subject of study in our laboratory. We have observed the following kinds of debris:

(1) The most common non-platelet particles we have observed are erythrocyte fragments. These have been referred to as "micro-

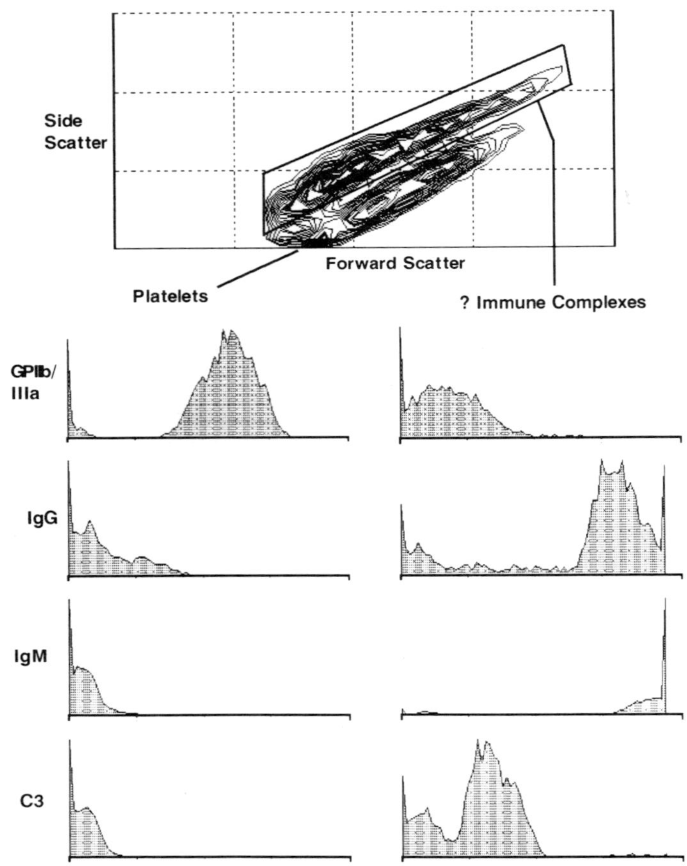

Fig. 2. Labeling pattern of putative immune complexes. At the top is shown the light scatter pattern of a sample which contains a large amount of "debris" which falls above the usual platelet cluster. The left panels show the labeling of the platelets, which are positive for GPIIb/IIIa, have low levels of IgG, and are negative for IgM and C3. The right panels show the labeling of the ? immune complexes which have little or no GPIIb/IIIa but have extraordinarily high levels of IgG, IgM, and C3. Each of the histograms has a four decade logarithmic horizontal scale. Thus the IgM levels of the ? immune complexes is nearly 10^4 that of the platelets.

spherocytes" (*2,11,29*). They are typically seen in patients who are severely thrombocytopenic (platelet count less than 20 000), and they are associated with immune thrombocytopenia rather than non-immune etiologies (*2*). They co-purify with platelets, but do not express platelet-specific glycoproteins such as GPIb and GPIIb/IIIa. They are best recognized because they do express the erythrocyte marker glycophorin. They frequently fall exactly in the middle of the platelet cluster as defined by light scatter. They are typically much less heterogeneous in size than platelets.

(2) We have also observed the presence in some patients of extremely heterogeneous particles which label very brightly with antibodies directed against immunoglobulins and complement components. These particles usually are seen in patients with an active immunological process, they extend from much smaller to much larger in size than normal platelets, and they usually have increased wide angle light scatter compared with platelets (Fig. 2). We have hypothesized that they may be immune complexes.

(3) In rare cases we have observed non-platelet particles in the size range of platelets which have leukocyte markers and may represent fragments of leukocytes. Our best characterized example was in a patient with very high count chronic lymphocytic leukemia.

(4) Finally, we and others have observed the presence of particles which do have platelet surface markers but which are much smaller than normal platelets (sub-micron). This "platelet dust" or platelet micro-particles have been observed both *in vitro* and *in vivo* (*9,10,25*).

Thus when studying platelets by any technique it is important to realize that purified platelets can be heavily contaminated by non-platelet particles with similar physical characteristics. Even in the flow cytometer some of these non-platelet particles can be difficult to distinguish. It is for this reason that nearly all of our platelet studies are based on identification of platelets by both physical characteristics, i.e. the typical platelet cluster, and immunological characteristics, i.e. labeling with either anti-GPIb or anti-GPIIb/IIIa.

V. PLATELET ACTIVATION MARKERS

Recently several groups have described immunological markers of platelet activation which may prove to be of considerable clinical utility. Although beyond the scope of this review, there is sufficient information published to strongly suggest that platelet activation is important in the pathogenesis of some forms of cardiovascular disease (*8*). Platelet activation *in vivo* has been detected primarily by measurement of the secreted products of activated platelets. These methods are very sensitive, but do not allow precise definition of the time or place at which activation is occurring. Thus the availability of immunological markers of platelet activation offers the promise of determining exactly when and where platelets become activated *in vivo* if the appropriate samples can be obtained.

In addition to the possible clinical applications, the research applications for such markers are too numerous to describe here. The one area of research which our laboratory has emphasized is the use of these new markers to obtain correlated information about the sequence of events occurring during platelet activation and aggregation as described below.

Table I is a summary of the immunological markers of platelet activation of which we are aware. In general, they fall into three categories.

First are those which are components of the membranes of platelet granules. These have the interesting property that they are not present on the surface of resting platelets, but become incorporated into the platelet membrane in the process of "exocytosis" of the granule contents when the granule membrane fuses with the platelet surface membrane. These are probably the most spectacular of the activation markers in that they switch from completely negative on resting platelets to strongly positive on platelets which have undergone the release reaction.

Table I

IMMUNOLOGICAL MARKERS OF PLATELET ACTIVATION

Antigen	Clone designation	References
Alpha granule membrane proteins GMP-140 (PADGEM)	S12, W40, KC4	(*4,17*)
Lysosomal proteins 53kD	2.28	(*19*)
Fibrinogen receptor (GPIIb/IIIa)	PAC1, 7E3	(*22,23*)
Fibrinogen	9F9	(*5*)
Thrombospondin	5G11	(*10*)
Factor Xa		(*26*)

They have the drawback however that they only appear on platelets which are fully activated in the sense that they have released the contents of their granules. It is quite likely that there are intermediate or more subtle levels of platelet activation which would be missed by these markers.

Second are those epitopes that result from an activation-related change on the platelet surface. The best example of this is PAC1 which binds to a new epitope of GPIIb/IIIa which appears at the time of platelet activation. Such a marker appears to reveal a more subtle, perhaps earlier stage in platelet activation. However, because the antigen is one which is at least potentially present on all platelets the change from resting to activated tends to be less dramatic, and, at least in the case of PAC1 the epitope is easily affected by fixation. A number of monoclonal antibodies to GPIIb/IIIa have, to varying degrees, the property of binding more strongly to activated than to resting platelets.

Finally, there are several antigens which appear on the activated platelet because they bind to receptors on the platelet surface. These include fibrinogen itself, thrombospondin, and factor Xa. The levels of these substances on activated platelets tend to be low, and the amount can be influenced by factors external to the platelet itself, but the presence or absence of such substances can be of considerable interest because of the functional consequences of their binding.

In addition to these there are several non-immunological techniques which have been applied to the flow cytometric study of platelet activation. These include the measurement of Ca^{2+} ion flux using fluorescent dyes that are sensitive to Ca^{2+} concentration (*13*), and the use of dyes such as mepacrine which enter platelet-dense granules and then are released upon activation (*6*). Both of these methods, although very powerful, require that the platelets be loaded with the dye prior to activation and thus, although useful for *in vitro* research, do not lend themselves to the measurement of *in vivo* platelet activation.

VI. PLATELET ACTIVATION AND AGGREGATION

The protocol we have been using in most of our studies of platelet activation involves immediate fixation of whole blood followed by labeling as described above. In the course of experiments in which platelet agonists were added to the blood prior to fixation we observed that two dramatic changes were taking place in the platelets. First, the platelets became positive for the activation marker (GMP-140 in this case). Second, the size distribution and light scatter of the platelets changed in such a way as to suggest that the platelets were aggregating. This led us to a series of experiments in which we showed that the flow cytometer was capable of measuring not only the expression of the activation markers but also the phenomenon of aggregation (*3*). This is illustrated in Fig. 3. Several interesting observations have come out of these studies. Platelet aggregation as measured on the flow cytometer has the same general properties as that measured on an aggregometer and we believe that the information is equivalent in terms of clinical usefulness. There are however several differences between the aggregation seen in our system and that seen in an aggregometer.

First, the aggregates that we see in whole blood do not become much larger than leukocytes, i.e. about 20 μm, whereas the aggregates that form in platelet-rich plasma in an aggregometer are macroscopic precipitates. This may be due to the lower effective concentration of platelets in whole blood and the presence of so many other particles which decrease the rate of growth of the aggregate.

Second, we find that most of the aggregation that occurs in whole blood is reversible by about 15 min at 37°C whereas that which occurs in an aggregometer is usually described as irreversible. This difference may be related to the size of the aggregates that form in whole blood although our experience suggests that a considerable amount of the aggregation seen in an aggregometer is also reversible if one follows the reaction long enough.

Third, we have convincing evidence for the existence of platelet–leukocyte aggregates as well as platelet–platelet aggregates. Preliminary experiments have suggested that activated platelets are most likely to bind to monocytes rather than to erythrocytes, lymphocytes, or granulocytes (Fig. 4). This phenomenon may be of importance in view of the recently described structural relationship between several cell–cell adhesion molecules and GMP-140 (*16*).

Fourth, our data suggest that the phenomena of activation (as measured by expression of GMP-140) and aggregation are quite distinct. In particular we have observed that platelets which do not express GMP-140 can participate in aggregation. The exact sequence of platelet activation in terms of expression of surface markers and the relation between changes on

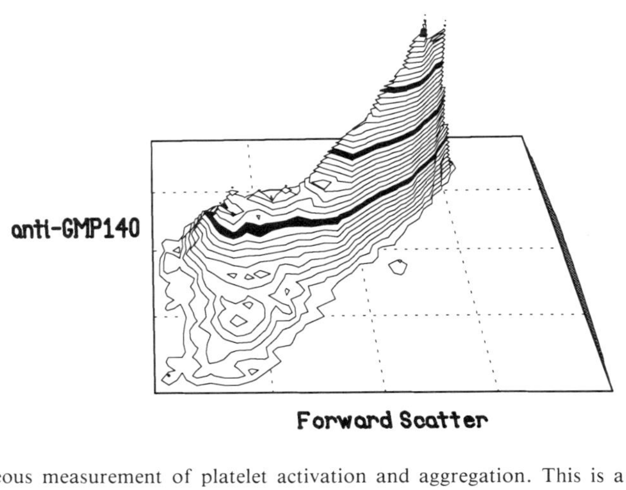

Fig. 3. Simultaneous measurement of platelet activation and aggregation. This is a flow cytometric analysis of platelets that have been induced to aggregate by addition of 10 μ*M* adenosine diphosphate (ADP) to heparinized blood. This data was obtained by fixing the blood after 8 min at 37°C. All of the particles in this analysis were GPIb positive. The horizontal axis is log forward scatter. The vertical axis is log fluorescence using phycoerythrin conjugated anti-GMP-140 to detect platelet activation. Within the platelet size range there are both activated and unactivated platelets. In addition platelet aggregates can be seen extending to the upper right. The peak of aggregates in the upper right are approximately 10–20 μm in diameter.

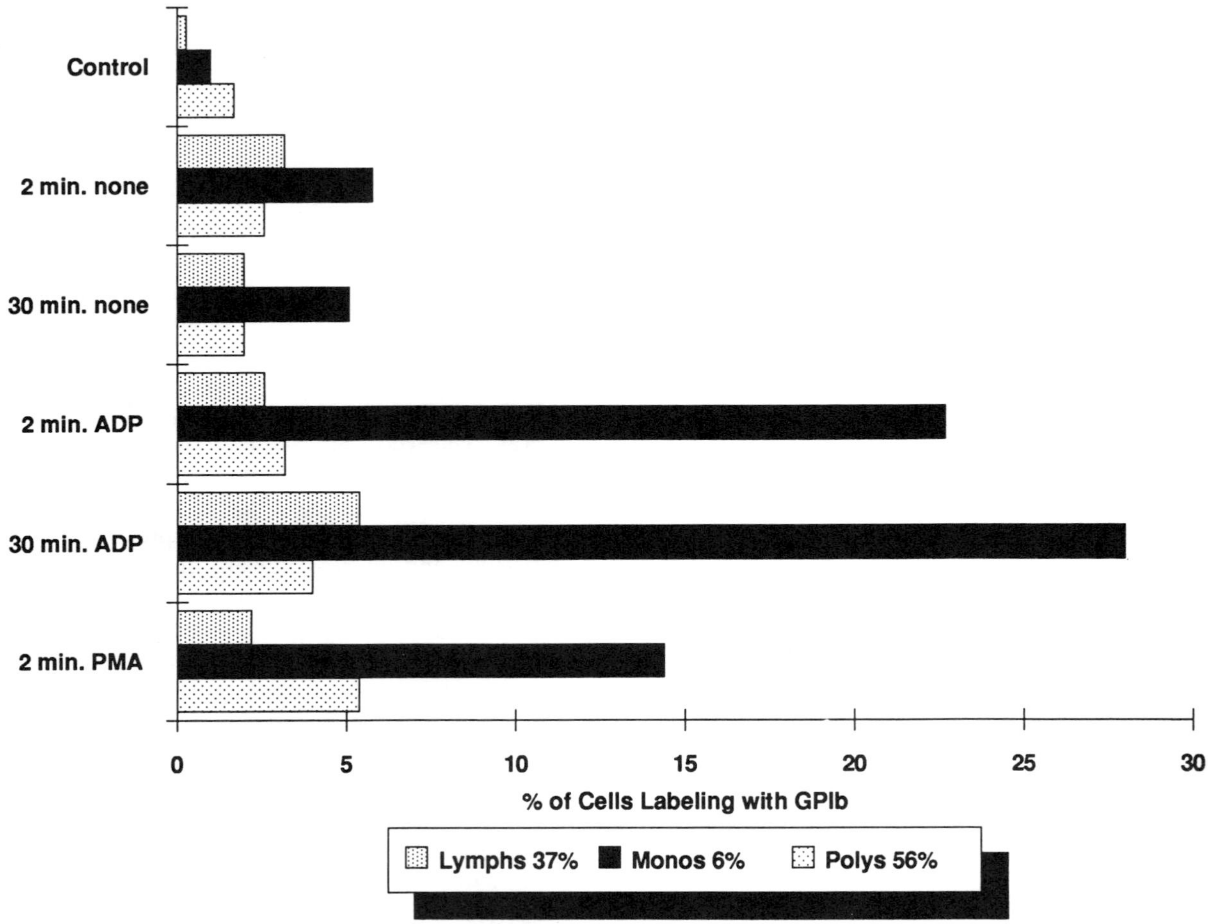

Fig. 4. Binding of platelets to leukocytes in whole blood. In this experiment heparinized whole blood samples were treated with either nothing, adenosine diphosphate (ADP), or phorbol myristate acetate (PMA) for the indicated times at 37°C. The blood was then fixed, and labeled so as to allow the measurement of the levels of binding of anti-GPIb to each of the major leukocyte classes. It can be seen that even in the absence of agonist there is some preferential binding of platelets to monocytes and that this is markedly increased by agents that cause platelet activation. Since monocytes constituted only 6% of the leukocytes in this sample, the results cannot be explained by random binding of platelets to other particles.

the surface and aggregation remain to be explored in detail. Other activation markers, such as the activation-dependent epitopes of GPIIb/IIIa, are likely to be directly involved in platelet aggregation, whereas perhaps molecules such as GMP-140 are involved in binding of platelets to other types of cells. Flow cytometry will play an important role in this work in the future.

VII. THE CLINICAL UTILITY OF PLATELET ACTIVATION MARKERS

There are a number of areas in which one might expect platelet activation markers to prove to have clinical utility. These can be broadly classified as those applications involving "biocompatibility" issues, platelet storage and transfusion, those involving assessment of cardiovascular disease, and assessment of platelet dysfunction. Although none of these areas has been extensively explored there are some data.

In the area of biocompatibility we are referring to the use of platelet activation markers to assess the degree of interaction between platelets and foreign surfaces such as those encountered in dialysis, cardiopulmonary bypass, vascular prostheses, etc. George *et al.* (*10*) have reported data using GMP-140 and thrombospondin as markers in patients undergoing cardiopulmonary bypass. They have shown no increase in either GMP-140 or thrombospondin but have noted an increase in platelet membrane microparticles suggesting that platelet fragmentation but not release is taking place. Our own laboratory has shown a small, but statistically significant, rise in the proportion of GMP-140-positive platelets during bypass, as have Nieuwenhuis *et al.* (*19*). We have also performed "closed loop" bypass experiments in which we observed an initial transient wave of GMP-140-positive platelet aggregates at the initiation of circulation, followed by significant rise in GMP-140-positive platelets over a period of several hours (*20*). We have suggested that the results can be best explained by proposing that platelets are initially activated and aggregate in response to the foreign surface, but that many of the platelet aggregates subsequently dissociate, resulting in a mixture of activated and apparently unactivated platelets exiting the bypass. As the activated platelets return to the *in vivo* circulation our data suggest that they have a very short survival time and are essentially "filtered" out so that they do not appear in large numbers in the circulation and do not accumulate during bypass. It is entirely possible, as discussed above, that the apparently unactivated platelets which are not expressing GMP-140 are nevertheless abnormal; they may be found to be activated when other markers are used, and they may very well be dysfunctional as has been observed by others (*12*,*27*,*28*).

In the area of platelet storage and transfusion, there is abundant evidence that stored platelets undergo progressive activation. This has been shown by examination of shape change, Ca^{2+} concentration (*7*), release of platelet granule contents (*18*), and in our laboratory by expression of GMP-140 (*21*). In addition, we have accumulated evidence suggesting that GMP-140-positive platelets have a very short survival in the circulation once transfused. Thus it seems likely that many of the platelets routinely transfused have become activated during storage and they may not contribute very much to improved hemostasis in the recipient. Further studies of the mechanism and kinetics of this process may contribute to more effective storage of platelets in the future.

There is every reason to think that platelets may be involved in several different types of cardiovascular disease, and that the ability to assess platelet activation, especially *in vivo*, may be very useful in these diseases. Platelet deposition may play a role in the primary events of atherosclerosis and obstruction leading to myocardial infarction and stroke (*8*). Platelets certainly play a role in the reocclusion of vessels after they are opened, either surgically or after angioplasty. Finally, it is likely that platelets contribute to the process of venous thrombosis as well.

There are several difficulties that one encounters in attempting to use platelet activation to assess the time course or magnitude of each of these cardiovascular events. First, even if every platelet passing through a damaged coronary artery were activated the activated platelets would be so massively diluted in the systemic circulation that they would be difficult to detect. Most authors put the detection limit of current flow cytometric methods at about 1–5% activated platelets (*14*,*24*), and this may be optimistic for measurements made outside of a controlled laboratory situation. Second, there is every reason to believe that activated platelets have a much shorter survival time in the circulation than resting platelets. Thus the activated platelets probably do not accumulate in the circulation as the disease process continues. For these reasons, with present technology, it seems it will be necessary to look for platelet activation at or near the site of the disease process. For example, in preliminary studies we have been able to demonstrate high levels of platelet activation in samples drawn directly from a diseased coronary artery although we were unable to detect any increase above background in samples taken simultaneously from the pulmonary artery.

Despite these difficulties, there may be an important role for measurements of platelet activation *in vivo*. For example, it might be possible to estimate the rate at which platelets are being activated as they cross a vascular prosthesis, and such a measurement might predict which prostheses are likely to fail. A similar line of reasoning might apply to studying platelet activation across coronary vessels that have been opened by thrombolytic agents or by mechanical means. If, after flow is restored, there is a high level of platelet activation it might mean that the endothelium is so damaged, or the flow pattern is such that a platelet thrombus may form and lead to reocclusion.

Obviously a great deal more work needs to be done to determine how, when, and where studies of platelet activation using markers will be clinically useful.

The final area in which measures of platelet activation may be of clinical utility is in the diagnosis of platelet dysfunction. Use has already been made of flow cytometric assays of platelet surface glycoproteins in the diagnosis of genetic platelet disorders (*15*), and in assessment of platelet modification by plasmin (*1*). It seems likely that, as we come to understand better the sequence of activation marker display on normal platelets, we will be able to use this information to detect subtle abnormalities of platelet function.

The recent availability of immunological markers of platelet activation, combined with the unique capability of flow cytometry to identify small subpopulations of platelets and to quantitate the levels of markers, now allows us to assess platelet function in an entirely new way. The clinical and research potential of these developments is only beginning to be explored. We have attempted to summarize the current status of this field with special emphasis on the use of flow cytometry. We look forward with enthusiasm to future progress in this area.

REFERENCES

1. Adelman, B., Michelson, A. D., Handin, R. I., and Ault, K. A. (1985). Evaluation of platelet glycoprotein Ib by fluorescence flow cytometry. *Blood* **66**, 423.
2. Ault, K. A. (1988). Flow cytometric measurement of platelet associated immunoglobulin. *Pathol. Immunopathol. Res.* **7**, 395–408.
3. Ault, K. A., Rinder, H. M., Mitchell, J. G., Rinder, C. S., Lambrew, C. T., and Hillman, R. S. (1989). Correlated measurement of platelet release and aggregation in whole blood. *Cytometry.* **10**, 448–455.
4. Berman, C. L., Yeo, E. L., Wencel-Drake, J. D., Furie, B. C., Ginsberg, M. H., and Furie, B. (1986). A platelet alpha granule membrane protein that is associated with the plasma membrane after activation. *J. Clin. Invest.* **78**, 130–137.
5. Coller, B. S. (1985). A new murine monoclonal antibody reports an activation dependent change in the conformation and/or microenvironment of the platelet glycoprotein IIb/IIIa complex. *J. Clin. Invest.* **76**, 101.
6. Corash, L., Mok, Y., and Rheinschmidt, M. (1986). Analysis of platelet function using subcellular and surface associated fluorescent probes. *In* "Applications of Fluorescence in the Biomedical Sciences", pp. 567–584. Alan R. Liss, New York.
7. Feinberg, H., Sarin, M. M., Batka, E. A., Porter, C. R., Miripol, J. E., and Stewart, M. (1988). Platelet storage: Changes in cytosolic Ca^{2+}, actin polymerization, and shape. *Blood* **72**, 766–769.
8. Fitzgerald, D. J., Roy, L., Catella, F., and Fitzgerald, G. A. (1986). Platelet activation in unstable coronary disease. *New Engl. J. Med.* **315**, 983–990.
9. George, J. N., Pickett, E. B., and Heinz, R. (1986). Platelet membrane microparticles in blood bank fresh frozen plasma and cryoprecipitate. *Blood* **68**, 307–309.
10. George, J. N., Pickett, E. B., Saucerman, S., McEver, R. P., Kunicki, T. J., Kieffer, N., and Newman, P. J. (1986). Platelet surface glycoproteins: Studies on resting and activated platelets and platelet membrane microparticles in normal subjects, and observations in patients during adult respiratory distress syndrome and cardiac surgery. *J. Clin. Invest.* **78**, 340–348.

11. Gottschall, J. L., Collins, J., Kunicki, T. J., Nash, R., and Aster, R. H. (1987). Effect of hemolysis on apparent values of platelet associated IgG. *Am. J. Clin. Pathol.* **87**, 218–222.

12. Harker, L. A., Malpass, T. W., Branson, H. E., Hessel, E. A., and Slichter, S. J. (1980). Mechanism of abnormal bleeding in patients undergoing cardiopulmonary bypass: acquired transient platelet dysfunction associated with selective alpha granule release. *Blood* **56**, 824–834.

13. Johnson, P. C., Ware, J. A., Cliveden, P. B., Smith, M., Dvorak, A. M., and Salzman, E. W. (1985). Measurement of ionized calcium in blood platelets with the photoprotein aequorin: Comparison with Quin 2. *J. Biol. Chem.* **260**, 2069–2076.

14. Johnston, G. I., Pickett, E. B., McEver, R. P., and George, J. N. (1987). Heterogeneity of platelet secretion in response to thrombin demonstrated by fluorescence flow cytometry. *Blood* **69**, 1401–1403.

15. Marti, G. E., Magruder, L., Schuette, W. E., and Gralnick, H. R., (1988). Flow cytometric analysis of platelet surface antigens. *Cytometry* **9**, 448–455.

16. Marx, J. L. (1989). New family of adhesion proteins discovered. *Science* **243**, 1144.

17. McEver, R. P., and Martin, M. N. (1984). A monoclonal antibody to a membrane glycoprotein binds only to activated platelets. *J. Biol. Chem.* **259**, 9799–9804.

18. Murphy, S., and Gardner, F. H. (1971). Platelet storage at 22°C: Metabolic morphologic, and functional studies. *J. Clin. Invest.* **50**, 370.

19. Nieuwenhuis, H. K., van Oosterhout, J. J. G., Rozemuller, E., van Iwaarden, F., and Sixma, J. J. (1987). Studies with a monoclonal antibody against activated platelets: Evidence that a secreted 53,000 molecular weight lysosome like granule protein is exposed on the surface of activated platelets in the circulation. *Blood* **70**, 838–845.

20. Rinder, H., Rinder, C., Mitchell, J., Forest, R., Hillman, R. S., and Ault, K. A. (1988). The kinetics of platelet activation and aggregation during cardiopulmonary bypass — a wave of early aggregation is dissociated from a progressive increase in platelet activation. *Blood* **72**, 307a.

21. Rinder, H., Murphy, M., Ault, K. A., Mitchell, J., Stocks, J., Slichter, S., and Hillman, R. S. (1988). Quantitation of platelet activation during storage: evidence that activated platelets fail to circulate following transfusion. *Blood* **72**, 283a.

22. Shattil, S. J., Hoxie, J. A., Cunningham, M. C., and Brass, L. F. (1985). Changes in the platelet membrane glycoprotein IIb–IIIa complex during platelet activation. *J. Biol. Chem.* **260**, 11107–11114.

23. Shattil, S. J., Motulsky, H. J., Insel, P. A., Flaherty, L., and Brass, L. F. (1986). Expression of fibrinogen receptors during activation and subsequent desensitization of human platelets by epinephrine. *Blood* **68**, 1224–1231.

24. Sharttil, S. J., Cunningham, M., and Hoxie, J. A. (1987). Detection of activated platelets in whole blood using activation dependent monoclonal antibodies and flow cytometry. *Blood* **70**, 307–315.

25. Sims, P. J., Faioni, E. M., Wiedmer, T., and Shattil, S. J. (1988). Complement proteins C5b–9 cause release of membrane vesicles from the platelet surface that are enriched in the membrane receptor for coagulation factor Va and express prothrombinase activity. *J. Biol. Chem.* **263**, 18205–12.

26. Tracy, P. B., and Mann, K. G. (1983). Prothrombinase complex assembly on the platelet surface is mediated through the 74,000 dalton component of factor Va. *Proc. Natl. Acad. Sci. USA*, **80**, 2380–2384.

27. Wenger, R. K., Lukasiewicz, H., Mikuta, B. S., Niewiarowski, S., and Edmunds, L. H. (1989). Loss of platelet fibrinogen receptors during clinical cardiopulmonary bypass. *J. Thorac. Cardiovasc. Surg.* **97**, 235–239.

28. Zilla, P., Fasol, R., Groscurth, P., Klepetko, W., Reichenspurner, H., and Wolner, E. (1989). Blood platelets in cardiopulmonary bypass operations. Recovery occurs after initial stimulation, rather than continual activation. *J. Thorac. Cardiovasc. Surg.* **97**, 379–388.

29. Zucker-Franklin, D., and Karpatkin, S. (1977). Red-cell and platelet fragmentation in idiopathic thrombocytopenic purpura. *New Engl. J. Med.* **297**, 517–523.

3

Pathological Conditions

3.1

Bacterial and Viral Infections

ROBERT BJERKNES

Department of Pediatrics and The Gade Institute
Department of Pathology, University of Bergen
Haukeland Hospital, N-5021 Bergen, Norway

I. INTRODUCTION

In spite of the development of public health measures, vaccines and antibiotics, infections continue to be a major health problem. Humans are constantly in contact with a multitude of microorganisms living in the external environment, as well as on the skin and mucosal surfaces. Only relatively few of these microorganisms are virulent to man (*49*). However, the health of humans is dependent on the ability to seek out and destroy invading bacteria, fungi, viruses and parasites (*68,75*). The human host defence consists of mechanical barriers (e.g. skin and mucosa), unspecific first line mechanisms (e.g. phagocytic cells and complement) and specific secondary immune responses (e.g. production of antibodies and killer lymphocytes) (*54,68,75*). The importance of the different components of host defence is illustrated by the high rate of infection in patients with breakage of barriers (e.g. burn patients) (*3*) and congenital or acquired immunodeficiency syndromes (*54,68,109*). The microorganisms apply various tactics to overcome the defence capabilities of the host, and common virulence factors are pili, capsules, enzymes, toxins and lipopolysaccarides (*26,75*). To be able to effectively prevent and control infections, further knowledge on the biology of microorganisms and of human host defence mechanisms, as well as their interactions, is needed.

FLOW CYTOMETRY IN HEMATOLOGY
ISBN 0-12-432940-3

Flow cytometry has proven a useful technology for studies on the pathogenesis of infectious diseases. The present chapter reviews flow cytometric investigations of microorganisms and host defence mechanisms. In addition, clinical studies on patients with infection are discussed. This chapter does not aim to be a thorough review on either microbiology or immunology, but intends to demonstrate the usefulness of flow cytometric methods for the study of infections (see also Chapters 3.2 and 3.3).

II. BACTERIAL INFECTIONS

A. Studies of Bacteria

Flow cytometry offers several possibilities for bacterial characterization (*19*). Measurements of DNA and RNA content following staining with various fluorochromes, as well as assessments of light scatter characteristics, have allowed detection and quantification of bacteria (*5,81,106,108,114*). In addition, staining with chromomycin A3 and Hoechst 33258, dyes which bind preferentially to DNA rich in guanine–cytosine and adenine–thymine, made Van Dilla *et al.* able to discriminate between various bacterial species (*114*). The capacity to detect bacteria following DNA staining has been applied to analyses of blood, urine, serous cavity fluids, bile, wound exudates and the fluid obtained by bronchoalveolar lavage (*22,29,69,114*). In one study as few as 10 *Escherichia coli* were detected per ml of blood (*69*).

The use of fluorochrome-labeled antibodies allows for more specific and discriminative identification of bacteria (*6,23,29,47,56,82,106*). Further, immunofluorescence techniques can be applied to study the binding of bacterial toxins to eukaryotic cells (*76*). Bacterial species and subspecies can also be detected and discriminated by flow cytometry and fluorescently labeled rRNA-targeted oligonucleotides (*1*).

Bacterial growth has been characterized by flow cytometry (*106,108,114*) and several studies have demonstrated the effects of drugs and antisera on bacterial replication (*71,106,108*). Steen *et al.* reported that chloramphenicol, erythromycin, doxycyclin and streptomycin, antibiotics which are all inhibitors of ribosomal protein synthesis, inhibited the cell cycle differently (*106*). Another study demonstrated that flow cytometry accurately assessed the effects of β-lactam antibiotics on the cell wall formation of *E. coli* (*71*).

In infections, and especially in sepsis and meningitis, rapid microbiological diagnosis, including assessment of antibiotic susceptibility, is essential for optimal treatment (*28*). Most of today's methods for bacterial analyses of clinical specimens are time consuming, and in the case of blood culturing it may take days before the bacteria have multiplied sufficiently to be detected. Extensive efforts have therefore been made to develop rapid assays for bacterial analysis. Applying a flow cytometric technique, Cohen and Sahar were able to detect bacteria in 43 clinical specimens in less than 2 hours (*22*). Their results agreed with conventional methods with a sensitivity of 74% and a specificity of 88%. Further, bacterial sensitivity to amikacin was demonstrated in 1 hour in 92% of 13 positive specimens (*22*). Thus, detection of bacteria and testing of susceptibility to antimicrobial agents should have the potential for becoming another clinical application of flow cytometry.

B. The Host Response to Bacterial Infections

The granulocytic and mononuclear phagocytes constitute the first line of cellular defense against a variety of bacteria (*54,75*). The relative importance of granulocytic and mononuclear phagocytes varies with the type of microorganism. Neutrophilic granulocytes play a primary role in controlling infections with obligate extracellular pathogens, whereas monocytes and macrophages are more important for the host defense against facultative and obligate intracellular pathogens

(*68,75*). The phagocytes migrate to the site of infection, and here the microorganisms are ingested, killed and degraded (*68,75*). The invasion of microorganisms and their degradation by phagocytes induces a complex of immunological reactions which lead to the production of specific immunoglobulins and the development of specific responder T lymphocytes (*75,84*). This immune response may provide protection against future infections, a principle which is taken advantage of when immunizing with vaccines (*68*).

a. *Phagocyte Functions*

Several flow cytometric studies on the interaction between bacteria and phagocytes from healthy individuals have been reported (*8–10,12,99,110,111*). With the use of various species of bacteria it has been documented that differences in microbial structure affect phagocytosis (Fig. 1) (*8*). The phagocytosis of capsular *Streptococcus pneumoniae* by neutrophilic granulocytes was significantly lower than that of acapsular types (Fig. 1). In addition, a noteworthy heterogeneity has been observed between neutrophils ingesting the capsular pneumococci (not shown). The bacterial surface structure also influences phagocyte phagolysosomal acidification and intracellular degradation (*8*).

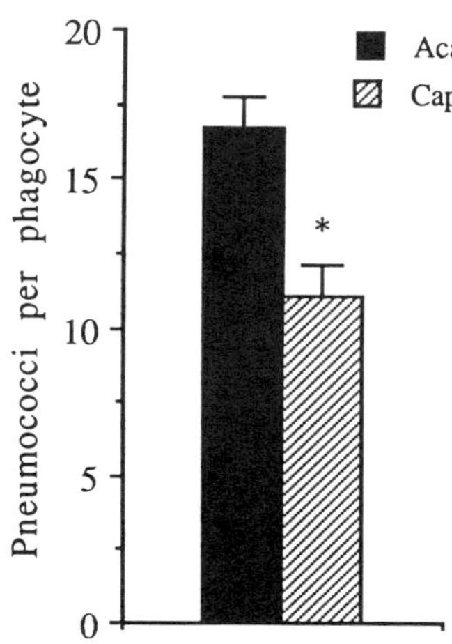

Fig. 1. Human neutrophilic granulocyte phagocytosis of capsular and acapsular pneumococci. The leukocytes (5×10^6/ml) were incubated with fluorescein isothiocyanate-labeled preopsonized bacteria (10^8/ml) for 10 min at 37°C before being analyzed by flow cytometry. The data are given as means ± SEM; $n = 8$; $^*p < 0.05$.

The many studies on neutrophilic granulocyte functions in patients with infections have reported both increased and normal, as well as decreased phagocytic capacity (*57,96,103*) and oxidative metabolism (*96,104*). However, these studies have in general used serum opsonized particles and techniques where bulk populations of phagocytes have been studied. The recognition of the existence of more or less independently functioning receptors for opsonins (IgG Fc receptors (mainly FcγRII and III) and complement receptors 1 (CR1) and 3 (CR3)) on phagocytic leukocytes (*58,113*), and significant functional heterogeneity between normal phagocytes (*34,94*), has demanded re-examinations of phagocyte functions in patients with infectious diseases.

We have examined the phagocytic capacity of neutrophilic granulocytes in 10 children (4–12 years) with bacterial pneumonia. At admission, a significant increased phagocytic capacity was demonstrated for patient neutrophilic granulocytes, as evaluated by the ingestion of IgG-coated *Steptococcus pneumoniae*, *Staphylococcus aureus* and *Haemophilus influenzae* (Table I). Neutrophilic granulocytes from four patients studied longitudinally showed a gradual decrease of phagocytic capacity to normal levels in 4–6 days (Fig. 2). The reduction of phagocytic capacity came concomitantly with clinical improvement. During the disease, the expression of patient neutrophilic granulocyte surface membrane FcγRII and FcγRIII was similar to that of healthy controls (not shown). Thus, the increased phagocytic capacity by neutrophilic granulocytes from patients with bacterial pneumonia was not due to increased levels of FcγRs, indicating up-regulation of other phagocytosis-related mechanisms.

In another study, adults with infections were shown to have subpopulations of neutrophils

Table I

NEUTROPHILIC GRANULOCYTE PHAGOCYTIC CAPACITY IN CHILDREN WITH BACTERIAL PNEUMONIA[a]

Bacteria	Number of bacteria per phagocyte[b]	
	Patients	Controls
Strep. pneumoniae	19.1 ± 1.6[c]	14.3 ± 1.3
Staph. aureus	18.4 ± 1.4	15.8 ± 1.7
H. influenzae	18.8 ± 1.5[c]	13.1 ± 1.3

[a] Neutrophilic granulocytes (5×10^6/ml) from 10 otherwise healthy children (4–12 years) were obtained at admission and incubated with fluorescein isothiocyanate-labeled IgG-coated bacteria (10^8/ml) for 10 min at 37°C and analyzed by flow cytometry (for details see (*11,17*)).
[b] The data are given as means ± SEM.
[c] $p < 0.05$.

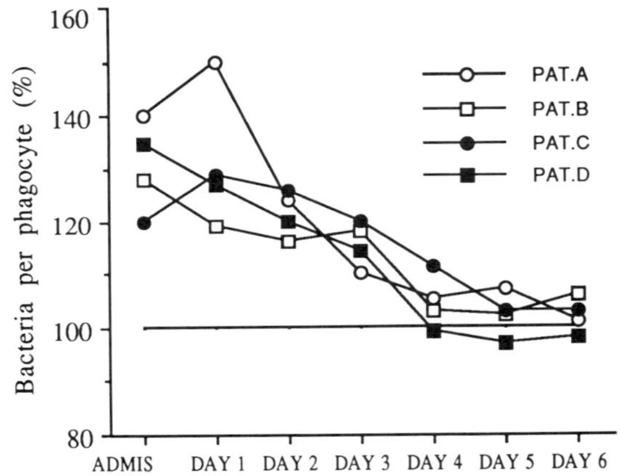

Fig. 2. Longitudinal measurements of the phagocytic capacity of neutrophilic granulocytes obtained from children with bacterial pneumonia. The leukocytes (5×10^6/ml) from four individuals (A–D) (otherwise healthy children; age 6–10 years) were incubated with fluorescein isothiocyanate-labeled preopsonized pneumococci (10^8/ml) and 10 min at 37°C before being analyzed by flow cytometry. The data are given as mean of two parallel samples.

with increased H_2O_2 production in their circulation (*7*). The fraction of primed neutrophilic granulocytes varied from 0% to 80%, and occurred in patients having infections with either Gram-positive or Gram-negative bacteria (*7*). We have observed similar responses in otherwise healthy children with bacterial pneumonias (not shown).

In patients with trauma who become septic, the situation is more complex. We have demonstrated that a burn injury itself induces significant reduction of H_2O_2 production by neutrophilic granulocytes (*17*). However, when these patients become septic, several patterns of responses are seen (*14*). Typical patterns include two populations of neutrophilic granulocytes at the onset of sepsis, of which one is primed, as compared to the situation before infection (*14*). Later in the septic period a population of neutrophilic granulocytes having an oxidative product formation even lower than before the onset of sepsis can be seen, suggesting deactivated or exhausted cells (*14*). Low oxidative responsiveness has also been demonstrated in neutrophilic granulocytes from gingival crevicular pockets in patients with periodontal disease (*66*). The low H_2O_2 production of these cells was attributed to either the large number of bacteria in the crevicular pockets or to specific inhibition by these bacteria (*66*).

In patients with burns and sepsis the neutrophilic granulocyte capacity of intracellular killing was significantly lower than in the same patients before the infections or compared to patients with no infectious complications (*14*). This may partly be attributed to altered capacity of H_2O_2 production. However, altered killing capacity may also be due to the significant changes of the oxygen-independent microbicidal mechanisms, e.g. phagolysosomal acidification (*15–17*).

The processes of chemotaxis, phagocytosis, oxidative product formation and degranulation all depend on the stimulation of membrane receptors. Binding of ligands to these receptors induces changes in the fluidity of the plasma membrane of leukocytes, and in one experimental flow cytometric study, changes in neutrophilic granulocyte membrane fluidity were demonstrated in rats infected with *E. coli* (*72*).

These changes were further shown to be dependent on an intact system of microtubules and microfilaments (*72*).

Following ingestion by phagocytic cells, microorganisms are degraded by the many digestive enzymes liberated into the phagosome during degranulation. In Whipple's disease, a rare infectious disorder, bacillary material is accumulated in the macrophages of affected tissues (*18*,*50*), suggestive of either a macrophage defect or an unusual structure of the infecting microorganism resistant to degradation. In longitudinal studies of a patient with Whipple's disease we were able to show a persisting defect of monocyte and macrophage intracellular degradation of bacteria and zymosan particles, demonstrating a mononuclear phagocyte dysfunction (*13*,*18*).

The influence of antimicrobial medication on phagocyte functions has been studied extensively by a variety of techniques (*43*,*55*,*67*). We have used flow cytometry to examine the effects of various antibiotics on the oxygen-dependent microbicidal mechanisms in human neutrophilic granulocytes. Clindamycin, an antibiotic which accumulates in neutrophilic granulocyte lysosomes, inhibited leukocyte H_2O_2 production at a concentration of 1.0 mM (Fig. 3). At therapeutic concentrations, i.e. 0.01 mM, the oxidative product formation was unaltered. In contrast, penicillin G and erythromycin did not influence neutrophilic granulocyte oxidative burst at any concentration tested (not shown). Hence, multiparameter flow cytometry may become an important supplement to already existing techniques for studies of the interaction of antimicrobials and leukocytes.

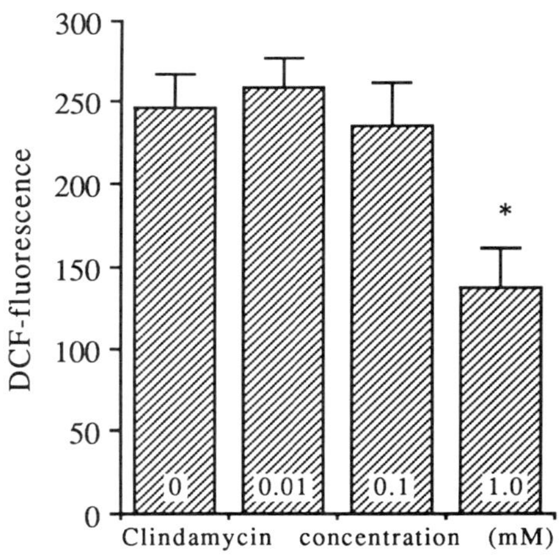

Fig. 3. The effect of clindamycin on the human neutrophilic granulocyte H_2O_2 production. The leukocytes (10^6/ml) were loaded with dichlorofluorescein diacetate (5 μM) and incubated with increasing concentrations of clindamycin. The leukocytes were then stimulated with phorbol myristate diacetate for 15 min and analyzed with flow cytometry. The H_2O_2 was estimated from the leukocyte dichlorofluorescein (DCF) fluorescence. The data are given as mean fluorescence ± SEM; $n = 8$; $^*p < 0.05$.

b. Opsonin Response following Disease and Vaccination

Opsonins, i.e. immunoglobulins and complement spilt products, mediate attachment of microorganisms to phagocytes and stimulate their ingestion (*54*,*75*,*84*). *Neisseria meningitidis* serogroup B is relatively resistant to killing by serum bactericidal antibodies and complement, but is susceptible to killing by phagocytes following opsonization (*87*). The opsonic response in patients with serogroup B meningococcal disease has been examined using FITC-labeled meningococci (*98–100*). The neutrophilic granulocyte phagocytosis increased by about 100% in the presence of convalescent sera obtained 6 weeks after admission, and remained higher than that of sera from patients with acute infections for the 3 years examined (*98*,*100*). Parallel ELISA analyses demonstrated increased levels of IgG_1, IgG_3, IgM and IgA to type-specific group B meningococcal outer membrane proteins in the convalescent sera (*98*,*100*,*101*). The effective opsonin-mediated neutrophilic granulocyte phagocytosis of group B meningococci documents an additional protection against meningococcal infections by phagocytes and serum opsonins to that of bactericidal antibodies. In addition, recent studies, using sera from patients with serogroup B meningococcal disease and bacteria of serogroups A, B, C, W, and Y, demonstrate significant cross reactivity between the different serogroups of the antibodies produced in response to infection (*39*).

The opsonic response has also been studied in sera from healthy volunteers immunized with a vaccine against *N. meningitidis* serogroup B. Vaccination induced a significantly increased opsonic activity as evaluated by opsonin-mediated ingestion of meningococci (*98,100*). These findings are consistent with earlier studies demonstrating that vaccination of individuals deficient in the late complement factors may shift the burden of host defence against meningococci from serum bactericidal antibodies to phagocytosis (*87*).

III. VIRAL INFECTIONS

An increasing knowledge on the multitude of viruses and their capacity of causing infections in humans has emerged during the last decades (*49,68*). Flow cytometry has aided the study of viruses and their interaction with host cells. In addition, flow cytometry has become increasingly important for the assessment of clinical diagnosis and prognosis of viral infections, of which infection with human immunodeficiency virus (HIV) probably is the most widely studied.

A. Studies of Single Viruses

Viruses are small infectious particles consisting of a nucleic acid genome (DNA or RNA), associated proteins, glycoproteins and in some cases a lipid membrane envelope (*49,68*). They range in size from about 20 to 400 nm, making the largest viruses comparable with small bacteria. Viruses replicate within host cells. These cells either die shortly after the release of virus (lytic infection) or keeps on producing virus for a long period without being killed (non-lytic infection, e.g. retrovirus) (*49*).

The limited sensitivity of most current commercial flow cytometers makes them unsuitable for the detection of viruses. However, both Hercher *et al.* (*42*) and Steen *et al.* (*107*) have been able to detect and quantify single viruses by measurements of light scatter using laboratory-built instruments. In addition, Hercher *et al.* showed that single T2 phages could be discriminated from reovirus capsids (*42*). Thus, if made sufficiently sensitive, flow cytometers could become useful instruments for the counting and differentiation of viruses. The discrimination of viruses may be further facilitated by the use of fluorochromes or fluorescent antibodies.

B. The Interaction between Viruses and Host Cells

a. *Virus Receptors*

While some viruses, e.g. herpes simplex virus (HSV) and measles virus (*31,68*), have a broad range of host cells, others, e.g. HIV and Epstein–Barr virus (EBV) (*27,68*), have a much narrower range of cells that they infect. EBV mainly infects and transforms B lymphocytes, causing among other clinical pictures infectious mononucleosis (*68*). The use of fluorescein-labeled (*51,73,118*) or biotinylated EBV (*46*) has allowed flow cytometric identification of virus-binding lymphocytes in heterogeneous cell populations. Using biotinylated EBV in combination with fluorescently labeled antibodies to surface-associated structures on B cells, Inghirami *et al.* demonstrated that EBV attached to all B cells, including those bearing surface μ, δ, γ and α immunoglobulin heavy chains, and those that were CD5 positive (*46*). EBV did not bind to T lymphocytes, natural killer (NK) cells or monocytes (*46*).

McCune and Volsky applied FITC-labeled virus to study EBV receptor (EBVR) turnover (*73*). They demonstrated that after stripping of EBVR from the cell surface, the reappearance of the receptors was dependent on protein synthesis (*73*). In addition, EBVRs were more rapidly regenerated in a lymphoma cell line than in normal lymphocytes (*73*). Others have shown that the EBVR is also the receptor for complement split product C3d (CR2, CD21)

33,84), and that EBV binds to an epitope on CR2 which is recognized by the monoclonal antibody (MoAb) OKB7 (*46*). Using this antibody and dual color flow cytometry. Harabuchi *et al.* demonstrated that the expression of EBVR varies with the maturation stage and the state of activation of B cells (*41*). Consistently, EBV fails to transform proliferating normal B cells, i.e. cells in the S and G_2/M phases of the cell cycle, an ability which is associated with disappearance of surface EBVR and the inability of virus to attach to these cells (*46,85*). Wells *et al.* demonstrated increased EBVR expression of two EBV carrying Burkitt lymphoma-derived cell lines during proliferation (*118*), whereas Volsky and Anderson have demonstrated deficiency in EBVRs of B lymphocytes from patients with preleukemia (*115*).

Fluorescently labeled viruses have also been applied in flow cytometric studies of the interaction of influenza virus (*53*), Rous sarcoma virus (*77*) and adult T cell leukemia virus (HTLV-I) (*97*) with various target cells. Sinangil *et al.* demonstrated receptors for HTLV-I on both lymphoid and non-lymphoid cells from various species, but only a few cell types were infectible by the virus (*97*). Thus, the occurrence of receptors and even binding of virus to the cell surface do not always imply that the cell will be infected by the virus.

b. *Identification and Characterization of Virus-infected Cells*

Flow cytometry has been applied in studies of host cell virus antigens (*2,4,24,30,36,40,44,48,52,59,60,78,80,117*), as well as host cell growth (*35,45,60*). In addition, virus-induced host cell fusion has been determined (*25*).

In 1973 Cram and Brunsting reported the detection of hog cholera virus within PK cells using an immunofluorescent assay and flow cytometry (*24*). Later similar immunofluorescent techniques were used to detect cells infected with simian virus 40 (SV40) (*35,44,48,60*), murine leukemia virus (*40*) or HSV type 2 (*4*). An alternative approach was reported by Leary *et al.*, who were able to detect HSV type 2-infected cells by an indirect immunoperoxidase technique and flow cytometry (*59*). Goolsby *et al.*, comparing another immunoenzymatic staining procedure, the glucose oxidase antiglucose oxidase technique, to indirect immunofluorescence staining, found that the immunofluorescence method was superior for the flow cytometric detection of HSV type 2-infected and transformed cells (*36*).

A problem with immunofluorescence techniques for the demonstration and quantification of intracellular viral antigens is background fluorescence due to autofluorescence and unspecific binding of the immunological reagents. If dual colour measurements are performed, overlapping emission from the second fluorochrome may also be a problem. The standardization of an assay for the determination of intracellular SV40 antigens has been published by Jacobberger *et al* (*48*). In addition to the usual standardization steps involved with such an assay, e.g. optimalization of antibody concentrations, temperature, cell number and washes, they demonstrated the importance of specific high affinity primary antibodies, affinity-purified secondary antibodies (goat anti-mouse) and the inclusion of normal goat serum (blocking secondary antibody) to obtain minimal background fluorescence (*48*). Using this method, allowing combined measurements of host cell cycle and expression of an early cytomegalovirus (CMV) antigen, Elmerdorf *et al.* showed that human diploid fibroblasts were susceptible to CMV infection in all stages of the cell cycle (*30*). The same procedure was applied to analyses of bronchoalveolar lavage fluids of patients suspected to have CMV infection. The data obtained indicated a clinical potential for such assays (*30*).

SV40 has been an important model system for viral replication, transformation and gene regulation (*60,70*). Several studies have shown that infection with SV40 causes host cells, both those permissive and non-permissive of viral replication, to cycle (*60,61*). Lehman *et al.* studied the expression of early viral tumour (T) and late viral (V) SV40 antigens by permissive CV-1 cells and related this to cell cycle stage, using a dual colour flow cytometric assay (*60*). The amount of the early T antigen per cell increased as the cells entered the successive stages of the cell cycle, and could be detected 6–12 h after infection (*60*). In contrast, the late V antigen could be detected immediately after infection, but appeared in late S and G_2/M phases of the cell cycle (*60*). In addition to the various viral antigens that can be identified inside infected host cells, host cells may also express virally encoded antigens on their surface which can be detected by flow cytometry and immunofluorescence (*2,52,78, 80,117*). Using monoclonal antibodies to rubella antigens the *in vivo* expression of these antigens on human leukocytes has been demonstrated (*80*). In naturally acquired rubella, antigens were detected between 1 and 13 days after the onset of illness, whereas the same antigens were found between 5 and 12 days after immunization (*80*). Monocytes expressed the rubella antigens more frequently than lymphocytes and granulocytes, and the expression was more transient and the proportion of cells exhibiting the antigens was lower following immunization than after natural infection (*80*).

c. *Antiviral Drugs*

A few reports demonstrate the usefulness of flow cytometric methods for the evaluation of compounds with activity against viruses (*86,91,92*). Rosenthal *et al.*, studying the DNA content and cell cycle of human diploid fibroblasts, demonstrated that changes in cellular DNA could be detected within 24 h after HVS-1 infection (*86*). Infection was recognized by broadening of the normal diploid and tetraploid peaks, and the presence of greater than 4n DNA staining cells. Inhibition of virus-induced DNA synthesis and maintenance of normal growth pattern was indication of antiviral activity (*86*). Toxicity of the compound was indicated by loss of the S phase and tetraploid cell populations. Using this technique, the potent anti-HSV activity of acyclovir was confirmed (*86*).

Another principle for the evaluation of the anti-viral effects of drugs has been reported by Schols *et al.* (*92*). They determined the cytopathic effect of HIV-1 on a human T4 lymphocyte cell line following fluorescein diacetate (FDA) incubation (*92*). Esterases of live T4 lymphocytes converted FDA to fluorescein, allowing flow cytometric discrimination and quantification of live and dead cells. Applying this method to HIV-1-infected cultures of cells treated with various concentrations of the anti-HIV compound azidothymidine (AZT), they obtained a virus-inhibitory dose response comparable to that obtained by conventional methods (*92*). Using a flow cytometric indirect immunofluorescence assay, the same authors studied the capacity of various anti-HIV agents to inhibit the binding HIV-1 to T4 lymphocytes (*91*). They reported that AZT did not influence attachment of virus to the cells, whereas suramin partially blocked HIV-1 binding at concentrations that fully protected T4 lymphocytes against cytopathic effects (*91*). Flow cytometry may, thus, prove to be a valuable tool in further studies of antiviral drugs.

C. The Host Response to Viral Infections

The host response to viral infections includes production of virus-specific antibodies, as well as a cellular immune response (*49,68,75*). Even though infections by a variety of viruses are followed by life-long immunity, the virus-specific antibodies produced are not always protective (*27,68*). The presence of such antibodies does, however, document past exposure to a virus or an ongoing infection (*49,68*). The altered number of immunocompetent cells may reflect effective host mechanisms to overcome the infections (*68,84*), but the changes of leukocyte populations may also reflect progression of the infection (*32,62,88*) and indicate virus-induced immunosuppression (*32,88*).

a. *Antibodies and Diagnostic Assays*

At least two different principles have been used in flow cytometric assays for the detection of viral antibodies (*74,93,102*). The first is to use fresh virus-infected cells, expressing virus-induced antigens on their surface, as substrate in a flow cytometric indirect immunofluorescence assay (*102*). Using this technique, Sligh *et al.* detected serum antibodies to HIV-1 with a sensitivity and specificity similar to that of Western blot (*102*), the most widely used confirmatory test for HIV-1 antibodies. The second principle is the use of fluorescent microspheres to which purified viral antigens are adsorbed (*74,94*). In these assays microspheres of different sizes allow for the simultaneous detection and quantification of antibodies to several antigens of the same virus (*93*) or antibodies to different viruses (*74*). McHugh *et al.* increased the sensitivity of their assay by using a biotin–streptavidin amplification procedure with phycoerythrin, and by combining two different sizes of microspheres they were able to detect antibodies to CMV and HSV with a sensitivity that was at least sixfold greater than that of enzyme immunoassays (*74*).

b. *Phagocyte Functions*

Viral infections may impair the natural defence functions of phagocytic cells, as well as their antigen-processing and antigen-presenting capacity (*88*). In a flow cytometric study of the neutrophilic granulocyte phagocytic capacity in patients with serologically verified influenza A virus infection, we demonstrated significant reduced ingestion of pneumococci and *Haemophilus influenzae* (Fig. 4). Others studying the pathogenesis of HIV infection have used flow cytometry to study monocyte HLA-DR expression (*20*) and neutrophilic granulocyte functions (*89*) in various stages of the disease.

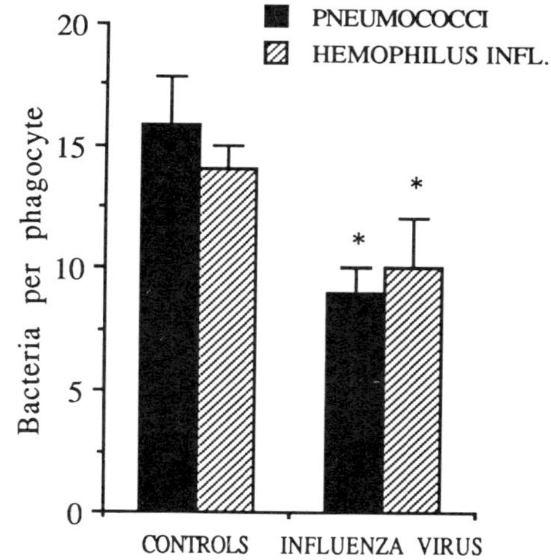

Fig. 4. The phagocytic capacity of neutrophilic granulocytes from patients with serologically verified infection with influenza virus type A. The leukocytes (5×10^6/ml) were incubated with fluorescein isothiocyanate-labeled preopsonized bacteria (10^8/ml) for 10 min at 37°C before being analyzed by flow cytometry. The data are given as means ± SEM; $n = 10$; $^*p < 0.05$.

c. *Leukocyte Subsets*

The leukocyte populations of peripheral blood, bronchoalveolar lavage fluid, exudate and tissue change numerically and functionally during infections (*27,32,37,38,62,63,65,79,83,90,120,121*). Multiparameter flow cytometry of leukocytes labeled by monoclonal antibodies has aided the study of host responses to infections, but most of this work has so far been limited to studies of lymphocytes.

In infectious mononucleosis, the EBV transforms and induces functional alterations of B cells (*68*). In addition, the infection is characterized by expansion of $CD8^+$ T cells and NK cells (*83*). These T cells express high levels of CD38 and HLA-DR, markers of T cell activation, whereas IL-2 receptors are lost or down-regulated in the acute phase (*90*). During convalence, the number of suppressor cells is reduced to normal, and the overall immune function is restored. In the CMV carrier status, significantly higher numbers of $CD3^+$, $CD4^+$,

CD8$^+$ and Leu7$^+$ cells than in controls have been demonstrated (*37,38*). Alterations of lymphocyte subsets have been used to discriminate between renal transplant rejection and CMV infection (*95*), and in one study an increased proportion of CD3$^+$ cells coexpressing IL-2R was found in 12 out of 14 patients with rejection, whereas an increased proportion of CD2$^+$ cells coexpressing HLA-DR was observed in all of 11 patients with CMV infection (*95*). Characteristic patterns of lymphocyte subset changes have also been reported in varicella, measles and hand–mouth–foot disease (*121*), as well as in hepatitis B virus carrier state (*120*) and in rhinovirus infection (*62*). In addition, the utility of lymphocyte immunophenotyping in monitoring patients with HIV infection is well established (*27,32,65*) (see also Chap. 3.2).

IV. FUNGAL AND CHLAMYDIAL INFECTIONS

Fungal cells are larger than bacteria making them even better suited for analyses by flow cytometry. Taguchi *et al.* applied light scatter measurements to differentiate *Exophila dermatitidis*, *E. monilia* and *E. jeanselmei*, the microorganisms causing phaeohyphomycosis (*112*). Flow cytometry has also been used to determine the β-galactosidase activity of *Saccharomyces cerevisiae* (*105,119*) and to study adenine biosynthetic mutants of the same yeast (*21*).

Candida albicans may cause invasive infections in patients with immunosuppression, e.g. diabetes mellitus, HIV, burn injuries or malignancies, or in patients having received antimicrobials altering the normal bacterial flora (*3,27,54,68*). Phagocytic cells are important for the defence against *C. albicans*, and neutrophilic granulocyte phagocytosis and intracellular killing of these fungi have been studied by flow cytometry (*11,14,16*). In patients with thermal injury and *C. albicans* sepsis, we have demonstrated that the capacity of neutrophilic granulocytes to kill *C. albicans* was significantly reduced as compared to the situation before the onset of sepsis (*11,14*). In another study, patients with *C. albicans* infections were shown to have subpopulations of neutrophils with increased H_2O_2 production in their circulation, similar to those with bacterial infections (*7*).

Chlamydiae are responsible for a large number of infectious diseases, e.g. trachoma and venereal infections (*68,75*). Chlamydial infections occur following specific attachment of the microorganisms to the host cell surface followed by entry into a phagocytic vacuole (*68*). Levitt *et al.* (*64*), and later Waldman *et al.* (*116*), using FITC-labeled anti-chlamydial monoclonal antibodies and flow cytometry, have studied the binding, internalization and growth of chlamydiae following interaction with various cell types.

V. CONCLUSION

Flow cytometry has already become an important tool in experimental studies of microorganisms and the protective mechanisms of host response. The steady increasing number of reports on simple reliable flow cytometric procedures for the diagnosis of infections suggests wide applications of this technology in clinical work in future.

REFERENCES

1. Amann, R. I., Binder, B. J., Olson, R. J., Chisholm, S. W., Devereux, R., and Stahl, D. A. (1990). Combination of 16S rRNA-targeted oligonucleotide probes with flow cytometry for analyzing mixed microbial populations. *Appl. Environ. Microbiol.* **56**, 1919–1925.

2. Anderson, J., Byrne, J. A., Schreiber, R., Patterson, S., and Oldstone, M. B. A. (1985). Biology of cloned cytotoxic T lymphocytes specific for lymphocytic choriomeningitis virus: Clearance of virus and in vitro properties. *J. Virol.* **53**, 552–560.

3. Artz, P., Moncrief, J. A., and Pruitt Jr., B. A. (1979), "Burns: A Team Approach." W. B. Saunders, Philadelphia.

4. Aurelian, L. (1982). Herpes simplex diagnosis: Antigen detection by ELISA and flow microfluorometry. *Diagn. Gynecol. Obst.* **4**, 375–388.

5. Baily, J. E., Fazel-Madjlessi, J., McQuitty, D. N., Lee, L. Y., Allred, J. C., and Oro, J. A. (1977). Characterization of bacterial growth by means of flow microfluorometry. *Science* **198**, 1175–1176.
6. Barnett, J. M., Cuchens M. A., and Buchanan, W. (1984). Automated immunofluorescent speciation of oral bacteria using flow cytometry. *J. Dent. Res.* **63**, 1040–1042.
7. Bass, D. A., Olbrantz, P., Szejda, P., Seeds, M. C., and McCall, C. E. (1986). Subpopulations of neutrophils with increased oxidative product formation in blood of patients with infection. *J. Immunol.* **136**, 860–866.
8. Bassøe, C.-F., and Bjerknes, R. (1985). Phagocytosis by human leukocytes, phagosomal pH, and degradation of seven species of bacteria measured by flow cytometry. *J. Med. Microbiol.* **19**, 115–125.
9. Bassøe, C.-F., and Bjerknes, R. (1984). The effect of serum opsonins on the phagocytosis of *Staphylococcus aureus* and zymosan particles measured by flow cytometry. *Acta Pathol. Microbiol. Immunol. Scand. Sect. C* **92**, 51–58.
10. Bassøe, C.-F., Solsvik, J., and Laerum, O. D. (1980). Quantitation of single cell phagocytic capacity by flow cytometry. *In* "Flow cytometry IV" (O. D. Laerum, T. Lindmo, and E. Thorud, eds), pp. 170–174. Universitetsforlaget, Bergen.
11. Bjerknes, R. (1984). Flow cytometric assay for combined measurement of phagocytosis and intracellular killing of *Candida albicans*. *J. Immunol. Methods.* **72**, 229–241.
12. Bjerknes, R., Bassøe, C.-F., Sjursen, H., Laerum, O. D., and Solberg, C. O. (1989). Flow cytometry for the study of phagocyte functions. *Rev. Infect. Dis.* **11**, 16–33.
13. Bjerknes, R., Laerum, O. D., and Ødegaard, S. (1985). Impaired bacterial degradation by monocytes and macrophages from a patient with treated Whipple's disease. *Gastroenterology* **89**, 1139–1146.
14. Bjerknes, R., and Vindenes, H. (1991). Impairment of human neutrophils in patients with burns and sepsis. Submitted for publication.
15. Bjerknes, R., and Vindenes, H. (1989). Neutrophil dysfunction after thermal injury: alteration of phagolysosomal acidification in patients with large burns. *Burns* **15**, 77–81.
16. Bjerknes, R., Vindenes, H, and Laerum, O. D. (1990) Altered neutrophil functions in patients with large burns. *Blood Cells* **16**, 127–143.
17. Bjerknes, R., Vindenes, H., Pitkänen, J., Ninneman, J., Laerum, O. D., and Åbyholm, F. (1989). Alterations of polymorphonuclear neutrophilic granulocyte functions in patients with large burns. *J. Trauma* **29**, 847–855.
18. Bjerknes, R., Ødegaard, S., Bjerkvig, R., Børkje, B., and Laerum, O. D. (1988). Whipple's disease: Demonstration of a persisting monocyte and macrophage dysfunction. *Scand. J. Gastroenterol.* **23**, 611–619.
19. Boye, E., and Løbner-Olesen, A. (1990). Flow cytometry: Illuminating microbiology. *New Biologist* **2**, 119–125.
20. Braun, D. P., Kessler, H., Falk, L., Paul, D., Harris, J. E., Blaauw, B., and Landay, A. (1988). Monocyte functional studies in asymptomatic human immunodeficiency disease virus (HIV)-infected individuals. *J. Clin. Immunol.* **8**, 486–494.
21. Bruschi, C. V., and Chuba, P. J. (1988). Nonselective enrichment for yeast adenine mutants by flow cytometry. *Cytometry* **9**, 60–67.
22. Cohen, C. Y., and Sahar, E. (1989). Rapid flow cytometric bacterial detection and determination of susceptibility to amikacin in body fluids and exudates. *J. Clin. Microbiol.* **27**, 1250–1256.
23. Cordery, M. C., Smith, I. M., and Mackenzie, N. M. (1986). Quantification by flow cytometry of specific binding of antibody to some serotypes of *Haemophilus pleuropneumoniae*. *Res. Vet. Med.* **41**, 277–278.
24. Cram, L. S., and Brunstig, A. (1973). Fluorescence and light-scattering measurements of Hg cholera-infected PK-15 cells. *Exp. Cell Res.* **78**, 209–213.
25. Cram, L. S., Forslund, J. C., and Jett, J. H. (1978). Quantitation of cell fusion by twenty-one strains of Newcastle disease virus using flow microfluorometry. *J. Gen. Virol.* **41**, 27–36.
26. Densen, P. and Mandell, G. L. (1980). Phagocyte strategy vs microbial tactics. *Rev. Infect. Dis.* **2**, 817–838.
27. De Vita, V. T., Hellman, S., and Rosenberg, S. A. (1985). "AIDS: Etiology, Diagnosis, Treatment, and Prevention." J. B. Lippincott Co., Philadelphia
28. Doern, G. V., Scott, D. R., and Rashed, A. L. (1982). Clinical impact of rapid antimicrobial susceptibility testing of blood culture isolates. *Antimicrob. Agents Chemother.* **21**, 1023–1024.
29. Donnelly, C. W., and Baigent, G. J. (1986). Method for flow cytometric detection of *Listeria monocytogenes* in milk. *Appl. Environ. Microbiol.* **52**, 689–695.
30. Elmendorf, S., McSharry, J., Laffin, J., Fogleman, D., and Lehman, J. M. (1988). Detection of an early Cytomegalovirus antigen with two-color quantitative flow cytometry. *Cytometry* **91**, 254–260.
31. Evans, A. S. (1984). "Viral Infections in Humans: Epidemiology and Control." Plenum, New York.
32. Fahey, J. L., Taylor, J. M. G., Detels, T., Hofmann, B., Melmed, R., Nishanian, P., and Giorgi, J. V. (1990). The prognostic value of cellular and serologic markers in infection with human immunodeficiency virus type 1. *New Engl. J. Med.* **322**, 166–172.

33. Frade, R., Barel, M., Ehlin-Henriksson, B., and Klein, G. (1985). gp140, the C3d receptor of human B lymphocytes, is also the Epstein–Barr virus receptor. *Proc. Sci. USA* **82**, 1490–1493.
34. Gallin, J. I. (1980). Human neutrophil heterogeneity exists, but is it meaningful? *Blood* **63**, 977–983.
35. Gershey, E. L. (1979). Simian virus 40 — host cell interaction during lytic infection. *J. Virol.* **30**, 76–83.
36. Goolsby, C., Gay, H., Docherty, J. J., and Todd, P. (1988). Flow cytometric detection of Herpes simplex virus type 2 and transformed cells by immunoenzymatic and by indirect immunofluorescence staining. *Cytometry* **9**, 126–130.
37. Gratama, J. W., Kluin-Nelemans, H. C., Langelaar, R. A., Den Ottolander, G. J., Stijnen, T., D'Amaro, J., Torsensma, R., and Tanke, H. J. (1988). Flow cytometric and morphologic studies of HNK1 + (Leu7 +) lymphocytes in relation to cytomegalovirus carrier status. *Clin. Exp. Immunol.* **74**, 190–195.
38. Gratama, J. W., Langelaar, R. A., Oosterveer, M. A. P., Van der Linden, J. A., Den Ouden-Noordermeer, A., Naipal, A. M. I. H., Visser, J. W. M., De Gast, G. C., and Tanke, H. J. (1989). Phenotypeic study of CD4 + and CD8 + lymphocyte subsets in relation to cytomegalovirus carrier status and its correlate with pokeweed mitogen-induced B lymphocyte differentiation. *Clin. Exp. Immunol.* **77**, 245–251.
39. Guttormsen, H. K., Bjerknes, R., Lehmann, V., Naess, A., Halstensen, A., Sørnes, S., and Solberg, C. O. (1991). Cross-reactive serum opsonins following meningococcal disease. Submitted for publication.
40. Hand Jr., R. E., Tennant, R. W., Yang, W.-K., and Lavelle, G. C. (1978). Immunofluorescent analysis of murine leukemia virus-infected cells by flow microfluorometry. *J. Immunol. Methods* **23**, 175–186.
41. Harabuchi, Y., Koizumi, S., Osato, T., Yamanaka, N., and Kataura, A. (1988). Flow cytometric analysis of Epstein-Barr virus receptor among the different B-Cell subpopulations using simultaneous two-color immunofluorescence. *Virology* **165**, 278–281.
42. Hercher, M., Mueller, W., and Shapiro, H. M. (1979). Detection and discrimination of individual viruses by flow cytometry. *J. Histochem. Cytochem.* **27**, 350–352.
43. Holmes, B., Quie, P., Wildhorst, D. B., Pollara, B., and Good, R. A. (1966). Protection against phagocytized bacteria from the killing action of antibiotics. *Nature* **210**, 1131–1132.
44. Horan, M., Horan, P. K., and Williams, C. A. (1975). Quantitative measurement of SV40 T-antigen production. *Exp. Cell Res.* **91**, 247–252.
45. Horan, P. K., Jett, J. H., Romero, A., and Lehman, J. M. (1974). Flow microfluorometry analysis of DNA content in chinese hamster cells following infection with Simian Virus 40. *Int. J. Cancer* **14**, 514–521.
46. Inghirami, G., Nakamura, M., Balow, J. E., Notkins, A. L., and Casali, P. (1988). Model for studying virus attachment: Identification and quantitation of Epstein-Barr virus-binding cells by using biotinylated virus in flow cytometry. *J. Virol.* **62**, 2453–2463.
47. Ingram, M., Cleary, T. J., Price, B. J., Price III, R. L., and Castro A. (1982). Rapid detection of *Legionella pneumophila* by flow cytometry. *Cytometry* **3**, 134–137.
48. Jacobberger, J. W., Fogleman, D., and Lehman, J. M. (1986). Analysis of intracellular antigens by flow cytometry. *Cytometry* **87**, 356–364.
49. Jawetz, E., Melnick, J. L., and Adelberg, E. A. (1978). "Review of Medical Microbiology," 13th edn. Lange Medical Publications, Los Altos.
50. Keren, D. F. (1981). Whipple's disease: A review emphasizing immunology and microbiology. *CRC Crit. Rev. Clin. Lab. Sci.* **14**, 75–108.
51. Khelifa, R., and Menezes, J. (1982). Use of fluoresceinated Epstein–Barr virus to study Epstein–Barr virus–lymphoid cell interactions. *J. Virol.* **41**, 649–656.
52. Kimura, T., Tanimura, E., Yamamoto, N., Ito, T., and Ohyama, A. (1985). The quantitative kinetic study of Dengue viral antigen by flow cytometry. I. An *in vitro* study. *Virus Res.* **2**, 375–389.
53. Kimura, Y., Yokochi, T., Miyadai, T., Yoshida, K., Yokoo, J., and Matsumoto, K. (1985). Characterization of a porcine kidney cell line resistant to Influenza virus infection. *J. Virol.* **53**, 980–983.
54. Klebanoff, S. J., and Clark, R. A. (1978). "The Neutrophil: Function and Clinical Disorders." Elsevier/North-Holland, Amsterdam.
55. Klempner, M. S., and Styrt, B. (1981). Clindamycin uptake by human neutrophils. *J. Infect. Dis.* **144**, 472–479.
56. Kornman, K. S., Patters, M., Kiel, R., and Marucha, P. (1984). Detection and quantitation of Bacteroides gingivalis in bacterial mixtures by means of flow cytometry. *J. Periodont. Res.* **19**, 570–573.
57. Krukowski, Z. H., and Smith, G. (1983). A unifying concept of polymorphonuclear neutrophil function in acute bacterial infection. *World J. Surg.* **71**, 424–429.
58. Lambris, J. D. (1988). The multifunctional role of C3, the third component of complement. *Immunol. Today* **9**, 387–393.
59. Leary, J. F., Notter, M. F. D., and Todd, P. (1976). Laser flow cytophotometric immunoperoxidase detection of Herpes simplex virus type 2 antigens in cultured human cells. *J. Histochem. Cytochem.* **24**, 1249–1257.
60. Lehman, J. M., Laffin, J., Jacobberger, J. W., and

Fogleman, D. (1988). Analysis of Simian virus 40 infection of CV-1 cells by quantitative two-color fluorescence with flow cytometry. *Cytometry* **9**, 52–59.

61. Lehman, J. M., Mauel, J., and Defendi, V. (1971). Regulation of DNA synthesis in macrophages infected with Simian virus 40. *Exp. Cell Res.* **67**, 230–233.

62. Levandowski, R. A., Ou, D. W., and Jackson, G. G. (1986). Acute-phase decrease of T lymphocyte subsets in rhinovirus infection. *J. Infect. Dis.* **153**, 743–748.

63. Levandowski, R. A., Weaver, C. W., and Jackson, G. G. (1988). Nasal-secretion leukocyte population determined by flow cytometry during acute rhinovirus infection. *J. Med. Virol.* **25**, 423–432.

64. Levitt, D., Zable, B., and Bard, J. (1986). Binding, ingestion, and growth of *Chlamydia trachomatis* (L_2 Serovar) analyzed by flow cytometry. *Cytometry* **7**, 378–383.

65. Liu, C.-M., Muirhead, K. A., George, S. P., and Landay, A. L. (1989). Flow cytometric monitoring of human immunodeficiency virus-infected patients. *Am. J. Clin. Pathol.* **92**, 721–728.

66. Loesche, W. J., Robinson, J. P., Flynn, M., Hudson, J. L., and Duque, R. E. (1988). Reduced oxidative function in gingival crevicular neutrophils in periodontal disease. *Infect. Immun.* **56**, 156–160.

67. Mandell, G. L. (1973), Interaction of intraleukocyte bacteria and antibiotics. *J. Clin. Invest.* **52**, 1673–1679.

68. Mandell, G. L., Douglas Jr., R. G., and Bennett, J. E. (1985). "Principles and Practice of Infectious Diseases." Wiley and Sons, New York.

69. Mansour, J. D., Robson, J. A., Arndt, C. W., and Schulte, T. H. (1985). Detection of *Escherichia coli* in blood using flow cytometry. *Cytometry* **6**, 186–190.

70. Martin, R. G. (1981). The transformation of cell growth and transmogrification of DNA synthesis by Simian virus 40. *Adv. Cancer Res.* **34**, 1–68.

71. Martinez, O. V., Gratzner, H. G., Malinin, T. I., Ingram, M. (1982). The effect of some β-lactam antibiotics on *Escherichia coli* studied by flow cytometry. *Cytometry* **2**, 129–133.

72. Masuda, M., Komiyama, Y., Nishikado, H., Kuriki, H., Egawa, H., and Murata, K. (1989). Change in membrane fluidity of rat neutrophils accompanying *Escherichia coli* inoculation. *J. Leuk. Biol.* **46**, 169–74.

73. McCune, B., and Volsky, D. J. (1984). Application of flow cytometry to studies on the distribution, persistence, and regeneration rate of Epstein–Barr virus receptors. *Cytometry* **5**, 523–529.

74. McHugh, T. M., Miner, R. C., Logan, L. H., and Stites, D. P. (1988). Simultaneous detection of antibodies to Cytomegalovirus and Herpes simplex virus by using flow cytometry and a microsphere-based fluorescence immunoassay. *J. Clin. Microbiol.* **26**, 1957–1961.

75. Mims, C. A. (1982). "The Pathogenesis of Infectious Disease." Academic Press, New York.

76. Notter, M. F. D., and Leary, J. F. (1985). Flow cytometric analysis of tetanus toxin binding to neuroblastoma cells. *J. Cell. Physiol.* **125**, 476–484.

77. Notter, M. D. F., Learly, J. F., and Balduzzi, P. C. (1982). Adsorbtion of Rous sarcoma virus to genetically susceptible and resistant chicken cells studied by laser flow cytometry *J. Virol.* **41**, 958–964.

78. Oldstone, M. B. A., Fujinami, R. S., Tishon, A., Finney, D., Powell, H. C., and Lampert, P. W. (1983). Mapping of the major histocompatibility complex and viral antigens on the plasma membrane of a measles virus-infected cell. *Virology* **127**, 426–437.

79. Oppenshaw, P. J. M. (1989). Flow cytometric analysis of pulmonary lymphocytes from mice infected with respiratory syncytial virus. *Clin. Exp. Immunol.* **75**, 324–328.

80. O'Shea, S., Mutton, D., and Best, J. (1988). In vivo expression of rubella antigens on human leucocytes: Detection by flow cytometry. *J. Med. Virol.* **25**, 297–307.

81. Pauu, A. S., Cowles, J. R., and Oro, J. A. (1977). Flow microfluorometric analysis of *Escherichia coli*, *Rhizobium meliloti*, and *Rhizobium japonicum* at different stages of the cell cycle. *Can. J. Microbiol.* **23**, 1165–1169.

82. Phillips, A. P., and Martin, K. L. (1983). Immunofluorescence analysis of bacilllus spores and vegetative cells by flow cytometry. *Cytometry* **4**, 123–131.

83. Reinherz, E. L., O'Brien, C., Rosenthal, P., and Schlossman, S. F. (1980). The cellular basis for viral-induced immunodeficiency: Analysis by monoclonal antibodies. *J. Immunol.* **125**, 1269–1274.

84. Roitt, I. M., Brostoff, J., and Male, D. K. (1989). "Immunology" Churchill Livingstone, London.

85. Roome, A. J., and Reading, C. L. (1987). Frequency of B-lymphocyte transformation by Epstein–Barr virus decreases with entry into the cell cycle. *Immunology* **60**, 195–201.

86. Rosenthal, K. S., Hodnichak, C. M., and Summers, J. L. (1987). Flow cytometric evaluation of antiherpes drugs. *Cytometry* **81**, 392–395.

87. Ross, S. C., Rosenthal, P. J., Berberich, H. M., and Densen, P. (1987). Killing of Neisseria meningitidis by human neutrophils: Implications for normal and complement deficient individuals. *J. Infect. Dis.* **155**, 1266–1275.

88. Rouse, B. T., and Horohov, D. W. (1986). Immunosuppression in viral infections. *Rev. Infect. Dis.* **8**, 850–873.

89. Ryder, M. I., Winkler, J. R., and Weinreb, R. N. (1988). Elevated phagocytosis, oxidative burst, and

F-actin formation in PMNs from individuals with intraoral manifestations of HIV infection. *J. Aquir. Immun. Def. Syndr.* **1**, 346–353.

90. Schmidt, C., Burman, K., Moss, D. J., Ryan, J., Burrows, S. R., and Misko, I. S. (1989), Interleukin-2 receptors in infectious mononucleosis. *Immunol. Lett.* **23**, 139–142.

91. Schols, D., Baba M., Pauwels, R., and De Clercq, E. (1989). Flow cytometric method to demonstrate whether anti-HIV agents inhibit virion binding to T4+ cells. *J. Aquir. Immun. Def. Syndr.* **2**, 10–15.

92. Schols, D., Pauwels, R., Vanlangendonck, F., Balzarini, J., and De Clercq, E. (1988). A highly reliable, sensitive, flow cytometric/fluorometric assay for the evaluation of the anti-HIV activity of antiviral compounds in MT-4 cells. *J. Immunol. Methods* **114**, 27–32.

93. Scillian, J. J., McHugh, T. M., Busch, M. P., Tam, M., Fulwyler, M. J., Chien, D. Y., and Vyas, G. N. (1989). Early detection of antibodies against rDNA-produced HIV proteins with a flow cytometric assay. *Blood* **73**, 2041–2048.

94. Seligmann, B. E., Chused, T. M., and Gallin, J. I. (1984). Differential binding of chemoattractant peptide to subpopulations of human neutrophils. *J. Immunol.* **133**, 264–2646.

95. Siegel, D. L., Fox, I., Dafoe, D. C., Power, M., Asplund, M., Zellers, L., Barker, C. F., and Prystowsky, M. B. (1989). Discriminating rejection from CMV infection in renal allograft recipients using flow cytometry. *Clin. Immunol. Immunopathol.* **51**, 157–171.

96. Simms, H. H., Frank, M. M., Quinn, T. C., Holland, S., and Gaither, T. A. (1989). Studies on phagocytosis in patients with bacterial infections. *J. Clin. Invest.* **83**, 252–260.

97. Sinangil, F., Harada, S., Purtilo, D. T., and Volsky, D. J. (1985). Host cell range of adult leukemia virus. *Int. J. Cancer* **56**, 191–198.

98. Sjursen, H., Bjerknes, R., Halstensen, A., Naess, A., Frøholm, O., Rosenqvist, E., and Solberg, C. O (1987). Serum opsonins to group B meningococci. *Acta Pathol. Microbiol. Immunol. Scand. Sect. C* **95**, 283–289.

99. Sjursen, H, Bjerknes, R., Halstensen, A., Naess, A., Soernes, S., and Solberg, C. O. (1989). Flow cytometric assay for the measurement of serum opsonins to *Neisseria meningitidis* serogroup B, serotype 15. *J. Immunol. Methods* **116**, 235–243.

100. Sjursen, H., Bjerknes, R., Halstensen, A., and Solberg, C. O. (1987). Serum opsonins to group B meningococci in patients and vaccinated volunteers. *Cytometry* Suppl. 1, 38 (abstract).

101. Sjursen, H., Wedege, E., Rosenqvist, E., Naess, A., Halstensen, A., Matre, R., and Solberg, C. O. (1990). IgG subclass antibodies to serogroup B meningococcal outer membrane antigens following infection and vaccination. *Acta Pathol. Mirobiol. Scand.* **98**, 1061–1069.

102. Sligh, J. M., Roodman, S. T., and Tsai, C. C. (1989). Flow cytometric indirect immunofluorescence assay with high sensitivity and specificity for detection of antibodies to human immunodeficiency virus (HIV). *Am. J. Clin. Pathol.* **91**, 210–214.

103. Solberg, C. O., and Hellum, K. B. (1972) Neutrophil granulocyte function in bacterial infections. *Lancet* **ii**, 727–730.

104. Solberg, C. O., Kalager, T., Hill, H. R., and Glette, J. (1982). Polymorphonuclear leukocyte function in bacterial and viral infections. *Scand. J. Infect. Dis.* **14**, 11–18.

105. Srienc, F., Campbell, J. L., and Bailey, J. E. (1986). Flow cytometry analysis of recombinant *Saccharomyces cerviciae* populations. *Cytometry* **7**, 132–141.

106. Steen, H. B., Boye, E., Skarstad, K., Bloom, B., Godal, T., and Mustafa, S. (1982). Applications of flow cytometry on bacteria: Cell cycle kinetics, drug effects, and quantitation of antibody binding. *Cytometry* **2**, 249–257.

107. Steen, B. H., Lindmo, T., and Stokke, T. (1989). Differential light scattering detection in an arc lamp based flow cytometer. *In*: "Flow Cytometry; Advanced Research and Clinical Applications" (A. Yen, ed.), pp. 63–80. CRC Press, New York.

108. Steen, H. B., Skarstad, K., and Boye, E. (1986). Flow cytometry of bacteria: Cell cycle kinetics and effects of antibiotics. *Ann. N.Y. Acad. Sci.* **468**, 329–338.

109. Styrt, B. (1989). History and implications of the neutrophil glycoprotein deficiencies. *Am. J. Hematol.* **31**, 288–297.

110. Sveum, R. J., Chused, T. M., Frank, M. M., and Brown, E. J. (1986). A quantitative fluorescent method for measurement of bacterial adherence and phagocytosis. *J. Immunol. Methods* **90**, 257–264.

111. Szejda, P., Parce, J. W., Seeds, M. S., and Bass, D. A. (1984). Flow cytometric quantitation of oxidative product formation by polymorphonuclear leukocytes during phagocytosis. *J. Immunol.* **133**, 3303–3307.

112. Taguchi, H., Tanaka, R., Nishimura, K., and Miyaji, M. (1988). Application of flow cytometry to differentiate *Exophiala dermatitidis*, *E. moniliae*, and *E. jeanselmei* from each other. *Mykopathologia* **103**, 87–90.

113. Unkeless, J. C. (1989). Function and heterogeneity of human Fc receptors for immunoglobulin G. *J. Clin. Invest.* **83**, 355–361.

114. Van Dilla, M. A., Langlois, R. G., Pinkel, D., Yajko, D. and Hadley, W. K. (1983). Bacterial characterization by flow cytometry. *Science* **220**, 620–622.

115. Volsky, D. J., and Anderson, R. W. (1983). Deficiency in Epstein–Barr receptors on B-lymphocytes of preleukemia patients. *Cancer Res.* **43**, 3923–3926.

116. Waldman, F. M., Hadley, W. K., Fulwyler, M. J., and Schachter, J. (1987). Flow cytometric analysis of *Chlamydia trachomatis* interaction with L cells. *Cytometry* **8**, 55–59.

117. Wang, F.-I., Pang, V. F., and Hahn, E. C. (1988). Flow cytometric analysis of porcine peripheral blood leukocytes infected with Pseudorabies virus. *J. Leukocyte Biol.* **43**, 256–264.

118. Wells, A., Steen, H. B., Saefland, S., Godal, T., and Klein, G. (1981). A microassay for quantitatively detecting the Epstein–Barr virus receptor on single cells utilizing flow cytometry. *J. Virol. Methods* **3**, 127–136.

119. Wittrup, K. D., and Bailey, J. E. (1988). A single cell assay of β-galactosidase activity of *Saccharomyces cervisiae*. *Cytometry* **9**, 394–404.

120. Yamada, H., Shimizu, H., Taniguchi, O., and Okumura, K. (1988). Leu-1 (CD5) B cell subpopulation in patients with various liver diseases — special reference to hepatitis B carrier and to changes caused by prednisolone therapy. *Int. Arch. Allergy Appl. Imunol.* **87**, 409–416.

121. Yanase, Y., Tango, T., Okumura, K., Tada, T., and Kawasaki, T. (1987). A comparative study of alteration in lymphocyte subsets among varicella, hand-foot-and-mouth disease, scarlet fever, measles, and Kawasaki disease. *Microbiol. Immunol.* **31**, 701–710.

3.2

HIV Infection and AIDS

BETSY M. OHLSSON-WILHELM

Departments of Medicine and Microbiology & Immunology
The Pennsylvania State University College of Medicine
Hershey, PA, USA

ALAN LANDAY

Departments of Immunology/Microbiology and Pathology
Rush-Presbyterian-St Luke's Medical Center, Chicago, IL, USA

I. INTRODUCTION

The human immunodeficiency virus (HIV) has caused an epidemic of acquired immunodeficiency syndrome (AIDS) of global proportions. Although initially regarded as a disease of homosexual men, it is clear that HIV can be spread by shared needles, by transfusion of infected blood, and by heterosexual contact. It can also be transmitted vertically during the prenatal period. Infected individuals are often referred to the hematologist for evaluation of anemia, leukopenia or thrombocytopenia. The frequencies and types of cytopenias for homosexual men are shown as a function of disease stage in Fig. 1 (*57*).

HIV may infect a number of different cells

FLOW CYTOMETRY IN HEMATOLOGY
ISBN 0–12–432940–3

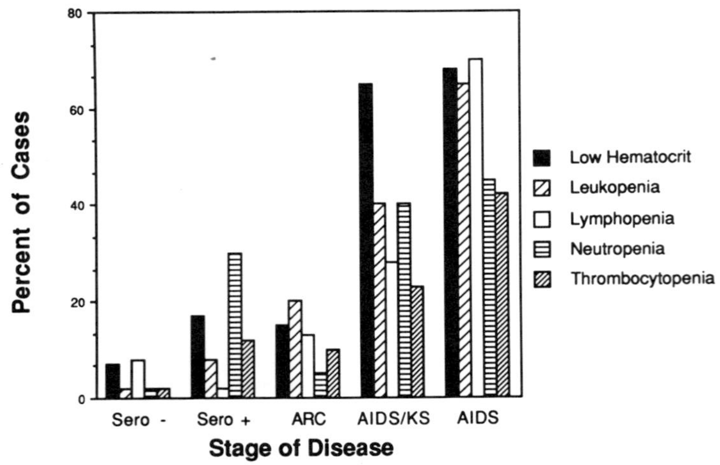

Fig. 1. Cytopenias in HIV disease. The percentage of cases showing a particular cytopenia is shown in groups of individuals representing different stages of HIV disease. Key: Sero – : seronegative asymptomatic homosexuals; Sero + : seropositive asymptomatic homosexuals; ARC: individuals with AIDS related complex; AIDS/KS: individuals with AIDS and Kaposi's sarcoma; AIDS: individuals with the acquired immunodeficiency syndrome. Data were taken from (*57*).

of the hematopoietic lineage. The major targets for HIV infection are the $CD4^+$ cells, both T cells and monocytes, since the CD4 molecule serves as a receptor for HIV (*23,51*). In addition, HIV can infect hematopoietic precursor cells ($CD34^+$) via Fc receptors (*10*) and other $CD4^-$ cells may be infected via an unidentified receptor which binds the fusion domain of the HIV envelope protein (*26*). Figure 2 summarizes changes in the peripheral blood lymphocyte subsets over the course of the disease. Progression is quite variable and

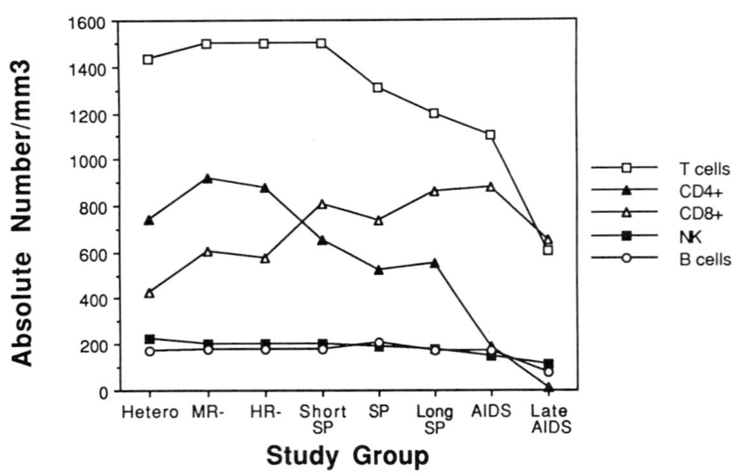

Fig. 2. Overview of lymphocyte subset changes. Key: Hetero: seronegative heterosexual men; MR – : homosexual men at moderate risk for HIV infection; HR – : homosexual men at high risk of HIV infection; Short SP: short-term seropositives; SP: seropositives; Long SP: long-term seropositives; AIDS: individuals with the acquired immunodeficiency syndrome; late AIDS: individuals late in the disease course. Data were taken from the Multicenter AIDS Cohort Study (*11,21*).

may require 10–20 years from infection until death.

Flow cytometric evaluation of hematopoietic cells allows assessment of immune system damage and prediction of the timing of disease progression. Such evaluation is useful in determining prognosis and in making therapeutic decisions — when to give antiretroviral therapy, e.g. zidovudine and when to initiate prophylaxis against *Pneumocystis carinii* pneumonia, e.g. aerosolized pentamidine or sulfamethoxazole/trimethaprim. This chapter will review flow cytometric tests which aid in the evaluation of hematopoietic cells from patients with HIV disease.

II. CHANGES IN THE PERIPHERAL BLOOD

A. Lymphocyte Subsets of Peripheral Blood

Evaluation of lymphocyte subsets in peripheral blood, especially $CD4^+$T lymphocytes, is of proven prognostic significance for individuals with HIV disease and provides an important part of the assessment of efficacy of antiretroviral therapy. Lymphocyte subsets are characterized by the expression of cell surface molecules defined by "cluster designations" ("CD—") (*24*). The subset of these antigens discussed in this chapter is given in Table I. Since most antigens are not lineage specific, monoclonal antibodies against them are reactive with more than one type of leukocyte. Multicolor flow cytometry is thus becoming essential for routine immunophenotyping.

Issues of accuracy, reproducibility, and precision in immunophenotyping are continually discussed by panels of experts such as the National Committee for Clinical Laboratory Standards (NCCLS). Methods involving staining of cells in whole blood followed by lysis of erythrocytes are strongly recommended. Consensus documents detailing methods are available (*33*). Accuracy of lymphocyte subset determination is best insured by performing a complete immunophenotypic profile and classifying all lymphocyte-like cells

Table I

LEUKOCYTE ANTIGENS USEFUL IN MONITORING HIV DISEASE

	Antigen
Components of basic panel	
CD3	T cell receptor complex
CD4	Class II MHC[a] receptor/HIV receptor
CD8	Class I MHC receptor
CD14	Monocyte Antigen
CD16	Fc_{gamma} RIII
CD19	B cell antigen
CD45	Leukocyte common antigen
CD56	N-CAM[b]
Other markers of interest	
CD11b	CR3 complement receptor
CD25	IL2 receptor
CD29	VLA β chain
CD38	46 kD glycoprotein
HLA-DR	MHC Class II
Leu8	Lymph node homing receptor

[a] MHC = major histocompatibility complex.
[b] CAM = cell adhesion molecule.

Table II

BASIC ROUTINE TWO-COLOR IMMUNOPHENOTYPING PANEL

Tube	Content[a]	Purpose	Comments
1.	IgG1/IgG2a[b]	Isotype control	Allows determination of non-specific binding
2.	CD45/CD14	Monocyte ($CD45^{+}_{intermediate}CD14^{+}$) and granulocyte ($CD45^{+}_{dim}CD14^{-}$) exclusion	Allows determination of fraction of cells in the lymphocyte analysis region which are lymphocytes ($CD45^{+}_{bright}CD14^{-}$). Allows determination of the fraction of total lymphocytes present in the lymphocyte analysis region.
3.	CD3/CD4	T helper cell ($CD3^{+}CD4^{+}$) determination	
4.	CD3/CD8	T suppressor cell ($CD3^{+}CD8^{+}$) determination	
5.	CD3/(CD16 + CD56)	Natural killer cell ($CD3^{-}[CD16+CD56]^{+}$) determination	
6.	CD3/CD19	B cell ($CD3^{-}CD19^{+}$) and total T cell ($CD3^{+}CD19^{-}$) determination[c]	

[a] Each tube contains two monoclonal antibodies (MoAb) directed against the designated antigen. The first MoAb is labeled with fluorescein; the second is labeled with phycoerythrin.

[b] Panel is chosen such that the first MoAb of each pair is an IgG1, and the second is an IgG2a. If MoAbs of different subclasses are used, the appropriate subclass matched control must be added or substituted.

[c] Total T cells ($CD3^{+}$) can be determined from tubes 3, 4, 5 and 6, giving an internal estimate of precision for each patient's panel.

as T cells, B cells, or NK cells. A typical, basic immunophenotyping panel is shown in Table II. Care must be exercised with AIDS samples, because of the frequent presence of nucleated red cells and alterations in light scatter parameters caused by certain antiretroviral agents. As increasing numbers of infected children are identified, two additional problems must be solved. First, reference values for normal, healthy children including neonates must be established. Initial data suggest a need for reference values for each 6-month period early in life. Second, since overall blood volume is small, a method requiring very little blood may be required. Use of a "cocktail" of monoclonal antibodies (*28*) as shown in Fig. 3 may therefore prove attractive.

a. *Declining $CD4^{+}$ Lymphocytes*

Figure 2 shows a major, selective depletion of $CD4^{+}$ lymphocytes during HIV disease (*13*). This decline begins during the first year after infection and proceeds more rapidly in the months preceding the development of full blown AIDS. Rates of decline of $CD4^{+}$ cells in AIDS patients are even faster, probably due to increased HIV replication. Therapeutic decisions are influenced by the absolute number of CD4-bearing cells as zidovudine is generally started at 500 $CD4^{+}$ cells/mm^3 and *Pneumocystis* prophylaxis at 200 $CD4^{+}$cells/mm^3. $CD4^{+}$ levels are highly prognostic for survival, especially late in HIV disease, and death occurs within 1 year in about 80% of untreated individuals with less than 10% $CD4^{+}$ lymphocytes. The percentage of $CD4^{+}$ lymphocytes is slightly more predictive for AIDS-free survival than is the absolute number (*50*), probably due to the difficulty in performing meaningful differentials on such lymphopenic blood samples.

Although preferential loss of particular $CD4^{+}$ subsets during certain stages of HIV

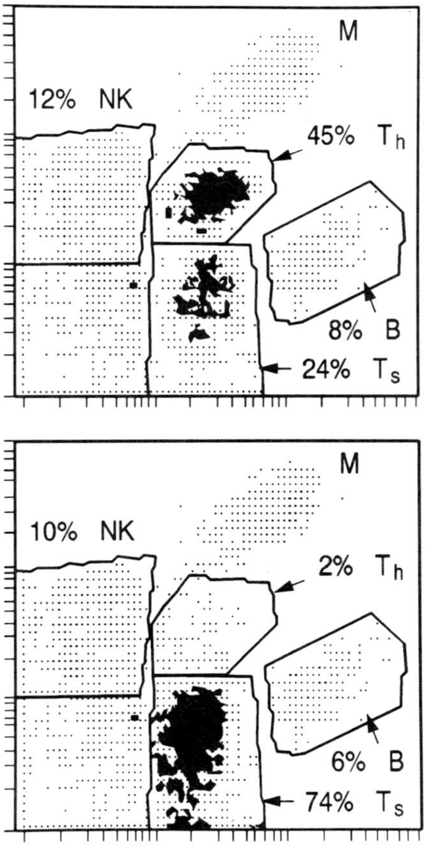

Fig. 3. Cocktail methods for small samples. The top panel represents a leukocyte-enriched preparation from a normal, healthy individual and the bottom panel a similar sample from an AIDS patient. Both were stained using a cocktail of antibodies (*28*). Unlabeled cells (bottom, left of each panel) are contaminating neutrophils or unlysed erythrocytes. M: monocytes, $CD14^+$; T_h: $CD3^+CD4^+$; T_s: $CD3^+CD8^+$; B: $CD19^+CD20^+$; NK: $CD16^+CD3^-$.

disease has been reported (*9,34*), other studies (*14,21*) which evaluated the $leu8^+$, $CD29^+$, and CD45RA subsets failed to support this. Physical sorting of $CD4^+CD29^+$ cells or $CD4^+CD45RO^+$ cells ("memory" T cells) and subsequent evaluation by viral culture, polymerase chain reaction (PCR), and functional assays reveals that these cells are preferentially infected compared to "naive" cells both *in vitro* and *in vivo* (*45*). This preferential infection may account for the failure of $CD4^+$ T cells from HIV-infected individuals to proliferate in response to soluble antigen, e.g. tetanus toxoid, *in vitro*.

b. *$CD8^+$ Lymphocytosis*

Total T lymphocyte number remains relatively stable during HIV disease until early symptoms of AIDS related complex (ARC) in spite of the dramatic decline in $CD4^+$ cells (Fig. 2). This stability results from a $CD8^+$ lymphocytosis, suggesting homeostatic control of total T cell number. Figure 4 shows the changes in the mean percentages of $CD8^+$ subsets as a function of disease progression seen in men participating in the Multicenter AIDS Cohort Study (MACS; *11,21*).

In general, activated subsets of CD8-bearing cells are seen to increase relative to the numbers found in uninfected individuals. $CD38^+CD8^+$ cells increase, beginning at seroconversion and remaining elevated throughout the disease course. Increases in this activated, immature $CD8^+$ subset have also been observed in individuals with active herpes virus infections (e.g. Epstein–Barr, cytomegalovirus) which decline as the infection is cleared. HLA-DR^+CD8^+ cells are relatively rare in HIV seronegative individuals (3–8%) but represent 15–30% of $CD8^+$ cells in HIV seropositive individuals. In the MACS study, this subset was higher in seropositive men prior to developing AIDS than in individuals with AIDS suggesting that at least some HLA-DR^+CD8^+ cells represent a protective component of the specific anti-HIV immune response, possibly cytolytic effector cells directed against HIV-infected cells (*12*). On the other hand, Stites *et al.* (*48*) regard increased numbers of these cells as a poor prognostic sign, implying that such cells may accelerate the $CD4^+$ decline or serve as a marker of HIV activity. Both $leu8^-CD8^+$ and $CD45RA^-CD8^+$ cells increase in individuals with HIV disease. Both antigens are lost upon maturation of naive cells to memory cells or upon activation of T cells *in vitro*. Thus the progressive increases in $leu8^-CD8^+$ and

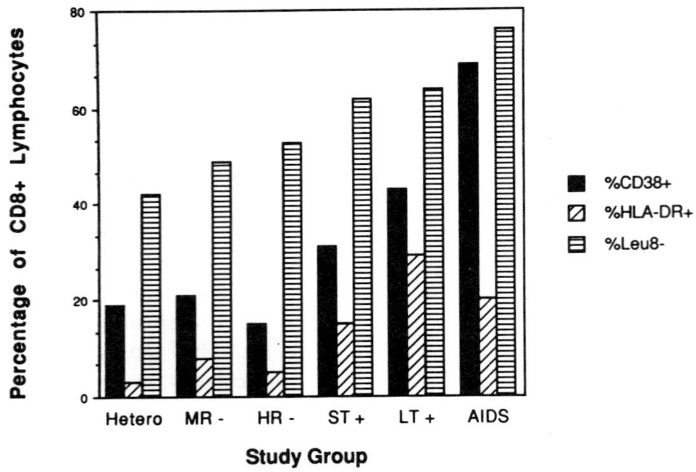

Fig. 4. Changes in $CD8^+$ subsets. Key: Hetero: seronegative heterosexual men; MR – : homosexual men at moderate risk for HIV infection; HR – : homosexual men at high risk of HIV infection; ST + : short-term seropositives; LT + : long-term seropositives; AIDS: individuals with the acquired immunodeficiency syndrome. Data were taken from the Multicenter AIDS Cohort Study (*11,21*).

$CD45RA^-CD8^+$ cells as HIV disease progresses support a role for activated $CD8^+$ lymphocytes in disease progression. $CD57^+CD8^+$ and $CD57^-CD8^+$ cells both increase in individuals with HIV disease compared to heterosexual controls (*12,27,39,55*). In individuals with late stage disease, a selective elevation in the $CD57^+CD8^+$ may accompany development of opportunistic infection (*27*). However, the significance of these increases is unclear, because HIV seronegative homosexuals also show a similar elevation when compared to heterosexual controls. This observation highlights the importance of defining the relevant control population when evaluating changes in lymphocyte subsets as a function of disease stage.

Information available from dual color surface immunophenotyping raises a number of questions, e.g. is there overlap between the HLA-DR^+CD8^+ cells and either the $CD38^+CD8^+$ subset or the $leu8^-CD8^+$ subset? Such a question is readily answered using a third color. Zeigler-Heitbrock *et al.* have used multicolor analysis to show that the HLA-DR^+CD8^+ subset is only partially overlapping with the $CD57^+CD8^+$ subset (*56*). The activation markers (HLA-DR, CD38, CD25, CD69) are expressed discordantly during HIV disease (*1*).

To summarize, full-blown AIDS is characterized by lymphopenia, low levels of $CD4^+$ lymphocytes, a low $CD4^+ : CD8^+$, and a high proportion of $CD38^+CD8^+$ cells.

c. *Gamma/delta-bearing T Cells*

Most (> 90%) peripheral blood T cells bear the α/β heterodimeric T cell receptor which is coexpressed with CD3 on the T cell surface. The rest express a second type of T cell receptor, the heterodimeric gamma/delta receptor (*35*). In uninfected individuals, such cells express neither CD4 or CD8 antigens although they do express CD3. In HIV-infected patients, these cells increase and can express varying levels of CD8 (*25,29*).

d. *Natural Killer Cells in Peripheral Blood*

Natural killer cells (NKs) are both phenotypically and functionally heterogeneous. They are the effector cells in antibody-dependent cellular cytotoxicity and are important in host defense against many pathogens. These large, granular lymphocytes express lineage independent markers, e.g. CD16 (FcγIII), CD56 (NCAM), CD57, or antigens characteristic of myeloid (CD11b) or T lymphocyte lineage (CD2, CD8). Overall levels of CD16$^+$ lymphocytes are comparable in seropositive homosexuals, symptomatic patients, and individuals with AIDS (*38*) even in the face of a major drop in NK function late in the HIV disease course. However, dual color analysis of CD56$^+$CD16$^+$ cells, the NK subset which mediates the most potent NK function, demonstrates that they are reduced in number. The dramatic decrease in NK function late in HIV disease is believed to result from impaired CD4$^+$ production of IL-2. *In vitro* function is restored by providing IL-2 (*3*). Selective depletion of low density CD8$^+$CD16$^+$ cells occur during HIV infection (*53*). No decrease in antibody-dependent cellular cytotoxicity (ADCC) has been observed (*37*). NK cells can be infected with HIV *in vitro* (*42*). Infection in this case is apparently mediated by a receptor other than CD4, possibly CD16.

e. *B Cells*

Values for both percentage and absolute number of B lymphocytes in samples from individuals with HIV disease lie on the low end of the normal range. Activated, immunoglobulin-secreting B cells are also seen (*30*).

B. Monocyte/Macrophage Subsets in Peripheral Blood

Demonstration of HIV infection of cells of the mononuclear phagocyte system in brain, lung, lymph nodes, spleen, and skin as well as peripheral blood has led to the hypothesis that such cells form a reservoir of virus within the body. Some of these cells bear CD4 and thus have an obvious receptor for the virus while others are apparently infected by a non-CD4-dependent mode of entry. Monocytes have one tenth the CD4 content of CD4$^+$ T lymphocytes. Although sufficient for HIV attachment and entry, this amount of CD4 is inadequate to support extensive viral glycoprotein-mediated cell to cell fusion as measured by the production *in vitro* of only limited numbers of multinucleated forms following infection. The short half-life of monocytes in peripheral blood prior to entering tissue has led to the speculation that it is the tissue macrophages which form the real viral reservoir.

Peripheral blood monocytes from a limited number of individuals with HIV disease have been sorted and examined for HIV proviral DNA using the polymerase chain reaction. In two of 10 HIV-seropositive, asymptomatic male homosexuals studied by Psallidopoulos *et al.* (*40*), integrated HIV-1 sequences were detected in the CD14$^+$ monocyte/macrophage population. More recently, Spear *et al.* (*47*) evaluated monocytes sorted on the basis of their CD14 marker from 12 seropositive asymptomatic individuals and eight individuals with AIDS. Only one asymptomatic individual had HIV-infected monocytes. In all individuals evidence for a much higher level of infection was found in the CD4$^+$ T lymphocytes.

Three different types of Fc receptors are found on monocytes. These include: (1) the FcγRI (CD64), responsible for ADCC; (2) the FcγRII (CD32), which mediates the monocyte-dependent mitogenic response of T cells and, (3) the FcγRIII (CD16), responsible for clearance of immune complexes. Defective FcR-mediated clearance of IgG-sensitized erythrocytes has been reported in patients with AIDS or ARC. There appear to be relatively normal numbers of FcγRI and FcγRII on monocytes from such individuals. In chimpanzee experimental models as well as in patients with ITP clearance of antibody-opsonized platelets has been successfully blocked with monoclonal anti-CD16 antibody (3G8) infusions (*5*). These observations support the role of the CD16 pathway in clearing immune complexes. Homsy *et al.* (*18*) have demonstrated that antibody to CD16

(3G8) efficiently blocks antibody-dependent enhancement of HIV infection. Their observations imply that FcγRIII mediates the uptake of HIV-enhancing antibody complexes into human macrophages. This finding conflicts with the observations of Takeda *et al.* (*49*) and Jouault *et al.* (*19*) showing that antibody-dependent enhancement of HIV infection of the monocytic cell line U937 seems to be related to the entry of anti-HIV immune complexes through FcγRI and FcγRII, as U937 does not express FcγRIII. One must remember that it requires many fewer molecules per cell (possibly only one?) to serve as a viral receptor than to give a good flow cytometric signal.

HLA Class II antigens play an important role in immune interactions of monocytes with both T and B lymphocytes. Surface expression of HLA-DR is decreased 10-fold in patients with AIDS but not ARC. Interferon-gamma *in vitro* restores HLA-DR levels to normal in monocytes from individuals with AIDS (*16*). The decrease in monocyte class II expression is correlated with a decrease in ability to serve as accessory cells. In contrast, other studies on monocytes from AIDS patients have reported an increased amount of class II as well as interleukin 2 receptor (CD25) (*31*). Resolution of the mechanism of control of class II expression will require additional studies. Alveolar macrophages from both AIDS and ARC patients have increased expression of both class II HLA-DR and -DQ antigens (*43*). More recently, Spear *et al.* (*46*) have examined the expression of CD4, CD13, CD11b, and HLA-DR on $CD14^+$ monocytes from asymptomatic individuals. $CD14^+$ monocytes from patients with circulating serum p24 antigen had reduced surface expression of CD4, CD11b, and HLA-DR.

To summarize, HIV infection of monocytes has been demonstrated by PCR evaluation of sorted monocytes from a few patient peripheral blood samples. Infection may proceed via a CD4-mediated or a CD16-mediated mode of entry. Reduced surface expression of CD4, CD11b and HLA-DR is associated with increased serum p24 antigen.

C. Neutrophils

Susceptibility to opportunistic microorganisms is characteristic of AIDS patients, possibly due to altered neutrophil function. Among the interesting surface molecules of neutrophils is CD16, described above. Two forms of this gene have been identified which differ in their cellular attachment chemistry. Neutrophil CD16 is anchored to the plasma membrane through a phosphatidyl inositol glycan moiety and can be released by treatment with phosphatidyl inositol specific phospholipase C or by incubation with chemotactic peptides. The CD16 characteristic of macrophage and NK cells has both transmembrane and cytoplasmic domains. Neutrophil CD16 expression is significantly decreased in AIDS patients (*4*). Dual color analysis of CD16, CD11b, and CD32 on neutrophils from individuals with HIV disease showed that 25% of such neutrophils were autofluorescent and $CD16^-$, compared with 3% in seronegative individuals. CD16 loss may be related to the presence and/or frequency of common bacterial infections in advanced AIDS. Serum CD16 increases in early seropositives and decreases in AIDS patients. Because of the different modes of attachment of the CD16 molecule to various lineages, it was speculated that the serum CD16 antigen arose by cleavage from neutrophils, possibly as a result of having encountered HIV. Of 73 HIV-infected people examined by Khayat *et al.* (*22*), 30 were examined by flow cytometry for CD16 expression on $CD3^+$ cells and on NK cells. No specific changes were found in either the number of $CD16^+$ cells or in the $CD56^+$ cells. There was, however, a statistical correlation between the absolute number of $CD16^+CD3^+$ cells and the modulated titers of soluble CD16 in sera from individuals with HIV disease.

III. HIV TARGETS IN BONE MARROW

A. Bone Marrow Progenitor Cells

HIV seropositive sera inhibit growth of CFU

in agar cultures if the source of the CFU is marrow from an individual with AIDS. CFU from marrows of HIV seronegative individuals are unaffected by the HIV seropositive sera (*8*). These studies imply that CFU in marrows from AIDS patients are infected with HIV. Folks and his colleagues (*10*) have demonstrated that bone marrow progenitor cells can be infected by HIV. These cells are devoid of CD4 and show neither T cell markers nor the specific monocyte enzyme alpha-naphthyl esterase. In these studies virus was assembled preferentially on the internal membranes of cells derived from the cultured marrow progenitors.

B. Megakaryocytes

HIV-associated thrombocytopenia can appear in subjects early in the course of HIV disease who are not anemic and whose bone marrows have none of the abnormalities typical of AIDS marrows. Initially low platelet counts were thought to be due to immune mechanisms, similar to those operative in chronic ITP which accelerate platelet destruction. However, HIV-associated thrombocytopenia is distinct from ITP or lupus erythematosus-associated thrombocytopenia because (1) no compensatory megakaryocytosis is apparent; (2) aberrations of bone marrow megakaryocytes seen at the ultrastructure level are distinct from those seen in immune thrombocytopenias from seronegative individuals; and (3) thrombocytopenia reverses rapidly after treatment with zidovudine (*2,20*). In addition, electron micrographs of megakaryocytes from these individuals reveal primary injury on the surface membrane and on the peripheral cytoplasmic zone where both blebbing and vacuolization are seen. No impairment of megakaryocyte differentiation is apparent (*58,59*). These observations together with the detection of viral RNA in the megakaryocytes from HIV-infected individuals (*58*) suggest that the direct infection of megakaryocytes may be responsible for the diminished platelet numbers.

IV. QUANTITATION OF VIRUS LOAD

Optimal monitoring of HIV-infected individuals requires quantification of viral load to delineate the natural history of the infection and to assess the effectiveness of new antiretroviral therapies. There are three components to viral load: cell-free virus, infected cells containing actively replicating virus and viral components, and cells containing proviral DNA. HIV-infected cells outside of the peripheral circulation, for example in the brain, only compound the difficulties associated with making the determination. One approach is to assume that the viral load in the peripheral blood will reflect the viral load of the entire body.

The first experiments addressing viral load (*15*) suggested that very few peripheral blood cells were infected by HIV. Using *in situ* hybridization techniques, viral RNA was found in 86% (six out of seven) of lymph nodes examined and in only 50% (seven out of 14) of peripheral blood samples examined. Labeled cells were seen at a very low frequency ($< 0.01\%$ of total mononuclear cells). Viral RNA was estimated at between 20 and 300 copies per cell. New methods of viral culture, e.g. those of the American AIDS Clinical Trials Group, result in measurement not only of cells actively replicating HIV but also of cells induced to produce virus from proviral DNA. Recent attempts to quantify free virus, as well as actively infected cells (*6,17*) have been successful. Interestingly, the number of infectious viral particles or infected peripheral blood cells found is substantially higher (one out of 400) than those estimated by Harper *et al*. Division of the mean number of infected cells ($2200/10^6$ cells for AIDS patients: range 50–100 000) by the fraction of CD4-bearing cells, assumed to be 10% of the lymphocytes, allows one to infer that 1 in 40 (2.5%) of the CD4-bearing lymphocytes could be infected.

The rapid rate of analysis of cells by flow cytometry (approaching 6000 cells/s) allows relatively rare events, such as the circulating HIV-infected cell, to be sought using this technology. As an example, it would take 16.7 min to evaluate 1 000 000 cells at 1000 cells per second. Thus it is potentially feasible to quantify HIV-infected cells present at 0.01% of the peripheral blood mononuclear cells, as suggested by Harper *et al.* (*15*).

A. Cell Sorting

Schnittman and colleagues (*44*) used cell sorting to obtain specific subsets of circulating lymphocytes from infected individuals based on their cell surface phenotype. They performed polymerase chain reaction assays on various dilutions of the sorted CD4-bearing cells. All PCR positivity was found in CD4-bearing cells and indications were that approximately 1% of the CD4-bearing cells contained HIV proviral DNA. Psallidopoulos *et al.* (*40*) performed similar experiments using samples from seropositive asymptomatic individuals. Again, the major population of infected cells was the $CD4^+$ lymphocyte, found in all 10 samples. Estimates indicated that between 1 in 100 and 1 in 40 000 $CD4^+$ cells were infected, depending on the individual. Again, a small signal was occasionally obtained from the $CD4^-$ lymphocytes. These investigators suggest a viral load in four of their asymptomatic individuals (1/100 $CD4^+$ cells) very similar to the burden found in individuals with AIDS. These results and others (*40,44,52*) suggest that the number of circulating HIV-infected cells is considerably higher than previously reported. Schnittman *et al.* contend that the CD4-bearing lymphocyte is the principal reservoir of HIV infection. Further, because of the intensity per cell of the CD4 signal used for cell sorting, as well as the distribution of intensities per cell, they suggested that the vast majority of the CD4-bearing cells had not down-regulated the CD4 receptor as a function of HIV infection. A similar conclusion was drawn by Walker *et al.* who found HIV associated with $CD4^+$ cells selected from PBMCs by panning (*54*).

B. Cell-associated p24 Antigen

Lymphocyte p24-FCA is a new flow cytometric test (*36*) which quantifies the number of circulating lymphocytes which contain actively replicating HIV (Fig. 5). The test relies for its specificity on a monoclonal antibody (MoAb) which binds to the HIV core antigen p24 (or p25). The assumption is made that the presence of cytoplasmic HIV p24 antigen indicates active viral protein synthesis. Binding to the MoAb is revealed by indirect immunofluorescence after fixation of the cells to allow entrance of the MoAb. The number of lymphocytes containing actively replicating HIV is quantified by flow cytometry and expressed as the number of lymphocytes which are p24-antigen positive ($p24^+$). In model systems this test detects one HIV-infected lymphoblastoid cell among 10^4 uninfected cells (*7*).

Comparison of asymptomatic, HIV seropositive, serum p24 antigen-negative individuals to patients with AIDS either untreated or receiving zidovudine therapy showed that the range of values for asymptomatic individuals was remarkably similar to that seen in untreated individuals with AIDS. Individuals receiving zidovudine showed a much lower range of values than did the other two study groups. Again, the number of circulating virus-infected lymphocytes as revealed by this assay was high, in some cases approaching 25% of all circulating lymphocytes. Calculation of the ratio of $p24^+$ lymphocytes to $CD4^+$ lymphocytes revealed some interesting facts. First, the ratio was quite low (< 0.5) in 53 out of 55 asymptomatic individuals, consistent with confinement of the infection to a limited number of the CD4-bearing lymphocytes. Second, two of the 55 who had ratios between 0.5 and 1.0 differed from the rest of the study group in that they progressed to AIDS within 2 weeks of testing.

Other investigators have also employed flow cytometry to quantify HIV-infected peripheral blood cells. McSharry and his colleagues (*32*) performed a cross-sectional study using

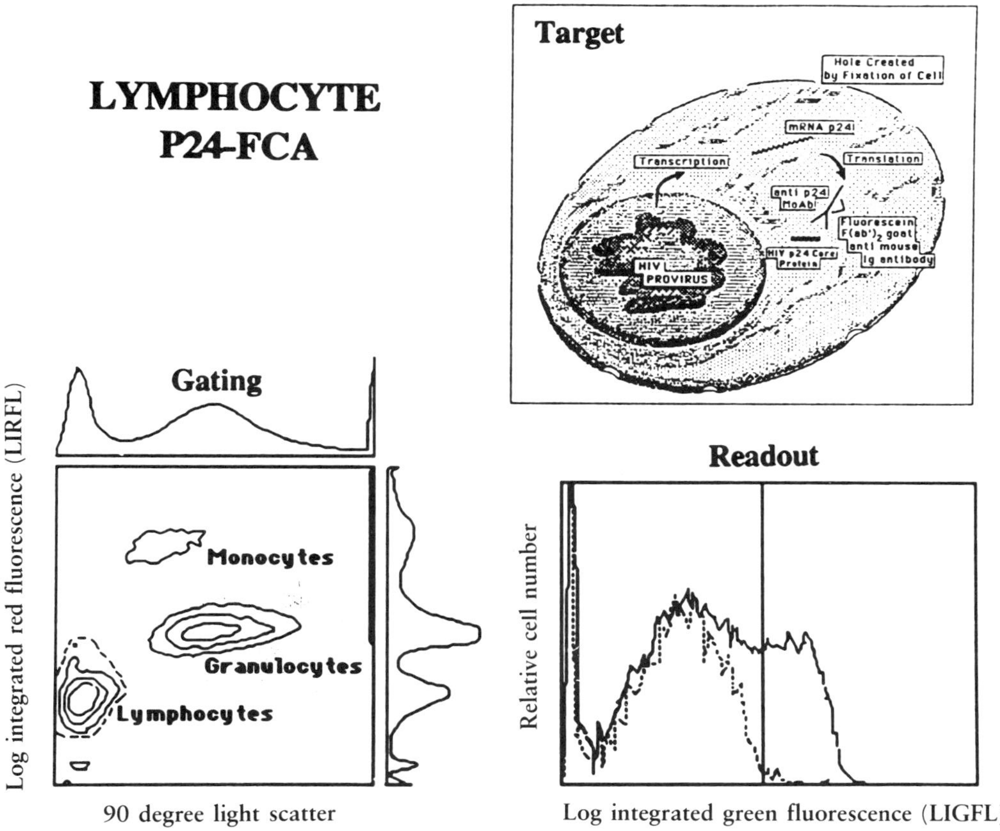

Fig. 5. Lymphocyte p24-FCA. Fixed, permeabilized HIV-infected lymphocytes are enumerated by lymphocyte p24-FCA. Binding of an HIV-p24-specific murine MoAb is revealed by indirect immunofluorescence and quantified by flow cytometry using a lymphocyte analysis region to identify cells with low 90° light scatter and low binding of a CD14-specific MoAb as lymphocytes (gating panel). Histograms for a seronegative control and an HIV-infected individual are shown.

PBMCs from individuals with HIV disease. Their approach differed slightly from that described above in that they measured binding of monoclonal antibodies directed against HIV as a function of the DNA content of the PBMCs. They also found quite high numbers (up to 24%) of cells which were positive for HIV-nef or HIV-p24. A significant correlation was demonstrated ($p = 0.0001$) between the CDC status of the patients and the percentage of p24 antigen positive mononuclear cells. No correlation was observed with serum p24 antigen.

C. Potential Usefulness of Direct Measures of Viral Load

Monitoring individuals with HIV disease should be aided by use of lymphocyte p24-FCA or other direct measures of cells actively replicating HIV. The current first line of defense, zidovudine, has some associated toxicity to bone marrow. These side effects continue to accrue as long as zidovudine is given. Since zidovudine inhibits HIV reverse transcriptase and is thus only active against the initial stages of HIV infection of cells, it would seem prudent to treat with the drug only when the viral infection is active. Thus use of this test may define appropriate treatment periods and, by allowing certain periods of recovery from unwanted side effects when the viral infection is relatively quiescent, may extend the period of time during which a given individual can be treated with zidovudine.

Another use of this technology will be to assay for active viral replication in individuals for whom the standard serotests are not meaningful, for example, infants born to HIV seropositive mothers who carry maternal antibody and individuals who may eventually receive an HIV vaccine. Preliminary data suggest that this technique may be able to detect infection in infants born to seropositive mothers (*41*), but longitudinal follow-up is needed for confirmation.

V. SUMMARY

Flow cytometric tests can aid substantially in hematologic evaluation of individuals with HIV disease. All components of the blood, bone marrow, and lymph nodes are amenable to analysis because of the easy preparation of single cell suspensions. Most of the work has focused on changes in lymphocyte subsets, particularly the relative numbers of CD4-bearing lymphocytes. Current therapeutic decisions are influenced by this determination. Specific subsets of the $CD8^+$ cells, notably those bearing CD38 or those which have lost the leu8 receptor, are of increasing interest as their prognostic significance is defined. In addition, erythroid and myeloid cell precursors, megakaryocytes, and platelets can be evaluated using flow cytometry. Lastly, tests under development may provide a direct measure of the number of HIV-infected cells.

ACKNOWLEDGEMENTS

The gracious assistance of Ms Lisa Doster in the preparation of this manuscript and Ms Elizabeth Steck in the preparation of the figures is gratefully acknowledged. One of us (B. M. O.-W.) was partially supported by a contract N01-HB-67023, Project 221924 from the National Heart, Lung, and Blood Institute and the other (A. L.) was supported in part by U01-AI25915 from the National Institute for Allergy and Infectious Disease.

REFERENCES

1. Bach, B. A., Campbell, D., Robertson, M. J. *et al.* (1989). T-cell activation markers (HLA-DR, CD38, CD25, CD69) are expressed discordantly in HIV infected individuals. *Blood* **74**, 339a.

2. Ballum, P., Belzberg, A., Devind, D. *et al.* (1988). Pathophysiology of HIV-ITP and the mechanism of the response to AZT. *Blood* **72**, 261a.

3. Bonavida, B., Katz, J., and Gottleig, M. (1986). Mechanism of defective NK cell activity in patients with AIDS and ARC 1. Defective trigger on NK cells for NKCH production by target cells and partial restoration by IL-2. *J. Immunol.* **137**, 1157–1163.

4. Boros, P., Gardos, E., Bekesi, O. *et al.* (1990). Changes in expression of FcRγRIII (CD16) on the neutrophils from HIV infected individuals. *Clin. Immunol. Immunopathol.* **54**, 281–189.

5. Clarkson, S. B., Bussel, J. B., Kimberly, R. P. *et al.* (1986). Treatment of refractory ITP with an anti-Fcγ receptor antibody. *New Engl. J. Med.* **314**, 1236–1239.

6. Coombs, R. W., Collier, A. C., Allain, J. P. *et al.* (1989). Plasma viremia in human immunodeficiency virus infection. *New Engl. J. Med.* **321**, 1626–1631.

7. Cory, J. M., Ohlsson-Wilhelm, B. M., Brock, E. J., *et al.* (1987). Detection of human immunodeficiency virus infected lymphoid cells of low frequency by flow cytometry. *J. Immunol. Methods* **105**, 71–78. **105**, 71–78.

8. Donahue, R. E., Johnson, M. M., Zon, L. I. *et al.* Suppression of *in vitro* haematopoiesis following HIV infection. *Nature* (*London*) **326**, 200–203.

9. Fletcher, M. A., Azen, S. P., Adelsberg, B. *et al.* (1989). Immunophenotyping in a multicenter study: the transfusion safety study experience. *Clin. Immunol. Immunopathol.* **52**, 38–47.

10. Folks, T. M., Kessler, S. W., Orenstein, J. M. *et al.* (1988). Infection and replication of HIV-1 in purified progenitor cells of normal human bone marrow. *Science* **242**, 919–922.

11. Giorgi, J. V., and Detels, R. (1989). T-cell subset alterations in HIV-infected homosexual men: NIAID multicenter AIDS cohort study. *Clin. Immunol. Immunopathol.* **52**, 10–18.

12. Giorgi, J. V., and Hultin, L. E. (1990). Lymphocyte subset alterations and immunophenotyping by flow cytometry in HIV disease. *Clin. Immunol.* **10**, 55–61.

13. Giorgi, J. V., Fahey, J. L., Smith, D. C. *et al.* (1987). Early effects of HIV on CD4 lymphocytes *in vivo*. *J. Immunol.* **138**, 3725–3730.

14. Giorgi, J. V., Nishanian, P. G., Schmid, I. *et al.* (1987). Selective alterations in immunoregulatory lymphocyte subsets in early HIV infection. *J. Clin. Immunol.* **7**, 140–150.

15. Harper, M. E., Marselle, L. M., Gallo, R. C. *et al.* (1986). Detection of T lymphocytes expressing human T-lymphotropic virus type III in lymph node and peripheral blood from infected individuals by *in situ* hybridization. *Proc. Natl. Acad. Sci. USA* **83**, 772–776.

16. Heagy, W., Kelley, V. E., Strom, T. B. *et al.* (1984). Decreased expression of human class II antigens on monocytes from patients with AIDS: increased expression with interferon gamma. *J. Clin. Invest.* **74**, 2089–2096.

17. Ho, D. D., Moughi, T, and Alam, M. (1989). Quantitation of HIV-1 in the blood of infected persons. *New Engl. J. Med.* **321**, 1621–1625.

18. Homsy, J., Meyer, M., Tateno, M. *et al.* (1989). The Fc and not CD4 receptor mediates antibody enhancement of HIV infection in human cells. *Science* **244**, 1357–1360.

19. Jouault, T., Chapuis, F., Olivier, R. *et al.* (1989). Opsonization, an infection pathway for HIV in monocytic cells. *AIDS* **3**, 125–133.

20. Karpatkin, S. (1988). Immunologic thrombocytopenic purpura in HIV-seropositive homosexuals, narcotic addicts and hemophiliacs. *Sem. Hematol.* **25**, 219–229.

21. Kaslow, R. A., Ostrow, D. G., Detels, R. *et al.* (1987). The multicenter AIDS cohort study: rationale, organization, and selected characteristics of the participants. *Am. J. Epidemiol.* **126**, 310–318.

22. Khayat, D., Soubrana, C., Andrieu, J. M. *et al.* (1990). Changes of soluble CD16 levels in serum of HIV-infected patients: correlation with clinical and biologic prognostic factors. *J. Infect. Dis.* **161**, 430–435.

23. Klatzmann, D., Barre-Sinoussi, F., Nugeyre, M. *et al.* (1984). Selective tropism of LAV for helper-inducer T-lymphocytes. *Science* **225**, 59–62.

24. Knapp, W., Rieber, P., Dorken, B. *et al.* (1989). Towards a better definition of human leucocyte surface molecules. *Immunol. Today* **10**, 253–258.

25. Landay, A., Ohlsson-Wilhelm, B. M., and Giorgi, J. V. (1990). Application of flow cytometry to the study of HIV infection. *AIDS* **4**, 479–497.20.

26. Levy, J. A. (1988). Human immunodeficiency viruses and the pathogenesis of AIDS. *J. Am. Med. Assoc.* **261**, 2997–3006.

27. Lewis, D. E., Puck, J. M., Babock, G. F. *et al.* (1985). Disproportionate expansion of a minor T cell subset in patients with lymphadenopathy syndrome and acquired immunodeficiency syndrome. *J. Infect. Dis.* **151**, 555–559.

28. Liu, C. M., Muirhead, K. A., George, S. P. *et al.* (1989). Flow cytometric monitoring of human immunodeficiency virus infected patients: Simultaneous enumeration of five lymphocyte subsets. *Am. J. Clin. Pathol.* **92**, 721–728.

29. Margolick, J. B., Carey, V., Rinaldo, C. *et al.* (1989). Development of antibodies to HIV-I is associated with an increase in circulating $CD3^+CD4^-CD8^-$ lymphocytes. *Clin. Immunol. Immunopathol.* **51**, 348–361.

30. Martinez-Maza, O., Grabb, E., Giorgi, J. V. *et al.* (1987). Infection with the human immunodeficiency virus (HIV) is associated with an *in vivo* increase in B lymphocyte activation and immaturity. *J. Immunol.* **138**, 3720–3724.

31. McCartney-Francis, N., Smith, P. D., Simon, G. *et al.* (1990). Expression of IL-2 receptors by monocytes from patients with AIDS and induction of monocyte IL-2 receptors by HIV *in vitro*. *J. Clin. Invest.* **85**, 192–199.

32. McSharry, J. J., Constantino, R., Robblano, E. *et al.* (1990). Detection and quantification of HIV-infected peripheral blood mononuclear cells by flow cytometry. *J. Clin. Micro.* **28**(*4*), 724–733.

33. National Committee for Clinical Laboratory Standards (1989). "Clinical Applications of Flow Cytometry Quality Assurance and Immunophenotyping of Peripheral Blood Lymphocytes." NCCLS Publication H-42P, Villanova, PA.

34. Nicholson, J. K. A., McDougal J. S., Jaffe, H. W. *et al.* (1985). Exposure to human T-lymphotropic virus type III/lymphadenopathy-associated virus and immunologic abnormalities in asymptomatic homosexual men. *Ann. Intern. Med.* **103**, 37–42.

35. Oettgen, H. C., Kappler, J., Tax, W. J. M. *et al.* (1984). Characterization of the two heavy chains of the T3 complex on the surface of human T lymphocytes. *J. Biol. Chem.* **259**, 12039.

36. Ohlsson-Wilhelm, B. M., Cory, J. M., Kessler, H. A. *et al.* (1990). Circulating HIV p24 antigen-positive lymphocytes: a flow cytometric measure of HIV infection. *J. Infect. Dis.* **162**, 1018–1024.

37. Ojo-Amaize, E., Nishanian, P. G., Heitjan, D. F. *et al.* (1989). Serum and effector-cell antibody-dependent cellular cytotoxicity (ADCC) activity remains high during human immunodeficiency virus (HIV) disease progression. *J. Clin. Immunol.* **9**, 454–461.

38. Plaeger-Marshall, S., Spira C. A., Giorgi, J. V. *et al.* (1987). Alterations in cytotoxic and phenotypic subsets of NK cells in AIDS. *J. Clin. Immunol.* **7**, 16–23.

39. Prince, H. E., Kreiss, J. K., Kasper, C. K. *et al.* (1985). Distinctive lymphocyte subpopulation abnormalities in patients with congenital coagulation disorders who exhibit lymph node enlargement. *Blood* **66**, 64–68.

40. Psallidopoulos, M. C., Schnittman, S. M., Thompson, L. M. *et al.* (1989). Integrated proviral human immunodeficiency virus type I is present in $CD4^+$ peripheral blood lymphocytes in healthy seropositives. *J. Virol.* **63**, 4626–4631.

41. Rich, K. C., Cory, J. M., Landay, A. L. *et al.* (1990). "Lymphocyte Associated p24 Antigen: A New Technique for Diagnosis of HIV Infection in Pediatric Patients", Vol. 3, p. 320. 6th Int. Conf. AIDS, San Francisco, CA.

42. Robinson, W. E., Mitchell, W. M., Chambers, W. H. *et al.* (1988). Natural killer cell infection and inactivation in vitro by the human immunodeficiency virus. *Hum. Pathol.* **19**, 535–540.

43. Roy, G., Rojo, N., and Ceyva-Cobian, F. (1987). Phenotypic changes in monocytes and alveolar macrophages in patients with AIDS and ARC. *J. Clin. Lab. Immunol.* **23**, 135–141.

44. Schnittman, S. M., Psallidopoulos, M. C., Lane, C. *et al.* (1989). The reservoir for HIV-1 in human peripheral blood is a T cell that maintains expression of CD4. *Science* **245**, 305–308.

45. Schnittman, S. M., Lane, H. C., Greenhouse, J. *et al.* (1990). Preferential infection of $CD4^+$ memory T cells by HIV-1: evidence for a role in the selective T-cell functional defects observed in infected individuals. *Proc. Natl. Acad. Sci. USA* **87**, 6058–6062.

46. Spear, G T., Kessler, H. A., Rothberg, L. *et al.* (1990). Decreased oxidative burst activity of monocyte from asymptomatic HIV-infected individuals. *Clin. Immunol. Immunopathol.* **54**, 184–191.

47. Spear, G. T., Ou, C.-Y., Kessler, H. A., Moore, J. L., Schochetman, G. and Landay, A. L. (1990). Analysis of lymphocytes, monocytes, and neutrophils from HIV-infected persons for HIV DNA. *J. Infect. Dis.* **162**, 1239–1244.

48. Sites, D. P., Casavant, C. H., McHugh, T. M. *et al.* (1986). Flow cytometric analysis of lymphocyte phenotypes in AIDS using monoclonal antibodies and simultaneous dual immunofluorescence. *Clin. Immunol. Immunopathol.* **38**, 161–177.

49. Takeda, A., Tuazon, C. U., and Ennis, F. A. (1988). Antibody-enhanced infection by HIV-1 via Fc receptor-mediated entry. *Science* **242**, 580–583.

50. Taylor, J. M. G., Fahey, J. L., Detels, R. *et al.* (1989). CD4 percentage, CD4 number and CD4:CD8 ratio in HIV infection: which to choose and how to use. *J. Acquir. Immun. Def. Synd.* **2**, 114–124.

51. Traunecker, A., Luke, W., and Karjalainen, K. (1988). Soluble CD4 molecules neutralize HIV-1. *Nature (London)* **331**, 84–86.

52. Ulrich, P. P., Busch, M. P., El-Beik, T. *et al.* (1988). Assessment of human immunodeficiency virus expression in cocultures of PBMCs from healthy seropositive individuals. *J. Med. Virol.* **75**, 1–10.

53. Vuillier, F., Bianco, N. E., Montagnier, L. *et al.* (1988). Selection depletion of low density $CD8^+$, $CD16^+$ lymphocytes during HIV infection. *AIDS Res. Hum. Retrovir.* **4**, 121–129.

54. Walker, C. M., Moody, D. J., Stites, D. P. *et al.* (1989). $CD8^+$ T lymphocyte control of HIV replication in cultured $CD4^+$ cell varies among infected individuals. *Cell Immunol.* **119**, 470–475.

55. Ziegler-Heitbrock, H. W. L., Schramm, W., Stachel, D. *et al.* (1985). Expansion of a minor subpopulation of peripheral blood lymphocytes ($T8^+$/$Leu7^+$) in patients with haemophilia. *Clin. Exp. Immunol.* **61**, 633–641.

56. Ziegler-Heitbrock, H. W. L., Stachel, D., Schlunk, T. *et al.* (1988). Class II (DR) antigen expression on $CD8^+$ lymphocyte subsets in acquired immune deficiency syndrome (AIDS). *J. Clin. Immunol.* **8**, 1–6.

57. Zon, L. E., Arkin, C., and Groopman, J. E. (1987). Hematological manifestations of HIV. *Br. J. Haematol.* **66**, 251–256.

58. Zucker-Franklin, D., and Cao, Y. (1989). Megakaryocytes of HIV-infected individuals express viral RNA. *Proc. Natl. Acad. Sci. USA* **86**, 5595–5599.

59. Zucker-Franklin, D., Termin, C. S., and Cooper, M. C. (1989). Structural changes in the megakaryocytes of patients infected with HIV-I. *Am. J. Pathol.* **134**, 1295–1304.

3.3

Detection and Analysis of Parasitic Infections

BARENDMONS AND CHRIS J. JANSE

Laboratory for Parasitology
University of Leiden,
Leiden, The Netherlands

I. INTRODUCTION

Parasitic protozoa which proliferate in the blood cause some of the most life-threatening infectious diseases in man. Malaria, caused by *Plasmodium* species, and trypanosomiasis, caused by *Trypanosoma* species, are among the top ten parasitic diseases defined by the World Health Organization (WHO).

Malaria causes about 1 million deaths and over 200 million infections per year. The situation is becoming worse every year, because of rapidly spreading drug and insecticide resistance of the parasite and the vector mosquitoes, respectively, and the absence of alternative control methods, like an effective vaccine, that can economically control this disease over vast endemic areas.

Sleeping sickness, caused by African trypanosomes, is less important than malaria in terms of mortality rates. However, every year tens of thousands of people fall victim to this disease. Moreover, African trypanosomes cause fatal anemia in domestic animals. In rural areas of South America up to 50% of people suffer from Chagas' disease, caused by American trypanosomes. Trypanosomiasis and

FLOW CYTOMETRY IN HEMATOLOGY
ISBN 0–12–432940–3

Chagas' disease are difficult to treat because of the scarcity of effective drugs.

Blood parasites are relevant for hematology because the morphology and number of different blood cells can be changed as a result of a parasitic infection. For example, infection with malaria parasites and trypanosomes can result in severe anemia. In addition, blood parasites pose a problem where blood transfusion is concerned.

There is an urgent need for the development of rapid, sensitive and reproducible techniques for the detection and the analysis of parasitic infections in blood; not only for human parasites but also for those infecting domestic animals, such as *Babesia* and *Theileria*. The classical, direct microscopical methods and the indirect serological methods to detect and count parasites are laborious and require skilled technical personnel. They also have the disadvantage of giving rather variable results.

Flow cytometry has been used in a number of cases to detect parasites in blood, but only recently has this technique shown itself to be a useful method in the field of parasitology. Possibilities for the application of flow cytometry in parasitology include: (1) rapid detection of parasites in blood samples; (2) rapid and reproducible counting of parasites; (3) species determination and stage determination; (4) quantitative measurement of DNA content and synthesis; (5) automated drugscreening; (6) separation of parasites from blood cells; (7) separation of different developmental stages of parasites; (8) separation of vital and dead parasites.

Reports on the use of flow cytometry for parasitological research are almost completely limited to *Plasmodium* and *Trypanosoma*. These parasites can easily be detected using DNA-specific fluorescent dyes such as ethidium bromide, Hoechst or DAPI. Remarkably little has been published on the use of species- or stage-specific monoclonal antibodies (MoAbs). The application of the flow cytometer is largely restricted to fundamental research, which is partly due to the high-tech status of the apparatus. Application of flow cytometry for field-oriented and clinical research has recently been initiated. Possible future applications for the technique in parasitology are manifold. However, a number of problems related to the specific needs for fixation and staining of the parasites have to be solved. It should be stressed that sample preparation and fixation procedures to be developed should be relatively simple and robust to allow application under primitive circumstances as generally found in the field stations in countries where parasitic diseases are endemic. The major blood parasites of man and the technical aspects relating to their detection and analysis will be discussed in this chapter.

II. PARASITES

Protozoan parasites of the genera *Plasmodium*, *Trypanosoma* and *Leishmania* occur in the circulating blood or within blood cells of humans. The first two genera will be more extensively described in this chapter.

Leishmania species are obligatory parasites, intracellular in mononuclear phagocytes in the skin and the reticulolymphatic tissues and are usually not found in the circulating blood.

Protozoan parasites of the genera *Theileria* and *Babesia* are essentially non-human parasites, although infection with *Babesia* is occasionally described in humans. These parasites are relevant to man because they are highly pathogenic to cattle, sheep and goats, resulting in important economic losses in the tropics. *Theileria* species live in nucleated cells in the lymphoid tissues. In the lymphocytes, merozoites are formed by schizogony, which are able to invade red blood cells (RBCs). Stages within the erythrocytes are infective to the vector, a tick (*32*). *Babesia* develops in erythrocytes, where it divides and also produces sexual forms infective to ticks (*32*).

A. *PLASMODIUM*

There are four species of human malaria parasites, *P. falciparum*, *P. vivax*, *P. ovale* and *P. malariae*, of which *P. falciparum* is the most hazardous. These parasites cause the disease malaria which is characterized by intermittent fever, severe anemia and, in sub-acute chronic cases, splenomegaly. In the case of a non-immune individual an infection with *P. falciparum* is often fatal, caused by severe complications, of which cerebral are most feared. Many non-human species of malaria parasites have been described which infect a wide range of vertebrates, ranging from reptiles, birds and rodents to non-human primates.

The life cycle of all species is essentially the same as far as the blood stages are concerned (*10*). After completion of the tissue phase, the parasites, called merozoites, invade the RBCs. Some species show a distinct preference for reticulocytes (*10*). During penetration the parasite works itself into the cell, taking a part of the erythrocyte membrane with it, which surrounds the intracellular parasite as a parasitophorous vacuole. Hence the parasite DNA is surrounded by four membranes (Fig. 1). Once inside the erythrocyte the parasite rapidly consumes the hemoglobin and, within 48–72 hours, the parasite divides via schizogony into 12–32 daughter merozoites. These merozoites are liberated by rupture of the erythrocyte membrane and are able to invade new RBCs. As a consequence of sequestration of the dividing forms (schizonts) of *P. falciparum* in capillary vessels, only the non-dividing forms of this parasite can be found in the peripheral blood. All the other species undergo the complete erythrocytic cycle in circulating RBCs, and therefore all the different developmental stages can be found. A variable proportion of the parasites in the bloodstream will differentiate into sexual cells, the gametocytes. These cells are able to infect the vector, a mosquito of the genus *Anopheles*, and are responsible for transmission of malaria.

The shape, membrane properties and metabolism of the infected RBC is different from that of the uninfected RBC (*7*). Different stages of the life cycle have specific antigens, which can also be exposed on the surface of the infected

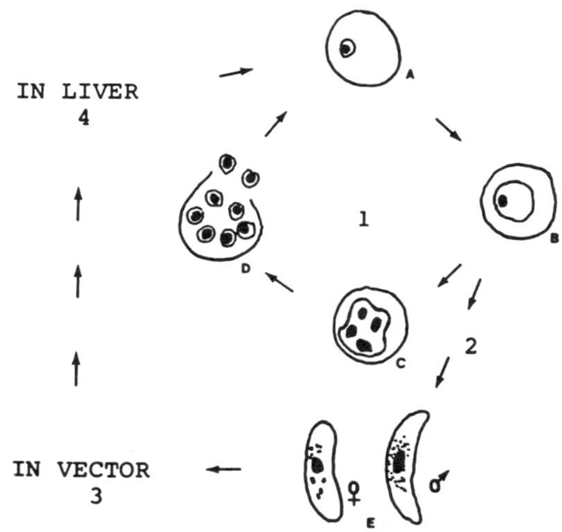

Fig. 1. Diagrammatic representation of the blood stages of the human malaria parasite *Plasmodium falciparum*. (1) Asexual multiplication in erythrocytes: (A) ringform (haploid); (B) trophozoite (haploid); (C) dividing schizont (polyploid); (D) mature schizont containing haploid merozoites. (2) Sexual development in erythrocytes: (E) male and female gametocytes, the precursor cells of the gametes. Gamete formation and fertilization occurs in the midgut of the vector mosquito (3). After meiotic and mitotic multiplication in the vector the parasites are injected into the blood stream by a mosquito bite. Parasites enter parenchymal cells of the liver (4) in which they multiply again by mitosis and within 1 week thousands of parasites per schizont are released into the blood stream.

cells. The stages of the life cycle which occur in the blood are haploid and their genome encompasses a complexity of about 2.5×10^7 base pairs. The DNA of malaria parasites is exceptionally rich in adenine (A) and thymine (T). Between 70% and 82% of the genome is comprised of A/T (*21*).

B. *Trypanosoma*

Many species of trypanosomes are known as parasites of vertebrates and invertebrates. In man, trypanosomes cause different diseases in Africa and in South America. The African trypanosomes, *T. brucei gambiense* and *T. brucei rhodesiense*, cause sleeping sickness, characterized by fever, anemia, delirium and finally coma (*43*).

Two distinct cell types can be seen, free in the bloodstream, the long slender forms which are correlated with a rising parasitemia, and the short stumpy forms which coincide with a decreasing parasitemia. The latter forms are believed to be responsible for infection of the vector, tsetse flies (*Glossina* spp.). Blood forms evade elimination by the immune system of the host by frequently changing their antigenic surface coat.

The South American parasite, *T. cruzi*, is concentrated intracellularly mainly in heart muscle and is scarcely present in the blood. The complications, known as Chagas' disease, can include anemia, cardiomyopathy, and meningoencephalitis. The blood stages are infective to bugs of the family *Triatomidae*, which act as the vector for transmission of the disease.

All the human trypanosomes appear in the blood as extracellular, motile cells. They are characterized by two distinct DNA-containing organelles, the nucleus and the kinetoplast (Fig. 2). African trypanosomes multiply in the blood by binary fission, whilst the multiplication of *T. cruzi* is largely restricted to the amastigote stage in systemic cells.

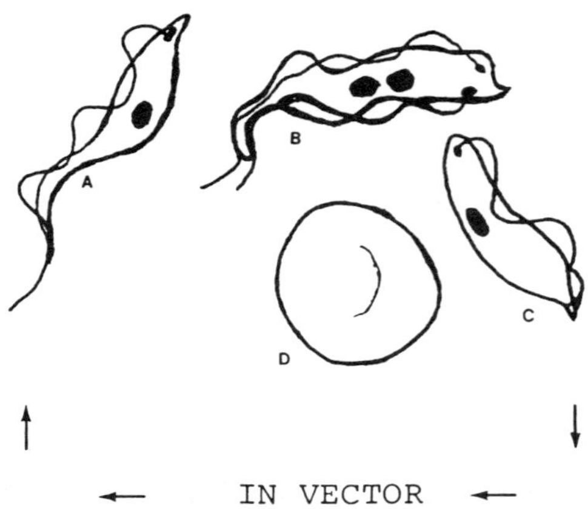

Fig. 2. Diagrammatic representation of the blood stages of African trypanosomes. (A) Long slender form; (B) dividing form; (C) short stumpy form; (D) erythrocyte. These parasites are free-living, motile stages and the erythrocyte is included for size reference. The short stumpy forms are believed to be responsible for the infection of the vector flies. After completion of the cycle in the fly, the parasites are injected in the new vertebrate host.

Trypanosomes are essentially diploid organisms with a genomic complexity of between 5 and 10×10^7 base pairs and guanine (G) and cytosine (C) content of about 45–55%.

III. PREPARATIVE ASPECTS OF PARASITE DETECTION BY FLOW CYTOMETRY

A. Fixation

General fixation procedures for eukaryotic cells generally apply also to parasites. Fixation is not only necessary to allow prolonged storage of cell suspensions, but also to make the targer molecules (DNA, RNA or proteins) accessible to a variety of dyes, which do not pass through the membranes of intact living cells. In the case of trypanosomes two membranes, the plasmalemma and the membrane of the nucleus, have to be permeabilized. In the case of *Plasmodium*, which are inside the erythrocyte, four membranes are hampering the penetration of the dye into the nucleus of the parasite.

Red blood cells infected with *Plasmodium*, *Babesia* and *Theileria* have been fixed with glutaraldehyde, (para)formaldehyde (see references in Table I) or dimethylsuberimidate (DMS) (personal observations). Both glutaraldehyde and (para)formaldehyde have the disadvantage that they leave the membranes of red blood cells relatively intact, which may hamper the penetration of larger fluorochromes, MoAbs or DNA probes. In addition, glutaraldehyde and (para)formaldehyde can have a significant quenching effect on the emission of fluorescence from certain DNA-bound fluorochromes (*5*).

In addition to the above-mentioned fixatives,

Table I

FLUORESCENT DYES USED FOR THE DETECTION AND ANALYSIS OF PARASITIC INFECTION (IMAGE = USED IN MICROFLUOROMETRIC STUDIES)

Dye	Specific for	Measurement	Parasite	References
Acridine orange	DNA/ RNA	Image	*Plasmodium*	(*40*)
		Flow	*Plasmodium*	(*11,12,19,46*)
		Flow	*Trypanosoma*	(*19,33*)
7-Amino- actinomycin-D	DNA (G/C)	Flow	*Trypanosoma*	R. F. Kooy, unpublished
Chromomycin-A3	DNA (G/C)	Flow	*Trypanosoma*	(*39*), R. F. Kooy, unpublished
		Image	*Trypanosoma*	(*34,35*)
DAPI	DNA (A/T)	Image	*Plasmodium*	(*18*)
		Image	*Trypanosoma*	(*34*)
		Flow	*Trypanosoma*	(*9*)
		Flow	*Plasmodium*	Personal observations
Ethidium bromide	DNA/RNA	Image	*Plasmodium*	(*4,22*)
		Flow	*Trypanosoma*	(*9,19*)
		Flow	*Plasmodium*	(*19*)
Feulgen/ Pararosaniline	DNA	Image	*Plasmodium*	(*22,23,25*)
		Image	*Trypanosoma*	(*2,28,47*)
Hoechst 33258	DNA (A/T)	Flow	*Plasmodium*	(*1,3,13,15,24,29,36*)
		Flow	*Babesia*	(*14*)
		Flow	*Theileria*	Personal observations
		Flow	*Trypanosoma*	(*28*)
Hoechst 33342	DNA (A/T)	Flow	*Plasmodium*	(*8,24*)
		Flow	*Trypanosoma*	(*9*)
Mithramycin	DNA (G/C)	Image	*Plasmodium*	(*37*)
		Flow	*Trypanosoma*	(*6*)
		Image	*Trypanosoma*	(*45*)
Propidium iodode	DNA/ RNA	Flow	*Plasmodium*	(*38*)
		Flow	*C. fasciculata*	(*16*)
Thiazole orange	DNA/ RNA	Flow	*Plasmodium*	(*31*)

trypanosomes have been fixed with ethanol or methanol (see references in Table I). These fixatives have the disadvantage of inducing cell aggregation. Clumping of cells can be diminished by adding glycerol or by vigorous stirring of the cell suspension while slowly increasing the ethanol/methanol concentration. Dvorak *et al.* (*6*) successfully used the detergent Triton X-100 at low pH as a fixative prior to mithramycin staining of trypanosomes. Triton X-100 does not lyse vertebrate cells if used in combination with a low pH buffer.

B. Staining of Parasites

a. *DNA and RNA Dyes*

Parasite detection has been carried out almost exclusively on the basis of fluorescent dyes, which are specific for nucleic acids. Table I gives a summary of the various dyes which have been employed for microfluorometry and flow cytometry of parasites. Many DNA-specific dyes show a preferential binding to either A/T or C/G combinations. The relative scarcity of guanine and cytosine in the genome of *Plasmodium* (18–30%) makes the dyes with a specificity for these bases less suitable, in contrast to trypanosomes, which have a much higher G/C content (45–55%).

The fluorescence of uninfected blood cells sometimes can play a role in the difficulties experienced with flow cytometric analysis of infected blood. The amount of DNA in white blood cells is about 50–200 times as high as in parasites. Leukocytes are therefore not the major problem in using DNA-specific dyes. Reticulocytes show fluorescence after staining with RNA-specific dyes, which can be in the same range as the fluorescence intensity of parasites (Fig. 3). Moreover, uninfected blood cells may show an aspecific fluorescence, which can hamper the detection of parasites.

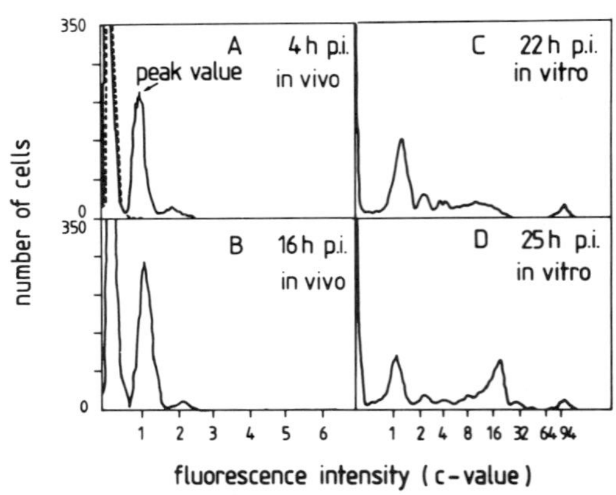

Fig. 3. The rate of DNA synthesis in *Plasmodium*. Frequency distributions of fluorescence values of erythrocytes infected with *Plasmodium berghei* at 4, 16, 22, and 25 hours postinfection (hpi) of the erythrocyte as measured by a FACS analyzer after Hoechst 33258 staining. At 4 hpi all parasites are haploid and do not synthesize DNA. The different peaks in (A) represent uninfected erythrocytes (see dotted line) and single and double infected cells. During development from 4 hpi (A) to 16 hpi (B), the parasites grow without synthesizing DNA. After that time, parasites start DNA synthesis (C). At 25 hpi (D) most parasites are schizonts containing 8–24 haploid daughter nuclei per erythrocyte. The haploid (1C) value is defined as the difference between the peak fluorescence value of single infected erythrocytes at 4 hpi and the mean background fluorescence. Cells with values higher than 64C are leukocytes, with a mean peak value of 115C. Panels A and B: linear gain setting, panels C and D: logarithmic gain setting.

When working with samples in which the red blood cells have been selectively lysed, platelets and Howell–Jolly bodies may interfere with parasite detection, especially when extremely low numbers of parasites are present (Fig. 4(A)).

Plasmodium

The A/T-specific DNA dyes, Hoechst 33258 and Hoechst 33342, give a relatively strong fluorescence with *Plasmodium* DNA after fixation of the parasites with glutaraldehyde or (para)formaldehyde or after vital staining of the parasites. The vast majority of the reported flow cytometric studies on malaria parasites have been carried out with these dyes.

Infected RBCs can be distinguished clearly from uninfected cells on the basis of Hoechst DNA fluorescence intensity (*1,3,13–15,24,36,37*). The related dye DAPI gives a comparable fluorescence intensity (FI) after fixation of malaria parasites with the above-mentioned fixatives (personal observations), although no reports on the use of DAPI for this purpose have been published since the initial report of Hyman and MacInnis in 1979 (*18*), who used the dye for microfluorometry. Both Hoechst 33258 and 33342 allow vital staining of malaria parasites in culture (*3,13,24*). After vital staining the parasites are infective to their hosts (*3,13*). However, only Hoechst 33258-stained parasites were able to develop normally *in vitro* after removal of the dye by washing (*24*).

After lysis of the RBCs, free parasites can be stained and counted by flow cytometry, allowing detection of very low parasitemias (see below).

The FI of Hoechst-stained parasites correlates well with the DNA content of the different developmental stages, which allows accurate monitoring of DNA synthesis during the intraerythrocytic cycle (*24*) (Fig. 3).

The FI of Hoechst-stained haploid stages varies less than 10% between different strains or isolates of several species (*26*, unpublished

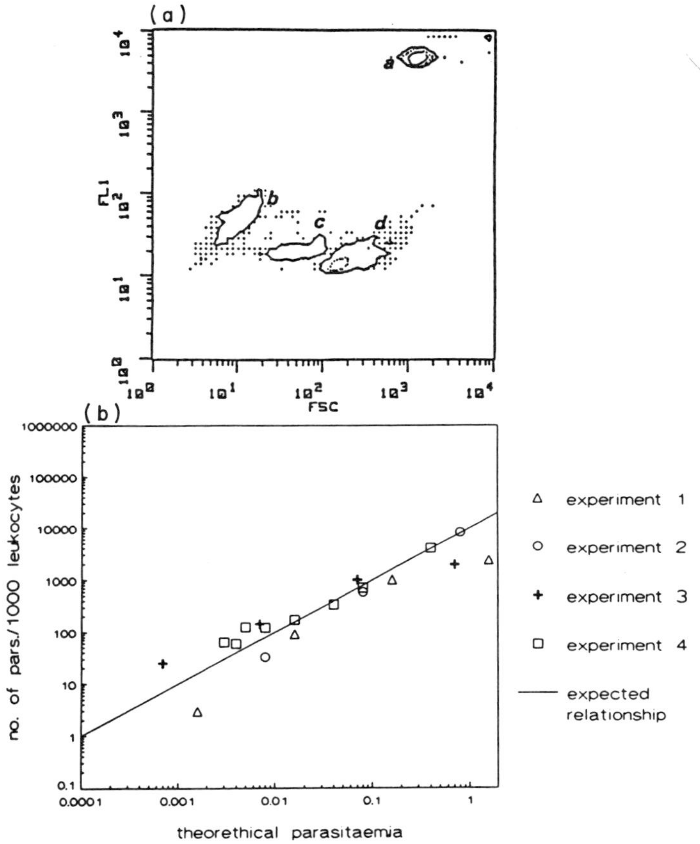

Fig. 4. The establishment of low numbers of *Plasmodium* parasites in blood samples. (A) Contour plot of different populations in a Hoechst-stained blood sample with a percentage of 0.02 of *Plasmodium berghei*-infected erythrocytes. The erythrocytes were selectively lysed before measurement. The X axis represents the forward angle light scatter and the *Y* axis the Hoechst DNA fluorescence. Four distinct populations can be distinguished. a = white blood cells, b = Howell–Jolly bodies and debris, c = haploid parasites, d = platelets. (B) The correlation between the calculated percentage of erythrocytes infected with *P. berghei* (theoretical parasitemia) in a blood sample and the parasitemia as measured by the FACS after Hoechst 33258 staining of the blood sample. In this experiment blood with a known parasitemia was serially diluted with uninfected blood. The erythrocytes were selectively lysed before measurement and the parasitemia is expressed per 1000 leukocytes.

observations). Between different species larger differences in FI exist. Compared to *P. falciparum*, *P. vivax* has a consistently lower FI, approximately 80%, whilst *P. malariae* seems to have a significantly higher FI of about 150% (*24* unpublished observations). These differences reflect differences in the total genome size and/or differences in G/C content of the DNA.

No reports have been published on the use of the G/C-specific DNA dyes chromomycin-A3, mithramycin and 7-amino-actinomycin-D. In our experiments, the FI of malaria parasites with these dyes was not sufficiently high to distinguish the parasitized RBC clearly from uninfected ones. This might be due partly to the fixation with glutaraldehyde or (para)formaldehyde, which may have a quenching effect on the FI (*5*). In addition, the low C/G content of *Plasmodium* DNA severely limits the number of specific binding sites. Mithramycin has been used for microfluorometry in combination with ethidium bromide, which stains both DNA and RNA, in order to enhance the fluorescent emission of DNA-bound ethidium bromide by energy transfer from excited mithramycin (*4*).

Ethidium bromide and propidium iodide have not been widely used for malaria research. A complicating factor with these dyes is the FI of RNA in reticulocytes, which can be in the same range as infected cells (personal observations). Saul *et al.* (*38*) however successfully used propidium iodide to detect and count *P. falciparum*-infected cells by flow cytometry. The dye, thiazole orange (*31*), originally introduced for the staining of RNA in reticulocytes, also allowed sensitive detection of *P. falciparum*-infected RBCs. The use of these dyes for counting infected cells is probably restricted to blood samples which contain very low numbers of RNA-containing reticulocytes. We were not able to accurately count the number of *P. berghei*-infected cells after staining with ethidium bromide and thiazole orange, due to reticulocyte fluorescence.

An advantage of ethidium bromide, propidium iodide and thiazole orange is that these dyes do not require activation by ultraviolet light, but can be excited by visible light at 488 nm, the optimum wavelength for argon ion lasers.

Acridine orange, which gives a green fluorescence with DNA and a red fluorescence with RNA, was used for flow cytometry on the rodent malaria parasite *P. yoelii* as early as 1977 (*19*). In that report, Jackson *et al.* described that they were not able to separate infected RBCs from uninfected RBCs. Other groups, however, analyzing *P. falciparum* parasites in mature RBCs, found a good correlation between microscopic counts of infected cells and flow cytometric counts using acridine orange (*11,12,46*). Relatively good staining specificity for nucleic acids occurred when cells were equilibrated with acridine orange in the presence of cations to reduce non-specific electrostatic interactions (46). Hare (*11*) used differences in red and green fluorescence of acridine orange to analyze the growth cycle of *P. falciparum* by flow cytometry.

Trypanosoma

In contrast to *Plasmodium*, the G/C-specific DNA dyes chromomycin-A3 and mithramycin have been most satisfactorily employed for the analysis of trypanosomes by flow cytometry (*6,39*, R. F. Kooy, unpublished observations). The higher DNA content and G/C content of trypanosomes probably explains the better results with these dyes in comparison with malaria parasites. Moreover, trypanosomes are extracellular parasites and therefore can easily be fixed with methanol or ethanol after separation of the parasites from blood cells. These fixatives may be more optimal with respect to quenching effects and penetration of membranes by the fluorochromes.

Flow cytometry has been used to establish the relative DNA content of different life cycle stages, including those occurring in the insect host, and to study DNA synthesis during the life cycle (*39*). Several African *Trypanosoma* species could be distinguished on the basis of chromomycin-A3 and mithramycin fluorescence intensity (*35*, R. F. Kooy, unpublished observations). Clones and strains of *T. cruzi* showed differences in FI of up to 40%, due to differences of the total DNA content (*6*).

7-Amino-actinomycin-D and Hoechst 33258 have also been successfully used to study DNA content and synthesis in trypanosomes by flow cytometry (*28*, R. F. Kooy, personal communication). The former can be excitated by visible light. To establish the DNA content of different stages of *Crithidia fasciculata*, also a kinetoplastid protozoan, parasites were stained with propidium iodide after RNAse treatment of the parasites (*16*). Acridine orange has been used to count trypanosomes accurately by flow cytometry (*34*). In one study DAPI, Hoechst 33258, ethidium bromide and acriflavine were used to study the uptake and the rate of accumulation of these dyes into multidrug-resistant clones of African trypanosomes (*9*).

Babesia and Theileria

Both *Babesia*- and *Theileria*-infected blood cells can be detected and counted by flow cytometry after vital staining with Hoechst 33258 (*14*) or after glutaraldehyde fixation and staining with this dye (personal observations). The FI with this dye is lower than that of *Plasmodium* (personal observations) but sufficient to separate parasitized from uninfected RBCs (*14*). These parasites probably have a smaller genome size than that of *Plasmodium*. No further detailed studies on these parasites are available.

b. Other Dyes

The literature on other staining methods is very fragmentary. Some dyes can be applied to study living cells, because they pass intact plasma membranes. Cationic fluorescent rhodamine dyes (rhodamine 123 and rhodamine 6G) accumulate in living malaria parasites (*27*). At an appropriate concentration the accumulation of these dyes reflects the membrane potential of the parasite. Thus one can apply rhodamine staining to test the viability of parasites. Rhodamine 6G has a fully reversible inhibitory effect on the development of *P. falciparum* in culture (*27*). Using these dyes separation of dead and living parasites by flow cytometry may be feasible.

DiOC1(3) (3,3′-dimethyloxacarbocyanine), also a dye responsive to cell membrane potential, has been used for vital staining of *P. vinckei* (*20*). Mature RBCs, immature RBCs and parasitized RBCs could be recognized and counted in the flow cytometer.

Detecting and counting parasites by flow cytometry after immunofluorescent labeling of parasites would open new perspectives for the analysis of parasitic infections (see below). There is, however, only one report of successful labeling of malaria-infected cells (*17*). Erythrocytes of mice infected with *P. yoelii* were analyzed for surface-bound immunoglobulins. The number of positive cells increased with rising parasitemia, but was always significantly higher than the number of infected RBCs.

IV. APPLICATIONS OF FLOW CYTOMETRY IN EXPERIMENTAL PARASITOLOGY

A. *Plasmodium*

Plasmodium-infected RBCs can easily be distinguished from uninfected RBCs on the basis of FI of DNA-specific dyes, both after glutaraldehyde fixation or vital staining of the parasites. Many studies showed a good correlation between manual counts of the percentage of infected cells in Giemsa-stained blood smears and flow cytometric counts of the percentage of infected cells after staining with Hoechst dyes (*1,3,14,15,24,36*).

The major advantage of flow cytometric analysis of blood cell suspensions is the speed and the high number of cells which can be measured. Flow cytometric counting is therefore more accurate from a statistical point of view and, when large numbers of samples have to be analyzed, less time consuming. For comparison the examination of 20 000 RBCs in a standard thin blood smear takes a skilled technician about 20 min. The collection of fluorescence data on 20 000 RBCs with a flow cytometer takes several seconds.

Assessment of the percentage of infected cells (the parasitemia) is less reproducible by flow cytometry when the parasitemia is less than 0.1%, due to background fluorescence of particles other than infected cells (personal observation). To detect and count very low numbers of parasites we therefore lyse the RBC in FACS lysing solution (Becton Dickinson, Mountain View, CA) before staining with Hoechst dyes, in order to get a higher concentration of parasites in the suspension to be measured. With this method we are able to detect parasitemias of 0.01–0.005% using blood samples of 50 μl from *P. berghei*-infected mice (Fig. 4(B)). In these experiments the number of Hoechst-stained leucocytes in the samples is used as a reference to calculate the percentage of RBCs which are infected. Some interference was however observed due to autofluorescence of platelets and aspecific binding of Hoechst to platelets causing overlap with the optical characteristics of malaria parasites. Therefore we subsequently improved our detection method by applying a combination of both Hoechst 33258 and propidium iodide. The latter dye does not fluoresce when excited with UV light, but it is excited by direct energy transfer from Hoechst. This energy transfer occurs only when both donor and acceptor molecules are within a very short distance of each other (25–35 Å). Using this dye combination, aspecific fluorescence can be eliminated by selecting the blue (Hoechst) fluorescing cells which are simultaneously emitting red fluorescence (propidium iodide). The detection limit with this technique is approximately 0.0004% parasitemia in blood of *P. falciparum*-infected patients (*42*). With this method the detection of malaria parasites is highly quantitative and its sensitivity is comparable to clinical diagnosis and even superior to routine microscopy in the field or other existing methods, including those using radioactively labeled DNA probes. Its simple sample collection and robust procedure make it a good alternative method especially suitable for follow-up studies and large-scale epidemiological surveys.

The FI of different developmental stages after Hoechst staining corresponds with the relative DNA content (*3,15,24*). Flow cytometric analysis of parasites during synchronized development in culture can therefore be used to give information about the time and rate of DNA synthesis. Different workers have used this method to quantitate the effect of drugs on the development (DNA synthesis) of malaria parasites (*24,37*). An example of assessment of the rate of DNA synthesis in *P. berghei*, a rodent malaria parasite, is given in Fig. 3.

Flow cytometry is thus a relatively simple and rapid method to assess the rate of DNA synthesis in *Plasmodium* in comparison with other methods, such as labeling of DNA with radioactive precursors or microfluorometry (*2,23,25*). In addition to the Hoechst dyes, acridine orange and thiazole orange can be used to follow both DNA and RNA synthesis of parasites in synchronized cultures (*12,31*).

Sorting of parasites or different developmental stages, based on differences in DNA content, makes it feasible to obtain parasites relatively free from other stages or uninfected blood cells (*3,15,24,36*). Moreover, vital staining and sorting would allow the selection of living cells, for example parasites with altered growth properties.

B. *TRYPANOSOMA*

To establish the DNA content of different life cycle stages and to quantitate DNA synthesis in *Trypanosoma* species, different DNA-specific dyes have been used (see above). As already stated, flow cytometry is a relatively simple and rapid method to assess the time and rate of DNA synthesis in parasites. In African trypanosomes, different species can be distinguished on the basis of fluorescence intensity after staining with G/C-specific DNA dyes (*34,35*). In contrast to *Plasmodium*, large differences in DNA content exist between clones and strains of *T. cruzi*. Flow cytometry can therefore be used to study dynamics of *T. cruzi* populations in which different clones with different DNA contents are present (*6*).

In trypanosomes flow cytometry has also been used to compare drug accumulation by drug-sensitive and multidrug-resistant clones (*9*). In this work the accumulation of several DNA- and/or RNA-specific dyes was measured. An advantage of this method is that data for single cells, not averages for populations, are obtained. Cells that take up less drug can be identified, quantitated and isolated by flow sorting.

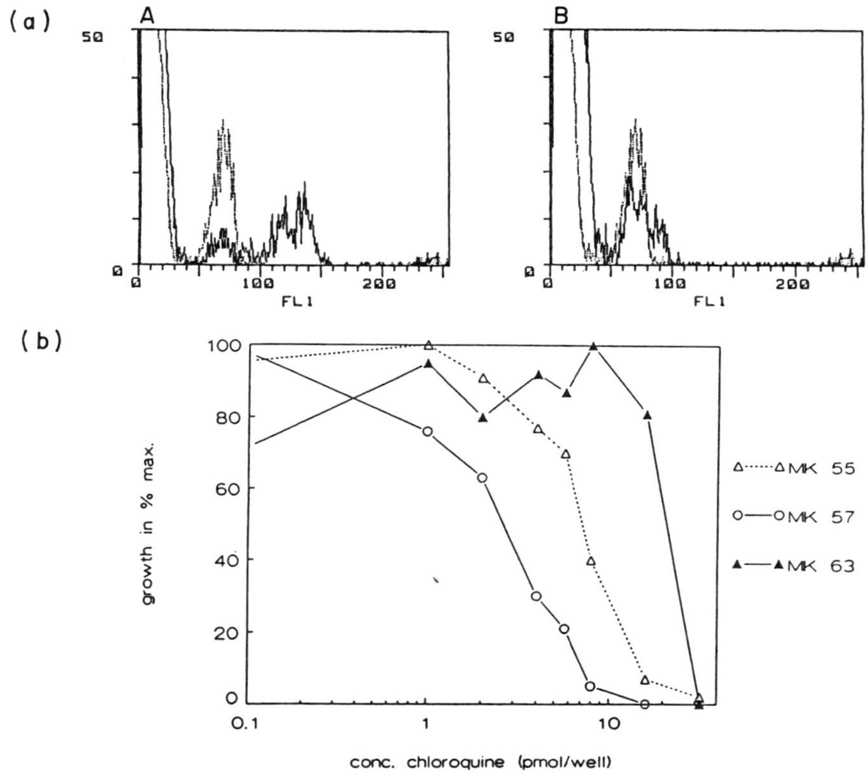

Fig. 5. Inhibition of the growth of *Plasmodium* by anti-malarial drugs. (a) Fluorescence histograms of erythrocytes infected with *P. falciparum* and stained with Hoechst 33258. The erythrocytes were obtained from two different wells (A, B) of a microtest culture plate. In this plate erythrocytes containing haploid stages of the parasite were incubated for 26 h under standard culture conditions in the presence of different concentrations of chloroquine. Panel A represents the analysis of the control well (containing no drug) at 0 (----) and at 26 (——) h of incubation at 37°C. Panel B represents the analysis of the well with the highest concentration of the drug at 0 (----) and 26 (——) h. This figure clearly shows the inhibitory effect of chloroquine on DNA synthesis in the parasite and the possibility to quantitate the effect of this drug on parasite growth. (b) The inhibitory effect of chloroquine on the growth of three different isolates of *P. falciparum* from Thailand. The growth is defined as the rate of DNA synthesis of the parasites in the presence of different concentrations of chloroquine as measured by the FACS (see (a)). The difference in susceptibility to chloroquine between the isolates is readily demonstrated.

C. *THEILERIA* AND *BABESIA*

Both *Theileria*- and *Babesia*-infected red blood cells can be distinguished from uninfected RBCs on the basis of fluorescence intensity after Hoechst staining (*14*, personal observations). Therefore it should be possible to use flow cytometry in experimental studies as described for *Plasmodium*.

V. APPLICATIONS OF FLOW CYTOMETRY IN CLINICAL STUDIES

Flow cytometry in parasitology has until now been limited to experimental studies. Based on the results of these studies, however, clinical or epidemiological applications of flow cytometry are possible, especially for malaria research.

When relatively large numbers of blood samples have to be screened for the presence of malaria parasites from populations in endemic areas, flow cytometry can be a more rapid and accurate method than microscopical examining and counting of parasites in Giemsa-stained blood smears (*1*,*42*). The fact that washing and centrifugation steps before fixation and staining with Hoechst dyes are not necessary (*1*, personal observations) makes flow cytometry attractive for large-scale assays. Fixation of the samples also permits assays to be stored and transported from the field and allows reference standards to be incorporated.

Because of the rapid spread of drug resistance among *P. falciparum* infections, the WHO has introduced an *in vitro* microtest for a better understanding of the epidemiology of drug resistance and for global monitoring of drug susceptibility. In this test *Plasmodium*-infected blood cells, suspended in culture medium, are incubated in a standard 96 well microtiter plate predosed with serial dilutions of the test drug. After 26–30 h incubation, parasite development is assessed in Giemsa-stained blood smears by counting the number of schizonts, i.e. the stages which synthesize DNA, followed by nuclear division. Flow cytometric reading of cells from the culture wells after Hoechst staining allows a rapid, accurate and detailed analysis of parasite growth as a function of drug concentration (*8*,*41*) (Fig. 5(a) and (b)). We have recently developed a method for the automatic reading of the microtests using the AutoMATE system (Becton Dickinson, Mountain View, CA) (*41*). In combination with standardized software to analyze and store the data, this method should greatly enhance the possibilities for worldwide, standardized monitoring of the spreading of drug resistance.

VI. FUTURE PROSPECTS

Flow cytometry is a useful tool in experimental analysis of *Plasmodium* and *Trypanosoma* infections, both for detection and counting of parasites and for analysis of DNA content and DNA synthesis. Until now most dyes used are DNA/RNA-specific dyes, which need ultraviolet or violet excitation. Further use of other fluorescent dyes, preferably those with excitation wavelengths other than UV, will facilitate the use of flow cytometry and enable the use of lower cost flow cytometers.

The contribution of flow cytometry to experimental and clinical studies will increase when immunocytochemical staining and detection of parasites is reproducible. Species-specific MoAbs would enable species determination of parasites with flow cytometry. Stage-specific MoAbs would be of use for both experimental and epidemiological purposes. For example the accurate and rapid counting of the percentage of sexual cells in *Plasmodium*-infected blood would be of great value for epidemiological transmission studies. In addition, combination of immunofluorescence and DNA-based fluorescence could enhance the specificity of parasite staining and hence the sensitivity of parasite detection by two-dimensional fluorescence analysis. For immunocytochemical detection of *Plasmodium*, *Babesia* and *Theileria* fixatives are needed which do not lyse the red blood cells but which allow penetration of the MoAbs and recognition of the antigens.

Automated measurement by flow cytometry and data analysis of the standard WHO microassay to test the drug susceptibility of *P. falciparum* would be of importance for worldwide drug testing, not only for monitoring of drug resistance, but also in the development of novel drugs. Given the accuracy and speed of the flow cytometric method for automated reading of drug tests (*41*) this might be one of the major future applications of flow cytometry. Detection of low numbers of malaria parasites is possible using flow cytometric analysis of blood samples. Other methods, alternative to classic microscopical reading of blood smears, involve the application of specific DNA or RNA probes. After a period of about 5 years of intensive research no DNA probes have been found which are consistently detecting parasite numbers lower than 1000 with a sufficient degree of specificity (*30*). In comparison, FACS-based diagnosis can reproducibly detect 0.0004% parasitemia in 50 μl of lysed blood (*42*). Apart from the non-radioactive procedure, the major advantage of flow cytometric analysis of blood samples is the speed and accuracy with which parasite numbers can be detected. Flow cytometry may prove very useful in the follow-up of drug treatment and the study of the impact of vaccination trials. For detection of extremely low parasitemias flow cytometry is not applicable. New and improved methods based on DNA and RNA probing are under way. Recently Waters and McCutchan described a new panel of oligonucleotide probes based on the small subunit ribosomal RNA genes from each of the four human malaria parasites. Hybridization to spotted parasites with the ribosomal RNA as a target can reveal the presence of fewer than 10 parasites after overnight exposure (*44*).

The future applications of flow cytometry in parasitological research can be manifold, provided that bench-top flow cytometers can be used. In the light of the impact of parasitic blood infections, in particular malaria, on morbidity and mortality in tropical countries, the further development and adaptation of flow cytometric techniques especially designed for parasitology deserves continuous attention.

ACKNOWLEDGEMENTS

The authors wish to thank Philip van Vianen, Frank Kooy, Hans Tanke and Hugo van der Kaay for useful comments, critical reading of the manuscript and communication of unpublished data. The development of flow cytometry in our group was partly sponsored by The Netherland's Ministry for Development Co-operation.

REFERENCES

1. Bianco, A. E., Battye, F. L., and Brown, G. V. (1986). *Plasmodium falciparum*: rapid quantification of parasitemia in fixed malaria cultures by flow cytometry. *Exp. Parasitol.* **62**, 275–282.

2. Borst, P., van der Ploeg, M., van Hoek, J. F. M., Tas, J., and James, J. (1982). On the DNA content and ploidy of trypanosomes. *Mol. Biochem. Parasitol.* **6**, 13–23.

3. Brown, G. V., Battye, F. L., and Howard, R. J. (1980). Separation of stages of *Plasmodium falciparum*-infected cells by means of a fluorescence-activated cell sorter. *Am. J. Trop. Med. Hyg.* **29**, 1147–1149.

4. Casaglia, O., Dore, E., Frontali, C., Ramoni, C., and Zenobi, P. (1985). Cytofluorometric analysis of DNA content in different life stages of *Plasmodium yoelli. Ann. Inst. Super. Sanita* **21**, 401–406.

5. Crissmann, H. A., Stevenson, A. P., Kissane, R. J., and Tobey, R. A. (1979). Techniques for quantitative staining of cellular DNA for flow cytometric analysis. *In* "Flow Cytometry and Sorting" (M. R. Melamed *et al.*, eds), pp. 243–262. John Wiley & Sons, New York.

6. Dvorak, J. A., Hall, Crane, M. S. J., Engle, J. C., McDaniel, J. P., and Uriegas, R. (1982). *Trypanosoma cruzi*: Flow cytometric analysis. I. Analysis of total DNA/organism by means of mithramycin-induced fluorescence. *J. Protozool.* **29**, 430–437.

7. Evered, D., and Whelan, J., eds (1983). "Malaria and the Red Cell." Pitman Books, London, p. 257.

8. Franklin, R. M., Brun, R., and Grieder, A. (1986). Microscopic and flow cytophotometric analysis of parasitemia in cultures of *Plasmodium falciparum* vitally stained with Hoechst 33342 — application to studies of antimalarial agents. *Z. Parasitenkd.* **72**, 201–212.

9. Frommel, T. O. (1987). Flow cytofluorimetric analysis of drug accumulation by multidrugresistant *Trypanosoma brucei brucei* and *T. b. rhodesiense. Mol. Biochem. Parasitol.* **26**, 183–192.

10. Garnham, P. C. C. (1966). "Malaria Parasites and Other Haemosporidia." Blackwell, Oxford.

11. Hare, J. D. (1986). Two-color flow-cytometric analysis of the growth cycle of *Plasmodium falciparum* in vitro: identification of cell cycle compartments. *J. Histochem. Cytochem.* **34**, 1651–1658.

12. Hare, J. D., and Bahler, D. W. (1986). Analysis of *Plasmodium falciparum* growth in culture using acridine orange and flow cytometry. *J. Histochem. Cytochem.* **34**, 215–218.

13. Howard, R. J., and Battye, F. L. (1979). *Plasmodium berghei*-infected red cells sorted according to DNA content. *Parasitology* **78**, 263–270.

14. Howard, R. J., and Rodwell, B. J. (1979). *Babesia rodhaini*, *Babesia bovis*, and *Babesia bigemina*: analysis and sorting of red cells from infected mouse or calf blood by flow fluorimetry using 33258 Hoechst. *Exp. Parasitol.* **48**, 421–431.

15. Howard, R. J., Battye, F. L., and Mitchell, G. F. (1979). Plasmodium-infected blood cells analyzed and sorted by flow fluorometry with the deoxyribonucleic acid binding dye 33258 Hoechst. *J. Histochem. Cytochem.* **27**, 803–813.

16. Hughes, D. E., Schneider, C. A., and Simpson, L. (1982). Isolation and characterization of drug resistant mutants of *Crithidia fasciculata*. *J. Parasitol.* **68**, 642–649.

17. Hunter, K. W., Finkelman, F. D., Strickland, G. T., Sayles, P. C., and Scher, I. (1980). Murine malaria: analysis of erythrocyte surfacebound immunoglobulin by flow microfluorometry. *J. Immunol.* **125**, 1.

18. Hyman, B. C., and Macinnis, A. J. (1979). Rapid detection of malaria and other bloodstream parasites by fluorescence microscopy with 4′6-diamidino-2-phenylindole (DAP1). *J. Parasitol.* **65**, 421–425.

19. Jackson, P. R., Winkler, D. G., Kimzey, S. L., and Fisher, F. M. (1977). Cytofluorograf detection of *Plasmodium yoelii, Trypanosoma gambiense*, and *Trypanosoma equiperdum* by laser excited fluorescence of stained rodent blood. *J. Parasitol.* **63**, 593–598.

20. Jacobberger, J. W., Horan, P. K., and Hare, J. D. (1983). Analysis of malaria parasite-infected blood by flow cytometry. *Cytometry* **4**, 228–237.

21. Janse, C. J., and Mons, B. (1987). DNA synthesis and genome structure of *Plasmodium*: a review. *Acta Leidens* **56**, 1–13.

22. Janse, C. J., van der Klooster, P. F. J., van der Kaay, H. J., van der Ploeg, M., and Overdulve J. P. (1986). Rapid repeated DNA replication during microgametogenesis and DNA synthesis in young zygotes of *Plasmodium berghei*. *Trans. Roy. Soc. Trop. Med. Hyg.* **80**, 154–157.

23. Janse, C. J., van der Klooster, P. F. J., van der Kaay, H. J., van der Ploeg, M., and Overdulve, J. P. (1986). DNA synthesis in *Plasmodium berghei* during asexual and sexual development. *Mol. Biol. Parasitol.* **20**, 173–182.

24. Janse, C. J., van Vianen, P. H., Tanke, H. J., Mons, B., Ponnudurai, T., and Overdulve, J. P. (1987). *Plasmodium* species: flow cytometry and microfluorometry assessments of DNA content and synthesis. *Exp. Parasitol.* **64**, 88–94.

25. Janse, C. J., Ponnudurai, T., Lensen, A. H. W., Meuwissen, J. H. E. Th., Ramesar, J., van der Ploeg, M., and Overdulve, J. P. (1988). DNA synthesis in gametocytes of *Plasmodium falciparum*. *Parasitology* **96**, 1–7.

26. Janse, C. J., Boorsma, E. G., Ramesar, J., van Vianen, Ph., van der Meer, R., Zenobi, P., Casaglia, O., Mons, B., and van der Berg, F. M. (1989). *Plasmodium berghei*: Gametocyte production, DNA content, and chromosome-size polymorphisms during asexual multiplication *in vivo*. *Exp. Parasitol.* **68**, 274–282.

27. Kato, M., Izumo, A., and Tanabe, K. (1987). Vital staining of *Plasmodium falciparum* with cationic fluorescent rhodamine dyes. *J. Parasitol.* **73**, 5.

28. Kooy, R. F., Hirumi, H., Moloo, S. K., Nantulya, V., Dukes, P., van der Linden, P. M., Duijndam, W. A. L., Janse, C. J., and Overdulve, J. P. (1989). Direct evidence of diploidy in metacyclic forms of African trypanosomes. *Proc. Natl. Acad. Sci. USA* **86**, 5469–5472.

29. Lamont, G., Saul, A., and Kidson, C. (1981). *Plasmodium falciparum*: assay of invasion of erythrocytes. *Exp. Parasitol.* **51**, 74–79.

30. Lanar, D. E., McLaughlin, G. L., Wirth, D. F., Barker, R. J., Zolg, W., and Chulay, J. D. (1989). Comparison of thick films, *in vitro* culture and DNA hybridization probes for detecting *Plasmodium falciparum* malaria. *Am. J. Trop. Med. Hyg.* **40**, 3–6.

31. Makler, M. T., Lee, L. G., and Recktenwald, D. (1987). Thiazole orange: a new dye for *Plasmodium* species analysis. *Cytometry* **8**, 568–570.

32. Mehlhorn, H. (1984). The piroplasms: life cycle and sexual stages. *Adv. Parasitol.* **23**, 37–103.

33. Mills, J. N., and Valli, V. E. O. (1978). A cytofluorometric method of counting trypanosomes. *Tropenmed. Parasit.* **29**, 95–100.

34. Muhlpfordt, H., and Berger, J. (1989). Characterization and grouping of *Trypanosoma brucei brucei, T. b. gambiense* and *T. b. rhodesiense* by quantitative DNA-cytofluorometry and discriminant analysis. *Trop. Med. Parasitol.* **40**, 1–8.

35. Muhlpfordt, H., Berger, J., and Glaser, N. (1985). Cytofluorometry as a method for the differentiation of trypanosomes. *Trop. Med. Parasitol.* **36**, 135–139.

36. Myler, P., Saul, A., Mangan, T., and Kidson, C. (1982). An automated assay of merozoite invasion of erythrocytes using highly synchronized *Plasmodium falciparum* cultures. *Aust. J. Exp. Biol. Med. Sci.* **60**, 83–89.

37. Rüssmann, L., Jung, A., and Heidrich, H. G. (1982). The use of percoll gradients, elutriator rotor elution, and mithramycin staining for the isolation and identification of intraerythrocytic stages of *Plasmodium berghei*. *Z. Parasitenkd.* **66**, 273–280.

38. Saul, A., Myler, P., Mangan, T., and Kidson, C. (1982). *Plasmodium falciparum*: automated assay of erythrocyte invasion using flow cytofluorometry. *Exp. Parasitol.* **54**, 64–71.

39. Shapiro, S. Z., Naessens, J., Liesegang, B., Moloo, S. K., and Magondu, J. (1984). Analysis by flow cytometry of DNA synthesis during the life cycle of African trypanosomes. *Acta Trop.* **41**, 313–323.

40. Sodeman, T. M. (1970). The use of fluorochromes for the detection of malaria parasites. *Am. J. Trop. Med. Hyg.* **19**, 40–42.

41. Vianen, Ph. P., Thaitong, S., Reinders, P. P., Engen, A. L. van, Keur, M., Tanke, H. J., Kaay, H. J. van der, and Mons, B. (1990). Automated flow cytometric analysis of drug susceptibility of malaria parasites. *Am. J. Trop. Med. Hyg.* **43**, 602–607.

42. Vianen, Ph. H., Thaitong, S., Engen, A. L. van, Keur, M. van der, Tanke, H. J., Kaay, H. J. van der, and Ons, B. (1991). Quantitative screening for low numbers of parasites in blood by flow cytometry. Submitted.

43. Vickerman, K. (1985). Developmental cycles and biology of pathogenic trypanosomes. *Br. Med. Bull.* **41**, 105–114.

44. Waters, A. P., and McCuthan, T. F. (1989). The rapid, sensitive diagnosis of the four human malarias. *Lancet* **8651**, 1343–1346.

45. Wells, J. M., Prospero, T. D., Jenni, L., and Le Page, R. W. F. (1987). DNA contents and molecular karyotypes of hybrid *Trypanosoma brucei*. *Mol. Biochem. Parasitol.* **249**, 103–116.

46. Whaun, J. M., Rittershaus, C., and Ip, S. H. C. (1983). Rapid identification and detection of parasitized human red cells by automated flow cytometry. *Cytometry* **4**, 117–122.

47. Zampetti-Boseler, F., Schweizer, J., Pays, E., Jenni, L., and Steinert, M. (1986). Evidence for haploidy in metacyclic forms of *Trypanosoma brucei*. *Proc. Natl. Acad. Sci. USA* **83**, 6063–6064.

3.4

Antibody-defined Surface Molecules of Normal and Malignant Hemopoietic Cells

WALTER KNAPP

Institute of Immunology
University of Vienna
Borschkegasse 8A, A-1090 Vienna, Austria

I. INTRODUCTION

The combination of flow cytometry as a very powerful technique in analytical cytology and the development of a great variety of monoclonal antibodies to leukocyte surface molecules has resulted in an ideal partnership providing us with excellent tools for the analysis of human hemopoietic cells.

Such monoclonal antibodies not only allow us today to define and characterize a variety of hemopoietic cell types by using flow cytometry, they also led to the discovery, functional characterization, purification, and cloning of a plethora of new leukocyte surface molecules.

Together with the tremendous enlightenment resulting from the success of this research, the vast variety of data produced has also created considerable confusion. It soon became apparent to everyone in the field that international co-operative efforts were essential to bring order out of this chaos, and to study all antibody-defined structures at the molecular level with all available reagents. The solution to this problem, the creation of an international

FLOW CYTOMETRY IN HEMATOLOGY
ISBN 0-12-432940-3

workshop series, turned out to be highly successful. In total, 89 different entities have meanwhile been defined in such workshops and given new internationally accepted cluster of differentiation (CD) designations (*41*). Antibodies of some of these specifities are already routinely being used for the flow cytometry analysis of human blood cells, others still need to be fully analyzed.

This review will concentrate on reagents presently available for the flow cytometric analysis of human hemopoietic cells with monoclonal antibodies. Aspects of flow cytometry instrumentation (*57*), DNA and RNA analysis (*5,30,59*) are being dealt with in recent reviews and/or in other chapters of this book.

II. ANTIBODY-DEFINED SURFACE MOLECULES

A. CD Molecules: Definition

CD designations are being assigned at regularly held international workshops on human leukocyte differentiation antigens (*8,40,53,61*) and are approved by the Nomenclature Committee of the International Union of Immunological Societies. The last such workshop was held in Vienna 1989 (*40*), the next one will be organized by Stuart Schlossman and will be located in the USA.

As a rule a CD designation is meant to describe a cluster of antibodies that identify the same molecular species. In this sense, the CD name, e.g. CD2, describes antibodies reacting with the 50 KDa transmembrane glycoprotein as expressed on resting human T cells. The CD designation can also be applied to describe the antigen but this should be made clear by using the terms CD antigen or CD molecule, in our example, f.i. CD2 molecule. The term anti-CD.. (f.i. anti-CD2) should not be used, however, for the description of CD antibodies, since this term may be confused with anti-idiotypes.

B. New CD Designations

At the recently held workshop in Vienna, in total 41 new CD designations have been assigned to groups of antibodies which define individual molecules on the surface of human immune cells (*40*). A considerable number of these clustered antibodies define new, not previously characterized membrane molecules of human blood cells and it can be anticipated that on the basis of that definition these molecules will soon be fully characterized in terms of their molecular and functional characteristics.

Other members of this panel of newly defined CD molecules represent well-known surface receptor structures like the transferrin receptor (CD71), the high affinity Fc-IgG receptor FcRI (CD64), several molecules of the VLA complex (CDw49b, CDw49d, CDw49f) and others. These molecules are not actually "new" but were, for the first time, defined on a collaborative basis with monoclonal antibodies from several laboratories.

Additional members of this panel of new CDs represent subdivisions of previously established clusters. So, for instance, CD2 was subdivided in CD2 and CD2R, CDw42 was divided into CD42a and CD42b, and CD45R was split up up into CD45RA, CD45RB and CD45RO.

The main characteristics of all presently defined CD clusters are listed in Table I. For further details the reader is referred to the proceedings of the international workshops on human leukocyte differentiation antigens and to several reviews (*8,17,36,37,39,40,41,53,61,62*).

III. IMMUNOPHENOTYPING OF BLOOD LYMPHOCYTES

A. Lymphocyte Differential

Immunophenotyping of blood lymphocytes is now probably the most widely used flow

Table I

CD ANTIBODIES/MOLECULES SO FAR ESTABLISHED IN INTERNATIONAL WORKSHOPS

CD design	Selected antibodies (submitter)	Main cellular reactivity	Other names of reactive component
CD1a	NA1/34 (McMichael, Oxford) T6 (Schlossman, Boston) Leu 6 (Evans, Buffalo)	Thymocytes (strong,), LHC, DC, B subset	
CD1b	NUT2 (Sagawa, Fukuoka) WM-25 (Bradstock, Sidney) 7C4 (Knapp, Vienna)	Thymocytes (moderate density), LHC, DC, B subset	
CD1c	M241 (Knowles, New York) 7C6 (Knapp, Vienna) L161 (Bernard, Villejuif)	Thymocytes (weak), LHC, DC, B subset	
CD2	9.6. (Hansen, Seattle) T11 (Schlossman, Boston) Leu 5 (Lanier, Mountain View)	T lineage	T11; SRBC receptor
CD2R	T11.3 (Reinherz, Boston VIT13 (Knapp, Vienna) D66 (Bernard, Nice)	Activated T cells	
CD3	T3 (Schlossman, Boston) UCHT1 (Beverley, London) Leu 4 (Evans, Buffalo)	T cells	CD3 complex
CD4	T4 (Schlossman, Boston) Leu3a (Evans, Buffalo)	T cell subset	T4, gp59
CD5	T1 (Schlossman, Boston) T101 (Royston, San Diego) UCHT2 (Beverley, London)	T, B cell subset	Tp67
CD6	T12 (Schlossman, Boston) T411 (Rieber, Munich)	T and B cell subset	T12
CD7	3A1 (Haynes, Durham) WT1 (Tax, Nijmegen)	T cell subset	
CD8	T8 (Schlossman, Boston) Leu 2a (Evans, Buffalo) UCHT4 (Beverley, London)	T cell subset	T8
CD9	BA2 (LeBien, Minneapolis) J2 (Ritz, Boston)	Pre-B, M, Plt	p24
CD10	J5 (Ritz, Boston) VILA1 (Knapp, Vienna) BA3 (LeBien, Minneapolis)	Lymphoid progenitor cells, cALL, Germ Ctr. B, G.	Neutral endopeptidase, gp100, CALLA
CD11a	TS2/16.1.1 (Springer, Boston) CRIS-3 (Vilella, Barcelona) MHM24 (McMichael, Oxford)	Leukocytes	Leukocyte function antigen-1 (LFA-1),
CD11b	Mac1 (Springer, Boston) Mo1 (Todd, Ann Arbor) VIM12 (Knapp, Vienna)	Monocytes, granulocytes, NK cells	C3bi Receptor, gp 155/95, Mac-1, Mo1
CD11c	S-HCL3 (Schwarting, Baltimore) Ki-M1 (Radzun, Kiel) 3.9 (Hogg, London)	Monocytes, granulocytes, NK cells, B cell subset, B cell prolymphocytic leukemia cells	gp150/95
CDw12	To be redefined	Monocytes, granulocytes	
CD13	My7 (Griffin, Boston) MCS-2 (Sagawa, Fukuoka) MoU28 (Winchester, New York)	Monocytes, granulocytes	Aminopeptidase N, gp150
CD14	UCHM1 (Beverley, London) Mo2 (Todd, Ann Arbor) MEM18 (Horejsi, Prague)	Monocytes	gp55

continued

Table I

(*Continued*)

CD design	Selected antibodies (submitter)	Main cellular reactivity	Other names of reactive component
CD15	My1 (Civin, Baltimore) VIMD5 (Knapp, Vienna) 1G10 (Bernstein, Seattle)	Granulocytes, some antibodies, also monocytes	3-FAL, X-Hapten, Lacto-N-fucopentaose III, SSEA
CD16	Bw209/2 (Kurrie, Marburg) VEP13 (Rumpold, Vienna) CLB/Fc Gran1 (v.d. Borne, Amsterdam)	NK, G, Mac.	FcγRIII
CDw17	GO35 (Thompson, Lexington) Hulym13 (McKenzie, Melbourne) T5A7 (Bernstein, Seattle)	Granulocytes, monocytes, platelets	Lactosylceramide
CD18	TS1/18.1 (Springer, Boston) MHM23 (McMichael, Oxford) M232 (Bernard, Paris)	Leukocytes broad	β-chain to CD11a,b,c
CD19	B4 (Nadler, Boston) HD37 (Dörken, Heidelberg)	B	Bgp95
CD20	B1 (Nadler, Boston) 1F5 (Clark, Seattle) 2H7 (Clark, Seattle)	B	Bp35
CD21	B2 (Nadler, Boston) HB-5 (Cooper, Birmingham)	B subset	C3d/EBV-Rec., CR2
CD22	HD39 (Dörken, Heidelberg) S-HCL1 (Schwarting, Berlin) To15 (Mason, Oxford)	Cytoplasm: pan B; surface: B subset	Bgp135
CD23	MHM6 (McMichael, Oxford) Blast 2 (Thorley-Lawson, Boston)	B subset	Low affinity Fcε receptor (FcεRII)
CD24	BA1 (LeBien, Minneapolis) VIBC5 (Knapp, Vienna)	B cells, granulocytes	
CD25	TAC (Waldmann, Bethesda) 7G7/B6 (Nelson, Bethesda) 2A3 (Maino, Mountain View)	Activated T and B cells and activated macrophages	Interleukin-2 receptor α, Tac-antigen
CD26	134-2C2 (Vilella, Barcelona) TS145 (Ueda, Nagoya) 4ELIC7 (Reinherz, Boston)	Activated T, B, and macrophages, liver canaliculi	Dipeptidyl-peptidase IV
CD27	VIT14 (Knapp, Vienna) CLB-9F4 (Tetteroo, Leiden) S152 (Bigler, New York)	T cell subset	
CD28	9.3 (Hansen, Seattle) Kolt2 (Sagawa, Kurume)	T cell subset	Tp44
CD29	4B4 (Schlossman, Boston) K20 (Boumsell, Paris)	Broad	Integrin β1 chain, platelet GPIIa
CD30	Ki-1 (Stein, Berlin) Ber-H2 (Schwarting, Berlin) HRS-4 (Pfreundschuh, Cologne)	Activated T and B cells, HRS cells	Ki-1 antigen
CD31	SG134 (Goyert, New York) TM3 (Ohto, Saitama)	Plt, M, G, B, (T)	Platelet GP IIa
Cdw32	CIKM5 (Pilkington, Melbourne) 2E1 (Tursz, Villejuif) IV.3 (Anderson, Columbus)	M, G, B, eosinophils	FcγRII receptor
CD33	My9 (Griffin, Boston) L4F3 (Bernstein, Seattle)	Monocytes, normal and malignant myeloid progenitor cells	gp67
CD34	My10 (Civin, Baltimore) BI-3C5 (Tindle, Brisbane) ICH3 (Levinsky, London)	Hematopoietic precursor cells	

continued

Table I

(*Continued*)

CD design	Selected antibodies (submitter)	Main cellular reactivity	Other names of reactive component
CD35	TO5 (Mason, Oxford) J3B11 (Kazatchkine, Paris)	Granulocytes (neutro. and eo.). monocytes, B-cells, some T/NK cells	CR1, C3b receptor
CD36	5F1 (Bernstein, Seattle) CIMeg1 (Pilkington, Melbourne)	M, Plt (B)	Platelet GPIV (GPIIIb)
CD37	HD28 (Dörken, Heidelberg) HH-1 (Funderud, Oslo)	B, (T), (M)	
CD38	T10 (Schlossman, Boston) HB7 (Tedder, Boston) GR7A4 (Garrido, Granada)	Plasma cells, thymocytes, activated T	T10
CD39	G28-10 (Ledbetter, Seattle) AC2 (Rowe, Birmingham)	B subset	
CD40	G28-5 (Ledbetter, Seattle) EA-5 (LeBien, Minneapolis)	B, carcinomas	
CD41	CLB-thromb/7 (Tetteroo, Leiden) VIPL1 (Knapp, Vienna)	Plt	GPIIb/IIIa
CD42a	FMC25 (Zola, Adelaide)	Plt	Platelet GPIX
CD42b	AN51 (McMichael, Oxford)	Plt	Platelet GPIb
CD43	G10-2 (Ledbetter, Seattle) G19-1 (Ledbetter, Seattle)	T cells, granulocytes (brain)	Leukosialin, sialophorin
CD44	F-10-44-2 (Dalchau, East Grinstead) GRHL1 (Garrido, Granada)	T cell, granulocytes, brain, RBC	Pgp-1, gp80-95
CD45	T29/33 (Trowbridge, San Diego) BMAC1 (Dalchau, East Grinstead)	Leukocytes	LCA, T200
CD45RA	2H4 (Schlossman, Boston)	T subset, B, G, M	Restricted T200, CD45 isoform
CD45RB	PD-7/26/16 (Mason, Oxford) This assignment was not based on Workshop studies, but on previously published data	T subset, B, G, M	Restricted T200, CD45 isoform
CD45RØ	UCHL1 (Beverley, Oxford) This assignment was not based on Workshop studies, but on previously published data	T subset, B, G, M.	Restricted T200, CD45 isoform gp 180
CD46	Hulym5 (E4.3) (McKenzie, Melbourne) J48 (Pesando, Seattle)	Leukocytes	Membrane cofactor protein (MCP)
CD47	BRIC126 (Anstee, Bristol) CIKM1 (Pilkington, Melbourne)	Broad	
CD48	WM68 (Henniker, Westmead) LO-MN25 (Ravoet, Brussels)	Leukocytes	
CDw49	CLB-thromb/4 (v.d. Borne, Amsterdam) Gi14 (Santoso, Giessen)	Plt, cultured T	VLA $\alpha 2$ chain, platelet GPIa
CDw49	B-5G10 (Hemler, Boston) HP2/1 (Sanchez-Madrid, Madrid)	M, T, B, (LHC), Thy	VLA $\alpha 4$ chain
CDw49f	GoH3 (Sonnenberg, Amsterdam)	Plt, (T)	VLA $\alpha 6$ chain
CDw50	101-1D1 (Vilella, Barcelona)	Leukocytes	
CD51	13C2 (Horton, London) NK1-M76 (Hogervorst, Amsterdam)	(Plt)	Vitronectin receptor (VNR) α chain
CDw52	Campath-1 (Waldmann, Cambridge) YTH66.9 (Waldmann, Cambridge)	Leukocytes	Campath-1

continued

Table I

(*Continued*)

CD design	Selected antibodies (submitter)	Main cellular reactivity	Other names of reactive component
CD53	MEM-53 (Horejsi, Praha) HD77 (Moldenhauer, Heidelberg) HI29 (Chen, Tiajin)	Leukocytes	
CD54	RR1/1 (Springer, Boston) LB-2 (Clark, Seattle) My13 (Civin, Baltimore)	Broad	ICAM-1
CD55	143-30 (Vilella, Barcelona) BRIC110 (Anstee, Bristol)	Broad	DAF (decay accelerating factor)
CD56	NKH1A (Ritz, Boston) NKH1 (Griffin, Boston) Leu19 (Lanier, Mountain View)	NK, activ. lymphocytes	NKH1, isoform of N-CAM
CD57	HNK-1 (Abo, Birmingham) L186 (Maino, Mountain View)	NK, T, B subset brain	HNK1
CD58	TS2/9 (Springer, Boston) BRIC5 (Anstee, Bristol)	Leukocytes, epithelium	LFA-3
CD59	MEM-43 (Horejsi, Praha) YTH53.1 (Waldmann, Cambridge)	Broad	
CDw60	M-T32 (Rieber, München) UM4D4 (Fox, Ann Arbor)	T subset	
CD61	VIPL2 (Knapp, Vienna) Y2/51 (Cordell, Oxford) CLB-thromb/1 (v.d. Borne, Amsterdam)	Plt	Integrin β3, VNR β chain, platelet GPIIIa
CD62	CLB-thromb/6 (v.d. Borne, Amsterdam) RUU-SP2.15.1 (Nieuwenhuis, Utrecht)	Plt activ.	GMP-140 or PADGEM
CD63	RUU-SP2.28 (Neuwenhuis, Utrecht) CLB-gran/12 (Fijnheer, Amsterdam)	Plt activ., M, (G, T, B)	
CD64	mAb32.3 (Guyre, Hannover) mAb22 (Guyre, Hannover)	M	FcRI, high affinity Fc-IgG receptor
CDw65	VIM2 (Knapp, Vienna) HE10 (Tursz, Villejuif)	G, (M)	Ceramide-Dodecasaccharide 4c
CD66	CLB-gran/10 (v.d. Schoot, Amsterdam) YTH71.3 (Waldmann, Cambridge)	G	
CD67	B13.9 (v.d. Schoot, Amsterdam) G10F5 (Thompson, Lexington) JML-H16 (Yokoyama, Kurume)	G	
CD68	Y2/131 (Mason, Oxford) EBM11 (Johansson, Glostrup) Ki-M7 (Radzun, Kiel)	Macrophages	
CD69	L78 (Buck, Mountain View) MLR3 (Cosulich, Genova) BI-Ac/p26 (Fiebig, Leibzig)	Activated B, T	Activation inducer molecule (AIM)
CDw70	Ki-24 (Stein, Berlin) HNE51 (Delsol, Toulouse) 200-4C3 (Poppema, Edmonton)	Activated T, B, Reed–Sternberg cells	Ki-24 antigen
CD71	T9 (Schlossman, Boston) 42.6.3 (Trowbridge, LaJolla) VIP1 (Knapp, Vienna)	Proliferating cells, macrophages	Transferrin receptor, T9 antigen
CD72	S-HCL2 (Schwarting, Berlin) BU-40 (Hardie, Birmingham) J3-109 (Pesando, Seattle)	B	

continued

Table I
(Continued)

CD design	Selected antibodies (submitter)	Main cellular reactivity	Other names of reactive component
CD73	1E9.28.1 (Thompson, LaJolla) AD2 (Gwin/Cooper, Birmingham)	B subset, T subset	Ecto-5′-nucleotidase (ecto-5′NT)
CD74	LN2 (Epstein, Frankfurt) BU-43 (Hardie, Birmingham)	B, M.	MHC class II assoc. invariant chain (ii)
CDw75	OKB4 (Ortho, Raritan) HH2 (Funderud, Oslo) EBU-141 (Gramatyki, Erlangen)	Mature B, (T subset)	
CD76	HD66 (Dörken, Heidelberg) CRIS-4 (Vilella, Barcelona)	Mature B, T subset	
CD77	38.13 (BLA) (Wiels/Tursz, Villejuif) 424/4A11 (Brodin, Lund)	Subset B	Globotriaosylceramide (Gb3), BLA
CDw78	Anti-Ba (MR11) (Kishimoto, Osaka) Leu-2 (MR11) (Lanier, Mountain View)	B, (M)	Ba

Abbreviations: Thy = thymocytes, B = B cells, T = T cells, M = monocytes, G = granulocytes, Plt = platelets, NK = NK cells, LHC = epidermal Langerhans cells. Cell types in brackets indicate weak reactivity.

cytometric analysis. Monoclonal antibodies selectively reactive with the various subsets of human lymphocytes can be obtained from various commercial sources and have proven useful for monitoring changes in luekemias, immunodeficiency diseases, autoimmune and other immunologically mediated diseases (*2,7,9,10,45,54,62,66,69,73,81*).

With the availability of fluoresceinisothiocyanate (FITC)- and phycoerythrin (PE)-labeled monoclonal antibodies staining of lymphocytes for the analysis of blood lymphocyte subsets has become very simple. Typically, 50 μl anticoagululated whole blood is incubated for 15 min with a combination of FITC- and PE-labeled antibodies. Afterwards, the red blood cells are lysed with lysing reagent (several proprietary lysing reagents are commercially available) followed by one washing step with phosphate buffered saline (PBS). Lymphocytes can then be analyzed or preferably first fixed with formaldehyde (0.5% in PBS) and analyzed the following day. Suitable marker combinations for the analysis of major lymphocyte subsets (*9,10,11,17,62*) are listed in Table II.

Table II
MAJOR BLOOD LYMPHOCYTE SUBSETS DISTINGUISHABLE WITH MONOCLONAL ANTIBODIES TO SURFACE MOLECULES

Subset designation	Marker profile
T lymphocytes	$CD3^+$
Helper/inducer T	$CD4^+$
Suppr./cytotox. T	$CD8^+$
Memory T	$CD3^+CD45R0^+$
Naive T	$CD3^+CD45RA^+$
TCR1 T	TCR gamma/delta$^+$
TCR2 T	TCR α/β^+
Activated T cells	$CD25^+CD3^+$ and/or HLA-D^+CD3^+
NK cells	$CD56^+CD3^-$
B cells	$CD19^+$
$CD5^+$B cells	$CD19^+CD5^+$

Typical flow cytometry profiles are shown in Fig. 1.

For details concerning the various procedural details the reader is referred to an excellent recent review on that topic (*44*) which also gives a detailed description of the safety precautions to be taken when evaluating potentially infectious blood specimens (see also Chap. 2.2).

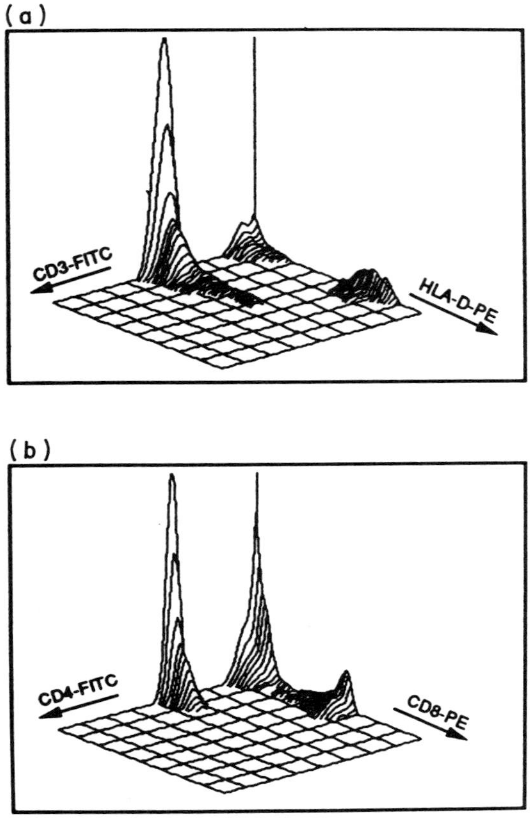

Fig. 1. Flow cytometric profiles of human peripheral blood lymphocytes stained with (a) CD3-FITC and anti-HLA-D-Phycoerythrin (PE) antibodies for the demonstration of T cells ($CD3^+$), B cells ($HLA\text{-}D^+CD3^-$) and activated T cells ($HLA\text{-}D^+CD3^+$) and (b) CD4-FITC/CD8-PE antibodies for the demonstration of helper/inducer T cells ($CD4^+$) and suppr./cytotoxic T cells ($CD8^+$).

B. Lymphocyte Activation

Under normal conditions the majority of circulating blood lymphocytes are in a resting state. Upon *in vitro* activation with antigens, stimulatory antibodies, lectins or cytokines, the surface marker profile of lymphocytes undergoes significant changes (*33,67*). Typically, activated lymphocytes begin to express "new" receptor molecules (activation antigens) and show alterations in the expression density of molecules also found on resting cells. Some of the more informative "activation antigens" recognized by CD antibodies which can be used to define the functional state of a given lymphocyte are listed in Table III. The kinetics of the appearance of CD2R epitopes during *in vitro* T cell activation are shown in Fig. 2.

C. Naive Versus Memory T Lymphocytes

Considerable interest has recently also been devoted to the phenotypic distinction between naive and memory T lymphocytes (*10*). A number of detailed studies have shown that these two functionally distinct T lymphocyte subsets can also be distinguished on the basis of their reactivity with certain monoclonal antibodies (*10,54,65*). The most clearcut distinction can be made with CD45RA antibodies (defining the naive T subset) and CD45RO antibodies (defining the memory T subset). Both antibody types are not T cell specific and have

Table III

MOLECULAR/FUNCTIONAL CHARACTERISTICS OF CD MOLECULES NEOEXPRESSED OR STRONGLY ENHANCED UPON ACTIVATION OF HUMAN T CELLS

CD designation (Workshop assignation)	
CD2R (4th Workshop, Vienna, 1989)	CD2R Molecule: same molecule as CD2 molecule, CD2R antibodies recognize CD2 epitopes unrelated to LFA3 binding site and neoexpressed on activated T cells.
	CD2R antibodies: mitogenic with CD2 antibodies, LFA3 molecules or PMA. Enhanced mitogenic effect in the presence of CD45 antibodies

continued

Table III

(*Continued*)

CD designation (Workshop assignation)	
CD25 (2nd Workshop, Boston, 1984)	55 kDa single-chain glycoprotein, also known as α-chain of the IL-2 receptor complex (by some authors also termed β-chain), Tac-antigen, low affinity IL-2 receptor. Forms together with the 75 kDa and possibly other chains the high affinity IL-2 receptor. CD25 molecules are also released from lymphoid cells and can be found in supernatants and in serum CD25 antibodies: some CD25 antibodies inhibit IL-2 binding and IL-2 dependent T and B cell proliferation
CD26 (2nd Workshop, Boston, 1984)	CD26 molecule: 120 kDa single chain transmembrane glycoprotein probably identical with the proteolytic enzyme dipeptidyl peptidase IV, a serine-type exopeptidase. Expression of CD26 molecules on T cells is up-regulated upon T cell activation CD26 antibodies: CD26 antibodies trigger lymphokine secretion, proliferation and cytotoxicity of pre-activated T cells
CD30 (3rd Workshop, Oxford, (1986)	CD30 molecule: 105 kDa single chain glycoprotein also known as Ki-1 antigen. Expressed on activated but not on resting T and B cells. Also expressed on Hodgkin and Sternberg Reed cells and on anaplastic large cell NHLs CD30 antibodies: no functional effects so far established.
CD38 (3rd Workshop, Oxford, 1986)	CD38 molecule: 45 kDa type II single chain transmembrane glycoprotein. The CD38 molecule has no significant homology with other known surface molecules. On the basis of its cellular distribution it has been speculated that CD38 may be a growth factor receptor of growth-related cell–cell adhesion molecule CD38 antibodies: CD38 antibodies can potentiate the proliferative response to IL-2 of IL-2-dependent murine natural killer cells.
CDw49b (4th Workshop, Vienna, 1989)	CDw49b molecule: 170 kDa glycoprotein, also known as VLA-α2 chain or platelet GPla; non-covalently associated with CD29 molecules in VLA-2 complexes, which functions as collagen receptor. Expressed upon long-term activation of T cells CDw49b antibodies
CD56 (4th Workshop, Vienna, 1989)	CD56 molecule: 220/135 kDa glycoprotein, heavily glycosylated isoform of N-CAM. Expressed on NK cells and long-term activated T cells CD56 antibodies: can inhibit NK/target cell interaction in certain system
CD69 (4th Workshop, Vienna, 1989)	CD69 molecule: dimeric phosphorylated glycoprotein of 28/34 kDa. Expressed within 3 hours after T cell activation. Molecule also known as activation inducer molecule (AIM) CD69 Antibodies: comitogenic with submitogenic concentration of PMA
CDw70 (4th Workshop, Vienna, 1989)	CDw70 molecule: molcule not as yet fully characterized. Expressed on Hodgkin and Sternberg Reed cells of Hodgkin disease, certain non-Hodgkin lymphoma cells and an activated T and B cells CDw70 antibodies; no functional effects so far established.

continued

Table III

(*Continued*)

CD designation (Workshop assignation)	
CD71 (4th Workshop, Vienna, 1989)	CD71 molecule: integral membrane glycoprotein composed of two SH-linked identical subunits of 95 kDa. Oriented in the membrane with the N-terminus inside the cell. Molecule better known as transferrin receptor. It mediates cellular iron uptake via internalization and recycling of transferrin. In addition to its role in delivering iron, the transferrin receptor may also be involved in the regulation of cell growth CD71 antibodies: several CD71 antibodies block transferrin binding and lead to an arrest in cell division and an accumulation of cells in S-phase

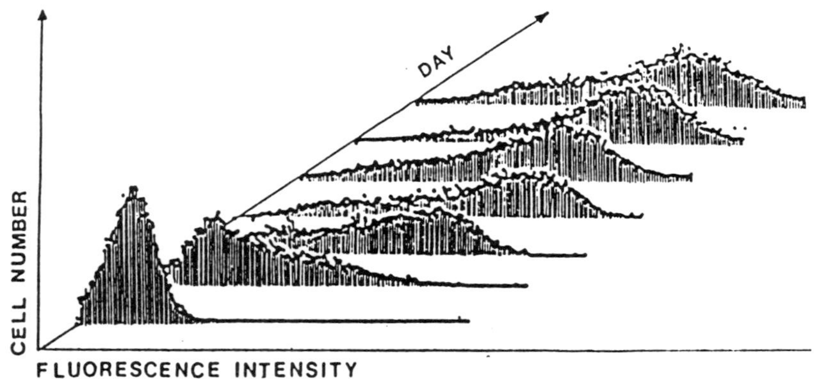

Fig. 2. Kinetics of the appearance of CD2R (antibody VIT13) reactive activation associated epitopes on human T cells upon activation with PHA.

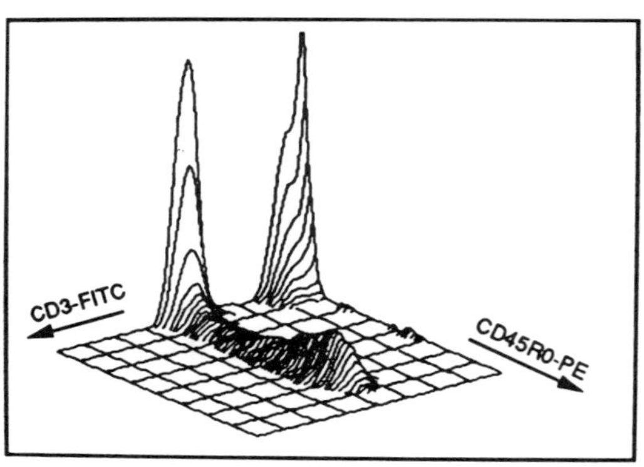

Fig. 3. Flow cytometric profiles of human lymphocytes double stained with FITC-labeled CD3 antibody UCHT1 and PE-labeled CD45R0 antibody UCHL1 (identifying memory T cells).

to be combined with a T cell-specific CD3 antibody. Flow cytometric profiles of such stainings for naive and memory T cells are shown in Fig. 3.

In newborns virtually all T cells are of CD45RA type and do not express CD45R0. Within increasing age the proportion of CD45R0 T cells steadily increases and reaches at an age of approximately 16 years the adult value of 30–60%.

IV. ANALYSIS OF BLOOD MONOCYTES AND GRANULOCYTES

Historically, the immunological analysis of surface molecules on phagocytic cells and the functional analysis of phagocytic cells by flow cytometry is rather young. During the last several years, considerable progress has been made, however, in this area (reviewed in *15,29,38,60*).

In contrast to the analysis of lymphoid cells where different subsets of lymphocytes are distinguished on the basis of their receptor repertoire, no solid evidence is available that human blood polymorphonuclear cells or monocytes are composed of individual and stable functional subsets that can be distinguished on the basis of their surface receptor repertoire. It rather seems that the functional state of a given cell determines what surface receptors are being expressed and at which density.

The flow cytometric analysis of monocytes and polymorphonuclear cells (PMN) is, therefore, mainly used for the relative quantitation of surface receptor molecules rather than for subset enumeration. Particularly for such quantitations, standardization of staining procedures and flow cytometry instrument alignment is mandatory (*52*).

Analysis of monocyte and PMN surface receptor expression density is of special interest in inflammatory disease states and for monitoring changes in conjunction with various therapeutic manipulations with cytokines (*52*).

Analysis of functional receptor structures is also of clinical relevance in suspected immunodeficiency states. Several laboratories could demonstrate, for instance, that complete or partial deficiency of CD11b (complement receptor CR3) expression is associated with severe functional defects of phagocytes (*4,42,76*). Other such defects occasionally encountered are deficiencies of FcRI (CD64) (*3*) or MHC class II (*69*) expression.

V. FLOW CYTOMETRIC ANALYSIS OF BONE MARROW SAMPLES

Quite in contrast to the enormous number of reports on flow cytometric analysis of blood samples from healthy individuals and patients with various diseases, less information is available for bone marrow samples. Most studies on bone marrow samples deal with the phenotypic analysis of blast cells in patients with overt leukemia. In these studies, usually only patients are included with 80% or more blast cells in the bone marrow. These blast populations are usually gated on the basis of forward and orthogonal light scattering and the percentage of antibody-reactive cells is reported.

Only few studies deal with the immunological analysis of normal or nearly normal bone marrow samples or of bone marrow samples from leukemia patients with minimal residual disease (*13,16,18,22,23,46,47,51,58,64,71*). One reason for that may be the technical difficulty of analyzing a heterogeneous cell mixture such as the bone marrow. It contains all maturation stages of myelomonocytic cells with different levels of autofluorescence and variable densities of Fc receptor expression leading to variable non-specific antibody binding. It also contains small lymphocytes and large blast cells which also differ in their non-specific immunoglobulin binding and autofluorescence densities.

Most investigators have tried so far to circumvent this problem by setting individual scatter gates for lymphocytes, blast cells and granulated myelomonocytic cells and then analyzing each gated population separately with a panel of monoclonal antibodies (*6,16,46,47, 63*). The problem with such scatter gate settings is the difficulty of standardizing these settings and the impossibility of comparing results between different laboratories using different instrumentations. One way to circumvent this problem is to use double or triple staining of strongly expressed informative surface molecules with a sufficiently wide separation of positive and negative cells. Unfortunately, some of the more interesting surface molecules are only weakly expressed, and particularly for the analysis of bone marrow cells improved reagents or reagent combinations are clearly needed.

VI. TYPING OF LEUKEMIA CELLS

Leukemia diagnosis classically was based on morphological and cytochemical criteria together with the clinical picture. Particularly for malignant proliferations of undifferentiated and of lymphoid cells it is often very difficult, however, to assign the cell lineage and differentiation stage of a given cell population. Typing of these cells with monoclonal antibodies considerably facilitates this process and gives additional information previously not available.

A. Acute Leukemias

In acute undifferentiated leukemias (AL) where blast cells can be difficult to classify on the basis of morphological criteria, typing of leukemic cells with monoclonal antibodies is of particular importance. Table IV gives a list of antibody specificities commonly used for the analysis of acute leukemia blast cells. For details concerning the results obtained with such analyses the reader is referred to a variety of review articles on this topic (*24,29,34,39,48,77,78,80*).

The prognostic value of immunological blast cell phenotyping is well established for the distinction between lymphoid and myeloid leukemias and for the subclassification of acute lymphoblastic leukemias. Not so well established as yet is the clinical significance of immunological AML subclassifications. Several studies on that question are presently ongoing and initial results have already been reported (*29,31,32,49,68*).

The first step in an immunological evaluation of leukemic blast cells is the definition of the direction (hemopoietic lineage) in which the malignant blast cells have differentiated. This is then followed by the definition of the differentiation stage within a given lineage at which the majority of blast cells are arrested. Some of the most frequently used marker combinations are shown in Table IV. Leukemic blast cells seem, in most instances, to be replicas of normal hemopoietic cells at an early stage of differentiation. Monoclonal antibodies known to selectively react with a given lineage of

Table IV

MARKER PROFILES OF BLAST CELLS IN PATIENTS WITH ACUTE LEUKEMIAS

Subset designation	Marker
T lineage	
Pro thymocyte	$CD7^+$, ($CD2^+$), TdT^+, $CD34^+$
Immature thymocyte	$CD7^+$, $CD2^+$, $CD5^+$, $cyCD3^+$, TdT^+
Common thymocyte	$CD7^+$, $CD2^+$, $CD5^+$, $cyCD3^+$, $CD1^+$
Mature thymocyte/Ly	$CD7^+$, $CD2^+$, $CD5^+$, $CD3^+$, $CD27^+$
B lineage	
Pro-B cell	$CD19^+$, $CD24^+$, $CD34^+$, TdT^+
Pre-B cell	$CD19^+$, $CD24^+$, TdT^+, $CD10^+$, $cy\mu^+$
Mature B cell	$CD19^+$, $CD24^+$, $CD20^+$, SmIg
Plasma cell	$CD38^+$, $CD45^-$, $cyIg^+$
Myeloid lineage	
AML blasts	$CD13^+$, $CD33^+$, $CDw65^+$ (heterogeneous expression). Mature monocytic forms $CD14^+$
Megakarioblasts	$CD61^+$, $CD41^+$
Erythroblasts	(Glycophorin A)

hemopoietic cells including precursor cells, can therefore be used for the designation of the predominant lineage and cell type involved in a leukemic process.

Ideally, such marker molecules should be expressed on all differentiation stages of a given lineage starting with the earliest detectable precursor cell committed to a given lineage down to the fully differentiated material and cell. Unfortunately, such ideal reagents which are absolutely lineage specific and fulfil all the above-mentioned requirements do not exist. It is, therefore, necessary to use a panel of different monoclonal antibodies and to evaluate the reaction pattern with these reagents rather than to rely on the reactivity of individual antibodies. Lineage decision must, as any other serological cell type characterization, always be based on a typical reaction pattern which includes both positive and negative reactions.

Increasingly, in addition to surface markers, intracytoplasmic markers such as, for instance, cytoplasmic CD3 or CD22 in immature T or B cells, are being analyzed by flow cytometry and supplement the armentarium for a correct cell lineage and cell type assignment (*14,37,82*).

The idea that leukemia cells in general are phenotypically and genotypically similar or identical to normal hematopoietic progenitors has recently been challenged by an increasing number of reports on atypical antigen expression and gene rearrangements in acute undifferentiated leukemias (*21,28,43,48*). In about 10% of AML cases, for example, blast cells coexpress myeloid antigens and TdT, whereas in normal populations this cellular phenotype has never been reported. In approximately the same proportion of AMLs immunoglobulin gene or T cell receptor gene arrangements can be observed. Strikingly frequent, as discussed above, are also coexpression of myeloid and thrombocytic or erythroid antigens in myeloid blast crisis of CML. Other frequent coexpressions such as the expression of the T cell-"specific" antigen CD7 on AML blasts and the expression of the "specific" myeloid antigens CD13 and CD33 on ALL blasts are more difficult to interpret, since the lineage specificity of these markers has not yet been fully explored in normal malignant hematopoiesis.

Greaves *et al.* (*28*) argue that many of the reported examples of infidelity are suspect on technical grounds. They further suggest that even the verifiable and consistent examples of apparent lineage infidelity also do not reflect actual misprogramming (leading to infidelity) but rather the existence of a transient phase of limited promiscuity of gene expression occurring also in normal bipotential or multipotential progenitors and able to be preserved as a relic in leukemic blast cell populations that are in maturation arrest.

B. Malignant Lymphomas

Lymphocyte surface membrane phenotype analysis has also become an accepted tool for the analysis of leukemic blood cells in patients with NHL (*20,70,72,74*). A first impression about the nature of proliferating lymphocytes can already be obtained from the basic lymphocyte differential panel (see above and Table II). It shows whether the circulating lymphocytes in a patient with lymphocytosis are predominantly of B ($CD19^+$) or of mature T lymphocyte ($CD3^+$) type or whether they are of NK cell nature ($CD56^+CD3^-$ or $CD16^+$). For a more in-depth analysis of the malignant cell type and for the detection of lower proportions of leukemic cells contaminating the normal lymphocyte population, more refined methods are necessary.

VII. QUALITY CONTROL AND STANDARDIZATION

With flow cytometry becoming a routine technology for the analysis of hemopoietic cells, it becomes mandatory to stress the development of quality control procedures for both cell preparation and labeling and the instrumentation. Successful flow cytometry analysis of hemopoietic cells is dependent upon sample preparation, fluorescent reagents, instrument calibration and accurate sample and data analysis. At present, several committees of the immunology, pathology and hematology societies in the USA and Europe are working on quality assurance guidelines. So far, to the author's knowledge, no official guidelines have been published, but preliminary studies and proposals are available (*12,19,25,44,55*).

VIII. RECENT TECHNICAL DEVELOPMENTS

Apart from the continuously growing list of antibody reagents becoming available for the analysis of hemopoietic cells, several technical developments have been reported which make life easier and can contribute to a further improvement of flow cytometric evaluations of hemopoietic cells.

Triple staining of hemopoietic cells with monoclonal antibodies conjugated with three different fluorochromes has for quite some time been a technical possibility. So far, the number of studies reported using triple staining techniques is relatively small, however. The main reason for that is probably the fact that rather expensive dual laser systems were needed initially for such studies. Recently, several studies have shown that triple stainings can also be performed with single laser systems (*26,35,56*). There can be no doubt that the possibility of using single laser systems, which are widely available and much easier to handle, will lead to a much more widespread use of triple staining methods for the analysis of hemopoietic cells.

One of the still existing problems in the analysis of triple stained cells with single laser systems is the loss of sensitivity. One way to circumvent this problem may be the use of fluorescent liposomes for enhanced fluorescence as proposed by Gray *et al.* (*27*).

At least for the time being, monoclonal antibodies labeled with fluorochromes are certainly the most straightforward reagents for the analysis of hemopoietic cells. The more we learn about the structures recognized by these antibodies and the natural ligands binding to these structures, the more it becomes feasible to use such natural ligands labeled with fluorochromes for the analysis of these previously antibody-defined structures. Recombinant technologies now allow the production of such ligands in large quantities and by using such reagents in combination with antibodies one cannot only identify the total immunologically reactive cell structure (as detected by antibody) but also quantify the functionally intact cell structure in terms of its binding capacity (as detected with the appropriate labeled natural ligand). Studies of that kind have already been reported, e.g. the flow cytometric analysis of interleukin-2 receptors with biotinylated human interleukin-2 (*75*), and the flow cytometric analysis of glucocorticoid receptors using monoclonal antibody and fluoresceinated ligand probes (*50*). Similar studies are, in principle, also possible for DNA–DNA and DNA–RNA interactions.

In summary, the future is bright and there seems to be virtually no limitation for the molecular analysis of hemopoietic cells with flow cytometric methods. With the development of new fluorochromes, fluorogenic enzyme substrates and powerful amplification techniques it will in the future be possible to extend the list of parameters which can either simultaneously or consecutively be measured in flow cytometry. It will also be possible to analyze cell structures expressed in very low densities (< 1000 molecules per cell) which are, at present, still difficult or impossible to be reliably analyzed.

REFERENCES

1. Alarcon, B., Reguiro, J. R., Arnaiz-Villena, A., and Terhorst, C. (1988). Familial defect in the surface expression of the T-cell receptor-CD3 complex. *New Engl. J. Med.* **319**, 1203–1208.

2. Alarcon, B., and Terhorst, C. (1990). Congenital T-cell receptor immunodeficiencies in man. *Immunodef. Rev.* **2**, 1–16.

3. Anderson, C. L. (1989). Human IgG Fc Receptors. *Clin. Immunol. Immunopathol.* **53**, 63–71.

4. Anderson, D. C. (1987). Leukocyte adhesion deficiency: An inherited defect in the Mac-1, LFA-1 and p150,95 glycoproteins. *Ann. Rev. Med.* **38**, 175–194.
5. Andreeff, M., Assing, G., and Cirrincione, C. (1986). Prognostic value of DNA/RNA flow cytometry in myeloblastic and lymphoblastic leukemia in adults: RNA content and S-phase predict remission duration and survival in multivariant analysis. *Ann. N.Y. Acad. Sci.* **468**, 387–406.
6. Andreoni, C., Rigal, D., Bonnard, M., and Bernaud, J. (1990). Phenotypic analysis of a large number of normal human bone marrow samples by flow cytometry. *Blut.* **61**, 271–277.
7. Bach, M. A., and Bach, J. F. (1981). The use of monoclonal anti-T cell antibodies to study T cell imbalances in human diseases. *Clin. Exp. Immunol.* **45**, 449–456.
8. Bernard, A., Boumsell, L., Daussett, J., Milstein, C., and Schlossman, S. F. (Eds) (1983). "Leucocyte Typing." Springer Verlag, Berlin.
9. Beverley, P. C. L. (1987). Human T cell subsets. *Immunol. Letters* **14**, 263–267.
10. Beverley, P. C. L. Merkenschlager, M., and Wallace, D. L. (1989). Identification of human naive and memory T cells. *In* "Progress Immunology VII" (F. Melchers, ed.), pp. 432–438. Springer Verlag, Berlin.
11. Brenner, M. B., Groh, V., Porcelli, S. A., Hochsteinbach, F., Band, H., Parker, C., Lanier, L., McLean, J., and Krangel, M. S. (1990). Structure and distribution of the human gd T-cell receptor. *In* "Leucocyte Typing IV" (W. Knapp, B. Dörken, W. R. Gilks, E. P. Rieber, R. E. Schmidt, H. Stein, and A. E. G. v.d. Borne, eds), pp. 1049–1053. Oxford University Press, Oxford.
12. Brown, M. C., Hoffman, R. A., and Kirchanski, S. J. (1986). Controls of flow cytometers in hematology and cellular immunology. *Ann. N.Y. Acad. Sci.* **468**, 93–103.
13. Campana, D., Sheridan, B., Tidman, B., Hoffbrand, A. V., and Janossy, G. (1986). Human leukocyte function-associated antigens on lymphohemopoietic precursor cells. *Eur. J. Immunol.* **16**, 537–542.
14. Campana, D., Thompson, J. S., Amlot, P., Brown, S., and Janossy, G. (1987). The cytoplasmic expression of CD3 antigens in normal and malignant cells of the T lymphoid lineage. *J. Immunol.* **138**, 648–655.
15. Civin, C. I. (1990). Human monomyeloid cell membrane antigens. *Exp. Haematol.* **18**, 461–467.
16. Civin, C. I., Banquerigo, M. L., Strauss, L. C., and Loken, M. R. (1987). Antigenic analysis of hematopoiesis. VI. Flow cytometric characterization of My-10-positive progenitor cells in normal human bone marrow. *Exp. Haematol.* **15**, 10–17.
17. Clark, E. A., and Ledbetter, J. A. (1989). Structure function, and genetics of human B cell-associated surface molecules. *Adv. Cancer Res.* **52**, 81–149.
18. Clark, P., Normansell, D. E., Innes, D. J., and Hess, C. H. (1986). Lymphocyte subsets in normal bone marrow. *Blood* **67**, 1600–1606.
19. Cunninghamrundles, S. (1989). Standardization of T-cell subsets enumerations. *J. Am. Med. Assoc.* **262**, 1526.
20. Deegan, M. J. (1989). Membrane antigen analysis in the diagnosis of lymphoid leukemias and lymphomas. *Arch. Pathol. Lab. Med.* **113**, 606–618.
21. Delwel, R., Bot, F., Knapp, W., and Löwenberg, B. (1990). VIM-2, candidate monoclonal antibody for purging autologous marrow grafts in acute myeloblastic leukemia (AML). *Bone Marrow Transplant.* **2**, 149–154.
22. Delwel, R., Touw, I., and Löwenberg, B. (1986). Towards detection of minimal disease: Discrimination of AML precursors from normal myeloid precursors using a combination of surface markers. *In* "Minimal Residual Disease in Acute Leukemia" (A. Hagenbeck, and B. Löwenberg, eds), pp. 68–75. Martinus Nijhoff, Dordrecht.
23. Dinndorf, P. A., Andrews, A. R. G., Benjamin, D., Ridgway, D., Wolff, L., and Bernstein, I. D. (1986). Expression of normal myeloid-associated antigens by acute leukemia cells. *Blood* **67**, 1048–1053.
24. Drexler, H. G., Gignac, S. M., and Minowada, J. (1988). Routine immunophenotyping of acute leukemias. *Blut* **57**, 327–339.
25. Edwards, B. S., Altobelli, K. K., Nolla, H. A., Harper, D. A., and Hoffman, R. R. (1989). Comprehensive quality assessment approach for flow cytometric immunophenotyping of human lymphocytes. *Cytometry* **10**, 433–441.
26. Festin, R., Björkland, A., and Tötterman, T. H. (1990). Single laser flow cytometric detection of lymphocytes binding three antibodies labelled with fluorescein, phycoerythrin and a novel tandem fluorochrome conjugate. *J. Immunol. Meth.* **126**, 69–78.
27. Gray, A. G., Morgan, J., Linch, D. C., and Huehns, E. R. (1989). Enhanced fluorescence in indirect immunophenotyping by the use of fluorescent liposomes. *J. Immunol. Methods* **121**, 1–7.
28. Greaves, M. F., Chan, L. C., Furley, A. J. W., Watt, S. M., and Molgaard, H. V. (1986). Lineage promiscuity in hemopoietic differentiation and leukemia. *Blood* **67**, 1–11.
29. Griffin, J. D. (1987). The use of monoclonal antibodies in the characterization of myeloid leukemias. *Haematol. Pathol.* **1**, 81–91.
30. Hiddemann, W., Wörmann, B., Ritter, J., Thiel, E., Göhde, W., Lahme, B., Henze, G., and Schellong, G. (1986). Frequency and clinical significance of DNA aneuploidy in acute leukemia. *Ann. N.Y. Acad. Sci.* **468**, 227–240.

31. Holowiecki, J., Lutz, D., Krzemien, S., Stell-Holowiecka, B., Graf, F., Kelenyi, G., Schrany, V., Callea, V., Brugiatelli, M., Neri, A., Magy, T., Ihle, R., Jagoda, K., and Rudzka, E. (1986). CD15 antigen detected by the VIM-D5 monoclonal antibody for prediction of ability to achieve complete remission in acute nonlymphocytic leukemia. *Acta Haematol.* **76**, 16–19.

32. Holowiecki, J., Lutz, D., Krzemien, S., Schranz, V., Jagoda, K., Holowiecka, B., Graf, F., Kelenyi, G., Callea, V., Brugiatelli, M., Berceanu, S., Ihle, R., Krawczyk, M., and Rudzka, E. (1989). The expression of CD15 antigen on myeloid leukemia cells — a new prognostic index for complete remission and for survival. *Folia Haematol.* **116**, 289–295.

33. Holter, W., Majdic, O., Liszka, K., Stockinger, H., and Knapp, W. (1985). Kinetics of activation antigen expression by *in vitro* stimulated human T lymphocytes. *Cell. Immunol.* **90**, 322–330.

34. Hooijkaas, H., Adriaansen, H. J., and Van Dongen, J. M. (1988). Detection of central nervous system involvement in patients with leukemia or non-Hodgkin's lymphoma by immunological marker analysis of cerebrospinal fluid cells. *In* "Immunological Approaches to the Classification and Management of Lymphomas and Leukaemias" (J. M. Bennett, and K. A. Foon, eds), pp. 149–171. Kluwer Academic Publishers, Amsterdam.

35. Horan, P. K., Slezak, S. E., and Poste, G. (1986). Improved flow cytometric analysis of leukocyte subsets: Simultaneous identification of five cell subsets using two-color immunofluorescence. *Proc. Natl. Acad. Sci. USA.* **83**, 8361–8365.

36. Horejsi, and Bazil (1988). Leukocyte antigen review. *Biochem. J.* **253**, 1–26.

37. Janossy, G., Couston-Smith, E., and Campana, D. (1989). The reliability of cytoplasmic CD3 and CD22 antigen expression in the immunodiagnosis of acute leukemia: a study of 500 cases. *Leukemia* **3**, 170–181.

38. Knapp, W. (1982). Monoclonal antibodies against myeloid differentiation antigens. *Blut* **45**, 301–308.

39. Knapp, W. (1987). Serology of myeloid leukemias. *Cancer Rev.* **8**, 66–90.

40. Knapp, W., Dörken, B., Gilks, W. R., Rieber, E. P., Schmidt, R. E., Stein, H., and v.d. Borne, A. E. G., (eds) (1989). "Leucocyte Typing IV." Oxford University Press, Oxford.

41. Knapp, W., Rieber, P., Dörken, B., Schmidt, R. E., Stein, H., and von dem Borne, A. E. G. (1989). Towards a better definition of human leucocyte surface molecules. *Immunology Today* **10**, 253–258.

42. Kohl, S., Springer, T. A., Schmalstieg, F. C., Loos, L. S., and Anderson, D. C. (1984). Defective natural killer cytotoxicity and polymorphonuclear leukocyte antibody-dependent cellular cytoxicity in patients with LFA-1/OKM-1 deficiency. *J. Immunol.* **133**, 2972–2978.

43. Köller, U., Haas, O. A., Ludwig, W. D., Bartram, C. R., Harbott, J., Panzer-Grümayer, R., Hansen-Hagge, T., Ritter, J., Creutzig, U., Knapp, W., and Gadner, H. (1989). Phenotypic and genotypic heterogeneity in infant acute leukemia. II. Acute nonlymphoblastic leukemia. *Leukemia* **3**, 708–715.

44. Landay, A. L., and Muirhead, K. A. (1989). Procedural guidelines for performing immunophenotyping by flow cytometry. *Clin. Immunol. Immunopathol.* **52**, 48–60.

45. Lane, H. C., and Fauci, A. S. (1985). Immunologic abnormalities in the acquired immunodeficiency syndrome. *Annu. Rev. Immunol.* **3**, 477–500.

46. Loken, M. R., Shah, V. O., Dattilo, K. L., and Civin, C. I. (1987). Flow cytometric analysis of human bone marrow: I. Normal erythroid development. *Blood* **69**, 255–263.

47. Loken, M. R., Shah, V. O., Dattilo, K. L., and Civin, C. I. (1987). Flow cytometric analysis of human bone marrow. II. Normal B lymphocyte development. *Blood* **70**, 1316–1324.

48. Ludwig, W. D., Bartram, C. R., HGarbott, J., Köller, U., Haas, O. A., Hansen-Hagge, T., Heil, G., Seibt-Jung, H., Teichmann, J. V., Ritter, J., Knapp, W., Gadner, H., Thiel, E., and Riehm, H. (1989). Phenotypic and genotypic heterogeneity in infant acute leukemia. I. Acute lymphoblastic leukemia. *Leukemia* **3**, 431–439.

49. Majdic, O., Bettelheim, P., Stockinger, H., Aberer, W., Liszka, K., Lutz, D., and Knapp, W. (1984). M2, a novel myelomonocytic cell surface antigen and its distribution on leukemic cells. *Int. J. Cancer* **33**, 617–623.

50. Marchetti, D., Van, N. T., Gametchu, B., Thompson, E. B., Kobayashi, Y., Watanabe, F., and Barlogie, B. (1989). Flow cytometric analysis of glucocorticoid receptor using monoclonal antibody and fluoresceinated ligand probes. *Cancer Res.* **49**, 863–869.

51. Martens, A. C. M., and Hagenbeck, A. (1985). Detection of minimal disease in acute leukemia using flow cytometry: studies in a rat model for human acute leukemia 1,2,3. *Cytometry* **6**, 342–347.

52. Maurer, D., Fischer, G., Majdic, O., Hinterberger, W., and Knapp, W. (1991). Ratio of complement over Fc-receptor expression is a sensitive parameter of granulocyte activation. *Ann. Hematol.* **62**, 135–140.

53. McMichael, A. J., Beverley, P. C. L., Cobbold, S., Crumpton, M. J., Gilks, W., Gotch, F. M., Hogg, N., Horton, M., Ling, N., MacLennan, I. C. M., Mason, D. Y., Milstein, C., Speigelhalter, D., and Waldmann, H., eds (1987). "Leucocyte Typing III." Oxford University Press, Oxford.

54. Morimoto, C., Steinberg, A. D., Letvin, N. L.,

Hagan, M., Takeuchi, T., Daley, J., Levine, H., and Schlossman, S. F. (1987). A defect of immunonoregulatory T cell subsets in systemic lupus erythematosus patients demonstrated with anti-2H4 antibody. *J. Clin. Invest.* **79**, 762–768.

55. Muirhead, K. A., Wallace, P. K., Schmitt, T. C., Frescatore, R. L., Franco, J. A., and Horan, P. K. (1986). Methodological considerations for implementation of lymphocyte subset analysis in a clinical reference laboratory. *Ann. N.Y. Acad. Sci.* **468**, 113–127.

56. Parks, D. R., Hardy, R. R., and Herzenberg, L. A. (1984). Three-color immunofluorescence analysis of mouse B-lymphocyte subpopulations. *Cytometry* **5**, 159–168.

57. Parks, D. R., and Herzenberg, L. A. (1989). Flow cytometry and fluorescence-activated cell sorting. *In* "Fundamental Immunology", 2nd edn, pp 781–802.

58. Pearl, E. R., Vogler, L. B., Okos, A. J., and Crist, W. M. (1978). B lymphocyte precursors in human bone marrow: an analysis of normal individuals and patients with antibody-deficiency states. *J. Immunol.* **120**, 1169–1175.

59. Pollack, A., Moulis, H., Prudhomme, D. L., Block, N. L., and Irvin III, G. L. (1986). Flow cytometric analysis of cell kinetic responses using measurements of correlated DNA and nuclear protein. *Ann. N.Y. Acad. Sci.* **468**, 55–66.

60. Radzun, H. J., Parwaresch, M. R., Stingl, G., and Knapp, W. (1989). Neoplasms of monocytes, macrophages and dendritic cells. *In* "Human Monocytes" (L. G. Asherson, and M. Zembala, eds), pp. 529–542. Academic Press, London.

61. Reinherz, E. L., Haynes, B. F., Nadler, L. M., and Bernstein, I. D., eds (1986). "Leukocyte Typing II" Springer Verlag, New York.

62. Ritz, J., Schmidt, R. E., Michon, J., Hercend, T., and Schlossman, S. F. (1988). Characterization of functional surface structures on human NK-cells. *Adv. Immunol.* **42**, 181–211.

63. Ryan, D., Kossover, S., Mitchell, S., Frantz, C., Hennessy, L., and Cohen, H. (1986). Subpopulations of common acute lymphoblastic leukemia antigen-positive lymphoid cells in normal bone marrow identified by hematopoietic differentiation antigens. *Blood* **68**, 417–425.

64. Ryan, D. H., Mitchell, S. J., Henessy, L. A., Bauer, K. D., Horan, P. K., and Cohen, H. J. (1984). Improved detection of rare CALLA-positive cells in peripheral blood using multiparameter flow cytometry. *J. Immunol. Methods* **74**, 115–128.

65. Sanders, M. E., Makgoba, M. W., and Shaw, S. (1988). Human naive and memory T-cells: reinterpretation of helper-inducer and suppressor-inducer subsets. *Immunol. Today* **9**, 195–199.

66. Schmidt, R. E. (1989). Monoclonal antibodies for diagnosis of immunodeficiencies. *Blut* **59**, 200–206.

67. Schnabl, E., Stockinger, H., Majdic, O., Gaugitsch, H., Lindley, I. J. D., Maurer, D., Hajek-Rosenmayr, A., and Knapp, W. (1990). Activated human T-lymphocytes express MHC class I heavy chains not associated with β2-microglobulin. *J. Exp. Med.* **171**, 1431–1442.

68. Schwarzinger, I., Valent, P., Köller, U., Marosi, C., Schneider, B., Haas, O., Knapp, W., Lechner, K., and Bettelheim, P. (1990). Prognostic significance of surface marker expression on blast of patients with the novo AML. *J. Clin. Oncol.* **8**, 423–430.

69. Seligmann, M. (1989). Primary immunodeficiencies: current findings and concepts. *In* "Progress in Immunology VII" (F. E. A. Melchers, ed.), pp. 509–518. Springer Verlag, Berlin.

70. Shackney, S. E. (1986). The use of flow cytometry in the diagnosis and biological characterization of the non-Hodgkin's lymphomas. *Ann. N.Y. Acad. Sci.* **468**, 170–177.

71. Shah, V. O., Civin, C. I., and Loken, M. R. (1988). Flow cytometric analysis of human bone marrow. IV. Differential quantitative expression of T-200 common leukocyte antigen during normal hemopoiesis. *J. Immunol.* **140**, 1861–1867.

72. Simon, R., Durrleman, S., Hoppe, R. T., Bonadonna, G., Bloomfield, C. D., Rudders, R. A., Cheson, B. D., and Berard, C. W. (1988). The non-Hodgkin lymphoma pathologic classification project: long-term follow-up of 1153 patients with non-Hodgkin's lymphomas. *Ann. Intern. Med.* **109**, 939–945.

73. Smolen, J. S., Chused, T. M., Leisersen, W. M., Reeves, J. P., Alling, D. W., and Steinber, A. D. (1982). Heterogeneity of immunoregulatory T cell subsets in systemic lupus erythematosus: correlation with clinical features. *Am. J. Med.* **72**, 783–790.

74. Stein, H., and Mason, D. Y. (1985). Immunological analysis of tissue sections in diagnosis of lymphoma, *In* "Recent Advances in Haematology IV" (A. V. Hoffbrand, ed.), pp. 127–169. Churchill Livingstone, Edingburgh.

75. Taki, S., Shimamura, T., Abe, M., Shirai, T., and Takahara, Y. (1989). Biotinylation of human interleukin-2 for flow cytometry analysis of interleukin-2 receptors. *J. Immunol. Methods* **122**, 33–41.

76. Taylor, G. M. (1987). Altered expression of lymphocyte functional antigen in Down syndrome. *Immunol. Today* **8**, 366–369.

77. Thiel, E. (1983). Monoclonal antibodies against differentiation antigens of lymphopoiesis. *Blut* **47**, 247.

78. Thiel, E., Kranz, B. R., and Rhagachavar, A. (1989). Prethymic phenotype and genotype of pre-T (CD7 + /ER)-cell leukemia and its clinical significance within adult acute lymphoblastic leukemia. *Blood* **73**, 1247–1258.

79. Van der Schoot, C. E., Daams, M., Pinkster, J., and Tetteroo, P. A. T. (1987). Phenotypic analysis of acute myeloid leukemia. *In* "Genotypic, Phenotypic and Functional Aspects of Haematopoiesis" (F. Grignani, M. F. Martelli and D. Mason, eds), pp. 51–58. Raven Press, New York.

80. Van Dongen, J. J. M., Adriaansen, H. J., and Hooijkaas, H. (1987). Immunological marker analysis of cells in the various hematopoietic differentiation stages and their malignant counterparts. *In* "Application of Monoclonal Antibodies in Tumor Pathology" (D. J. Ruiter, G. J. Fleuren, and S. O. Warnaar, eds), pp. 87–116. M. Nijhoff Publ., Dordrecht.

81. Van Dongen, J. J. M., Adriaansen, H. J., and Hooijkaas, H. (1988). Immunophenotyping of leukaemias and non-Hodgkin's lymphomas. Immunological markers and their CD codes. *Neth. J. Med.* **33**, 298–314.

82. Van Dongen, J. J. M., Krissansen, G. W., Wolvers-Tetteroo, I. L. M., Comans-Bitter, W. M., Adriaansen, H. J., Hooijkaas, H., Van We, E. R., and Terhorst, C. (1988). Cytoplasmic expression of the CD3 antigen as a diagnostic marker for immature T-cell malignancies. *Blood* **71**, 603–612.

3.5

Biological Characterization and Therapy Monitoring of Leukemia

MICHAEL ANDREEFF

From the Sections of Leukemia and Experimental Hematology
Division of Medicine
The University of Texas M. D. Anderson Cancer Center
1515 Holcombe Boulevard, Houston, TX 77030, USA

ISBN 0-12-432940-3

I. INTRODUCTION

The development of single cell analysis by Casperson in 1936 was a turning point in cell research: visual impressions became quantifiable biochemical measurements, and ever since, the molecular composition of the cell has been probed in order to understand the network of interlinked regulatory mechanisms. The subsequent technological development became a driving force in these studies which initially focused on nucleic acids and then on immunological and molecular targets. Flow cytometry became the preferred tool for the investigation of complex hematopoietic cell populations and has now achieved an undisputed role in the study of leukemia. Investigations of clonality, cell kinetics, immunologic phenotyping, and cellular pharmacology revised the classifications of luekemias and changed treatment strategies. The development of new probes, in particular monoclonal antibodies (MoAbs), to cell surface antigens, improved our understanding of cell phenotypes with regard to hematopoietic lineage and degree of differentiation. In parallel, the molecular techniques of gene sequencing and cloning, western (Protein), northern (RNA) and southern (DNA) blotting resulted in breathtaking advances and knowledge of the etiology, pathogenesis and pathophysiology of malignancies and of leukemias. These molecular techniques do not analyze single cells but average results for large cell populations, with the exception of the DNA polymerase chain reaction (PCR) and *in situ* hybridization (FISH) (*112*). By adapting molecular probes to flow cytometry, multiparameter molecular analysis of single cells should become a possibility.

This chapter will survey recent flow cytometric studies of human leukemia and summarize the applications of flow cytometric techniques in diagnosis, treatment and monitoring.

II. NORMAL HEMATOPOIESIS

The electronic impedance technique for determinations of cell number and volume, introduced by Coulter in 1956 (*54*), is still the most widely used application of flow cytometry in hematology. Refinement of this technique came with the development of hydrodynamic focusing, i.e. aligning cells in the center of the flow channel by differential pressures between sample and sheath fluid. In the present, multiparameter flow cytometers, cell number and cell size are determined by measurements of laser light scattered determined at small angles (0.5–2°). which is proportional to cell volume (*143*), and by scatter at 90° which is related to granularity. These two measurements alone, on unstained samples, allow the discrimination of debris/erythrocytes/platelets from lymphocytes, monocytes and granulocytes (*175*). Increased light scatter resolution and mathematical transformation of orthogonal light scatter signals allows discrimination of peripheral blood neutrophils, eosinophils, basophils, monocytes, granular and non-granular lymphocytes (*200*). Supravital staining with the metachromatic dye acridine orange (AO) was another approach that allowed automated blood cell differentials by flow cytometry (*138*).

The complex binding of AO to nucleic acids was extensively investigated by Darzynkiewicz, who introduced the technique of simultaneous quantitation of cellular DNA and RNA content (*207*), following the work of Andreeff and Haag (*17*) who described flow cytometric staining for RNA by AO. This procedure has been invaluable for determining cell cycle analysis and DNA ploidy determinations in hematopoietic tissues, as it allows the dissection of cells in different stages of the cell cycle (G_0, G_{1A}, G_{1B}, S and G_2M). The metachromatic dye AO binds differentially to RNA and DNA, resulting in red (RNA) and green (DNA) fluorescence. The mechanism of binding was described by Darzynkiewicz and Kapuszinski (*59*). In non-proliferating peripheral blood cells, differences were found in the RNA content of granulocytes B, T and non-B, non-T cells: granulocytes B and suppressor T cells had the lowest, and helper (T_μ) cells and rare progenitors had increasing RNA content ((*6*) and unpublished data). Enriched peripheral blood granulocyte progenitor cells are characterized by highly increased RNA content (*71*), which decreases with maturation, an observation that parallels the decrease in basophilia observed in Wright–Giemsa-stained cells. The AO two-step technique, which employs Triton-X to make cells permeable for AO, results in damage to the cytoplasmic membrane. Granulocytes in particular are affected, and are stripped of their cytoplasm and elongated by the shearing force of inside the flow channel. The resulting elongation of the nucleus can be detected by increased green pulse width (GPW). Cells characterized by high GPW have lost their cytoplasmic RNA and have a characteristic increase in integrated green (DNA) fluorescence, which can be misinterpreted as hyperdiploidy. Gating on pulse width parameter corrects for this artefact. Pulse width not only identifies granulocytes, but is also very useful for discriminating cell aggregates. Gating on pulse width is crucial for determining the true number of cells in G_2M and for excluding cell aggregates that mimic polypoidy (*4*).

Staining the small amounts of RNA in reticulocytes with pyronin Y allowed Tanke to automate the counting of these immature erythrocytes (*192*); acridine orange and thiazole orange have also been used for this purpose.

A further modification of the AO technique allows the measurement of double stranded (native) vs single stranded (denatured) nucleic acids in the intact cell or nucleus. As shown by Darzynkiewicz (*60*), changes in the sensitivity of DNA to heat or acid accompany cells as they traverse the different compartments of the cell cycle. This technique discriminates all proliferating and quiescent cell compartments. In particular and paradoxically, G_0 cell chromatin is more sensitive to denaturation than that of G_1 cells and mitotic cells are more sensitive than G_2 cells. This technique allows an excellent assessment of proliferation in the hematopoietic system.

The development of specific monoclonal antibodies (MoAbs) to cell surface antigens was probably the single most important factor in the development of FCM as a hematological–immunological technique. An overview of MoAbs and their cluster designations (CD) is given in Chapter 3.4. The initial antibodies to lymphocyte antigens have now been complemented by those directed against antigens of the erythroid, myeloid and megakaryocytic lineages (*77,78*). These MoAbs have become indispensible tools in leukemia classification. Examples are given below for the specific leukemia subtypes.

While flow cytometry techniques have been successful in identifying a variety of blood cells, the heterogeneity of bone marrow cells has until recently precluded an automated approach to bone marrow analysis. Differentiation antigens are expressed over rather prolonged periods of cell development and are not sufficiently restricted to match the accuracy of conventional morphological evaluation. The co-expression of several antigens, however, is tightly regulated and can be utilized in an improved assessment of cell lineage and differentiation. Such developmental schema have been established for lymphoid (*127*), erythroid (*128*) and myeloid (*92*) cell lineages. Loken *et al.* (*128*) have measured the expression of the transferrin receptor, of glycophorin A and of the HLe-1 antigen and were able to correlate proliferation and differentiation for the erythroid cell lineage (Fig. 1(a)). In the myeloid series, the earliest antigen identified so far is CD-34 (MY10), followed by CD-33 (MY9), CD-13 (MY7), MY8, CD-11 (Mo1, OKM-1) and CD-14 (MY4, Mo2). However, antigens like MY9 and MY7 are expressed on CFU-GM, myeloblasts, promyelocytes, myelocytes, metamyelocytes, monoblasts and monocytes and are not helpful in identifying discrete steps of cellular differentiation. In conjunction with other antigens, however, they contribute to better delineation of cellular differentiation. Terstappen *et al.* (*199*), have delineated the myeloid differentiation pathway by analysis of CD11b, CD15, CD16 and found that the morphologic changes during neutrophil maturation could be identified by combined immunofluorescence/light scatter FCM (Fig. 1(b)). Quantitative comparison of myeloid antigens by five-dimensional cytometry was used by Terstappen *et al.* (*196*), to identify the

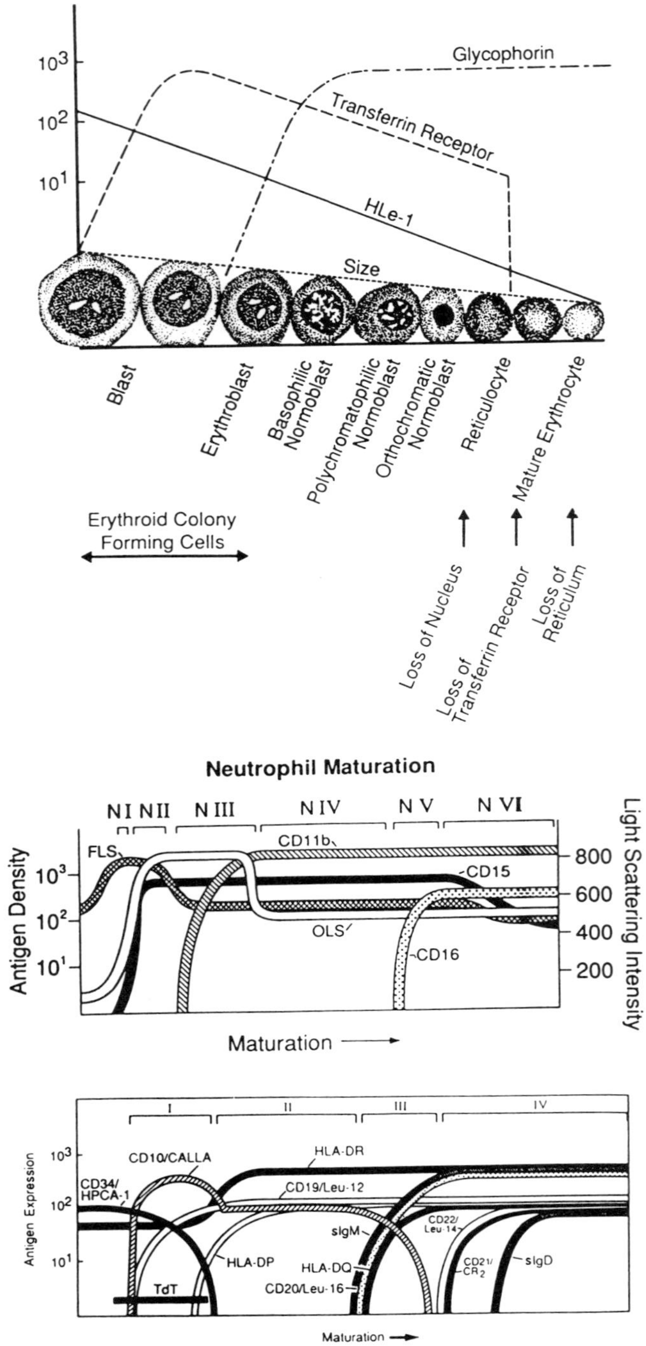

Fig. 1. (A) Schematic representation of erythroid cell maturation. Relative quantities of cell surface antigens are shown on top, morphology and colony formation on bottom (*128*). (B) Schematic representation of neutrophil maturation. Relative quantities of cell surface antigens are shown on left abscissa and light scatter on right abscissa, increased maturation on the ordinate. Six stages (NI–VI) can be discriminated. OLS = orthogonal light scatter, CD16 = MY23, CD11b = anti-Leu15, FLS = forward light scatter, CD15 = L16 (*201*). (C) Schematic representation of B cell maturation in normal human bone marrow. Four stages of maturation (I–IV) can be discriminated (*127*).

five types of peripheral blood cells (neutrophils, eosinophils, basophils, lymphocytes, monocytes). MoAbs CD11b, CD13, CD14, CD15, CD16, CD33, CD38, CD45, CD45R, anti-HLA-DR and anti-Leu8 were labeled with either fluorescin or phycoerythrin and measured in conjunction with forward and orthogonal light scatter. Loken *et al.* (*127*) used MoAbs to CD19, CD20, CD34, and CD10 to identify different steps in normal B cell development. $CD19^+$, $CD10^+$, $CD34^+$ cells comprised only 0.6% of marrow cells but contained the majority of terminal deoxynucleotidyl transferase (TdT^+) cells (Fig. 1(c)). HLA-Dr was expressed on all B lineage cells. During maturation cell surface IgM, CD20, and HLA-DQ were acquired at the same time and CD10 was lost when CD21 and CD22 were acquired. Normal plasma cells were identified by a similar approach (*198*). The frequency of these cells in normal bone marrow was 0.25% and early lymphoplasmacytoid (small, $CD22^+CD35^+sIgE^+$) cells could be discriminated from mature plasma cells (larger, $CD22^-CD35^-sIgE^-$). Unexpectedly, "myeloid-specific" antigens such as CD33 and CD13 and the "early B cell" antigen CD10 were also expressed.

DNA and DNA/RNA measurements in normal marrows are hampered by lineage heterogeneity. It can be demonstrated, however, that the cellular RNA content decreases when cells undergo differentiation. The RNA index, i.e. the cell cycle corrected RNA content of G_1 cells (*11*), was found to be 27.2 for normal myeloblasts and promyelocytes, 22.4 for myelocytes and 21.3 for metamyelocytes using normal bone marrow fractionated by a $1 \times G$ velocity sedimentation gradient. The number of cells in S phase also decreased with differentiation (*4*). The same was found in experimental cell systems when cells were treated with inducers of differentiation, including DMSO and TPA (*48,62*). A combined DNA/RNA/differentiation antigen approach would be useful for improved analysis of hematopoietic cell differentiation and proliferation. Ultimately, total cellular RNA measurements ought to be supplemented by quantitation of mRNA of cell cycle and differentiation dependent genes, including PCNA ("cyclin"), cdc2 (cyclin), Ki67, c-myc, c-fos, c-myb, c-ras, p53, and others. *In situ* hybridization techniques for specific mRNA should be adapted for flow cytometric quantitation in single cells.

Though FCM measurements of DNA can provide information regarding the number of cells in different cell cycle phases, the important cell cycle transit times were determined from autoradiographic data using labeled thymidine. A review of hematopoietic cell kinetics has recently been published (*3*). The adaptation of the BUdR technique to FCM has not yet provided complete kinetic analysis of normal hematopoietic cells, but the combination of techniques to identify restricted antigens and kinetic double labeling techniques like Dolbeare's BUdR/IUdR method (*120*) may finally result in a complete picture of the kinetics of developing hematopoietic cells. Pallavicini's elegant investigations of DNA content, DNA synthesis (BUdR) and agglutinin binding in hematopoietic cell populations point in this direction (*148*).

Many groups have attempted to identify the putative human hematopoietic stem cell and flow cytometry has become an important means to that end. Atzpodien *et al.* sorted human progenitor cells based only on light scattering properties and obtained 3- to 4-fold enrichment of CFU_{GM} and BFU_E (*27*). Lu *et al.* used MoAbs to CD34 and HLA-DR to enrich human marrow progenitor cells and achieved a cloning efficiency (CE) of up to 47% in fractions of cells expressing high CD34 and low DR antigens (*131*). By comparison, CE was 0.01% in CD34DR fractions. Lansdorp found that $CD34^+CD45RO^+$ cells were able to initiate hematopoiesis in long-term cultures (*123*). In the mouse system, the Stanford group succeeded in isolating a pluripotent hematopoietic stem cell by depletion of unwanted cells with magnetic immunobeads and MoAbs and subsequent cell sorting (*184*). The 1000-fold enriched CFUs were mostly in $G_{0/1}$ (>97%)

and only 30 of these cells were needed to rescue one-half of a group of lethally irradiated mice. Once the most immature, truly pluripotent "stem" cells are isolated, they will permit a much better understanding of hematopoietic development and of the mechanisms underlying luekemogenesis. They will also be used in the reconstitution of the human hematolymphoid system after ablative therapy, and in gene insertion therapy. In the human system a group of $CD34^+CD38^-$ progenitor cells was identified. This cell population encompassed only 1% of $CD34^+$ cells and 0.01% of all nucleated marrow cells (*197*). Sorting and culturing of these cells in the presence of cytokines resulted in a cloning efficiency of 25% and these cells had the capacity of self-renewal during five subsequent generations in the presence of cytokines (*197*).

III. Acute Myeloblastic Leukemia

A. Classification

The definition of acute leukemias is that of an accumulation of immature cells in the bone marrow "frozen" at an early stage of differentiation. This maturation arrest causes clinical consequences by way of quantitative or functional defects in hematopoiesis, manifested in depletion of erythrocytes, granulocytes/monocytes, and platelets. Anemia, infection and hemorrhage are the ensuring clinical consequences. The acute leukemias are classified by morphological and histochemical criteria as defined in the French–American–British (FAB) classification (*34,37*). They encompass the acute myeloid leukemias (M1 = myeloblastic without maturation, M2 = myeloblastic with maturation, M3 = promyelocytic, M4 = myelomonocytic, M5A = monocytic leukemia without maturation, M5B = monocytic with maturation), erythroblastic leukemias (M6), megakaryoblastic leukemias (M7) (*35*) and three subtypes of acute lymphocytic leukemia (L1, L2, L3). The reproducibility of this taxonomic system is not very good, but it provides at least morphology-based criteria that are used universally. The clinically important discrimination between AML and ALL can be achieved with a high degree of accuracy by well-trained hematologists, although FAB M1 and L2 cases can easily be confused. Treatment protocols and clinical results differ greatly for AML and ALL. ALL in children has a rate of complete remissions (CR) in excess of 90%, and 70% of children are long-term survivors. In adults the CR rate is in the 70–80% range, and 30–35% of patients achieving remission survive 5 years or longer (*15*). The response rates are lower in patients with AML, with only 10% of adults surviving 5 years (*51*). In AML, the FAB classification is of no prognostic importance, with the possible exception of M3 which may be associated with longer remission duration (RD) (*51*). Age, in all studies, is the main prognostic variable. Cytogenetics has recently emerged as another major prognostic factor (*73*). Unfortunately, no progress has been made over the past 15 years regarding survival of patients with acute leukemia (*50*).

FCM has greatly contributed to the unraveling of the heterogeneity of AML. In particular, nucleic acid content measurements have identified DNA stemlines (aneuploidy and multiploidy) and facilitated investigations of cell cycle kinetics under steady state conditions and following chemotherapy and experimental treatment with cytokines.

B. Cell Cycle Kinetics

The classical autoradiographic studies demonstrate a high rate of production of myeloblasts in AML. Conceptionally, it is important to understand that increased proliferation, decreased differentiation or decreased cell loss can all result in the accumulation of immature blast cells that characterize leukemia. Since it is very difficult to estimate cell loss in patients, the relative contributions of the different mechanisms are not known. The average T_s (DNA synthesis time) is estimated between 11 and 29 hours (*68*), but the range reported in the literature varies from 6 to over 40 hours (*52*). Cell cycle times determined autoradiographically vary from 49 to 95 hours, but by eliminating quiescent cells a probably more accurate generation time of 24 hours can be estimated. The size of the quiescent compartment is difficult to determine since cells are constantly entering and leaving it. Continuous labeling with radiolabeled DNA precursors in patients demonstrates the presence of 1–12% unlabeled (i.e. quiescent) cells even after 8–21 days of labeling (*168*). An alternative approach to the determination of the growth fraction is the estimation of the percentage of cells having DNA polymerase and primer template DNA (TTP index) (*122*). A recalculation of the published data shows a mean TTP index of 42.6% for AML and of 28.2% for ALL (*3*). The pretreatment bone marrow labeling index for myelocyte luekemia is approximately comparable to that of normal myeloblasts. Flow cytometric determinations of S phase (DNA synthesizing) cells vary: 5.1 ± 3.8% (*11*), 6.9 ± 3.6% (*22*), 11.3% (*108*), 5.06 ± 3.70 (*5*), and 9.9% (*167*). Variability in results can be attributed to: (1) sample error, i.e. quality of the marrow aspirate; (2) DNA dyes used; (3) coefficient of variation of measurement; (4) DNA histogram analysis program used. A major source of error is the composition of the sample studied: the more peripheral blood cells present in the "marrow" sample, the lower the measured percentage of cells in S phase. This is due to dilution with non-proliferating normal and low-proliferating leukemic blood cells. Consequently, S phase is consistently higher in marrow biopsies as compared to (diluted) marrow aspirates and peripheral blood samples in the same patient (*69,98,99*). Several methods were developed by Hiddemann *et al.* (*99*) and Holdrinet *et al.* (*107*) to subtract the peripheral blood component as estimated by the number of erythrocytes (or hemoglobin) which is presumed to be present in the blood but not in the true marrow compartment.

The development of antibodies to bromodeoxyuridine (BUdR) by Gratzner (*85,86*) and others (*67,211*) greatly facilitates the identification of S phase cells. The method allows the enumeration of the number of cells synthesizing DNA and also the calculation of T_S (*176,218*). Raza *et al.* (*159,161*) developed a double labeling technique for leukemic cells by combining *in vivo* BUdR exposure with *ex vivo* tritiated thymidine labeling and autoradiography. Short-term intravenous infusion of BUdR into patients with leukemia was used for *in vivo* labeling: the number of S cells in biopsies was 20%, in aspirates 9%. Double labeling experiments revealed T_S values from 4 to 49 hours (medium 17 hours), and T_C (cell cycle time) values from 16 to 292 hours (median 76 hours). The surprisingly short T_S values estimated in some samples could possibly be related to changes in the rate of DNA synthesis *ex vivo* and these results require confirmation by *in vivo* labeling techniques using BUdR and IUdR. Nevertheless, the extreme heterogeneity of cell cycle parameters appears to be of clinical significance, as patients with very short and with very long T_C had a poor response to therapy (*147,159*). Preisler *et al.* (*155*) reported that low pretreatment S phase was correlated with poor response in AML patients treated with high-dose ARA-C. The prognostic importance of pretreatment S phase in AML is uncertain as other studies have found no predictive value and may have to be reinvestigated using the BUdR technique. Raza *et al.* (*160*) reported that long T_c was associated with prolonged remission duration in AML which was interpreted as slower regrowth after remission, of leukemic cells with longer cycle times. A limitation of this approach is the lack of estimation of the number of G_0 cells by the BUdR pulse techniques used, which could be remedied by continuous infusion of BUdR in an attempt to label all cells entering S phase.

A very elegant technical modification was recently developed by Crissman *et al.* (*55*) and Kubbies *et al.* (*119*), who determined BUdR incorporation not by antibody binding, but by the quenching of Hoechst 33258 fluorescence. In this assay BUdR prevents denaturation of DNA and may permit the simultaneous determination of incorporated BUdR, of cellular

RNA content and of antibodies labeled with fluorescein isothiocyanate. If sensitive enough, this technique would allow cell kinetic measurements of antibody-specified cell lineages in the hematopoietic system. In a related approach BUdR suppressed the green (DNA) fluorescence of acridine orange-stained stimulated lymphocytes (*58*) and thus allowed the discrimination of cycling and non-cycling cells.

Terry *et al.* (*195*) have published a method to obtain potential doubling times from a single measurement of BUdR and DNA. The relative movement of BUdR-labeled cells allowed for computation of T_S. We have used this principle to determine T_{Dpot} and T_S for leukemic cell populations *in vivo*: patients were infused with BUdR for 1 hour and a single marrow sample was obtained after 4 hours (Fig. 2). A hyperdiploid DNA stemline (DI:1.4) can be distinguished from diploid cells. The BUdR positive aneuploid cells have moved from early towards mid-S allowing the calculation of $T_S = 40$ h and $T_{Dpot} = 10.2$ days. The diploid, putatively normal cells have only very few cells in S, which have not moved, indicating very long T_S and/or kinetic quiescence.

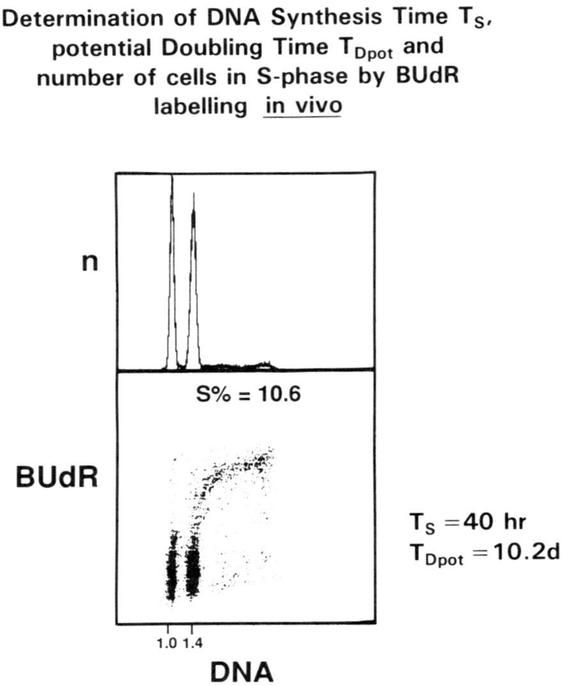

Fig. 2. BUdR (200 mg/m^2) was infused over 1 h and bone marrow was obtained after 4 h. BUdR was stained with IU4, and DNA with propidium iodide. T_S and T_{Dpot} were calculated according to the method of Terry *et al.* (*195*).

No significant cell kinetic differences were found among the different FAB subgroups (*5*), except in one series in which promyelocytic leukemia was found to have the lowest and erythroleukemia the highest S phase (*132*). Unclassified leukemia (M0) also had a high S phase compartment.

The introduction of acridine orange flow cytometry, first for RNA measurements (*17*) and subsequently for simultaneous measurements of DNA and RNA (*11*), added a new dimension to FCM of leukemia (*10,12*). The method allowed discrimination of AML from ALL (*26*), based on calculation of RNA index (RNA index = mean RNA content of $G_{0/1}$ cells of sample × 10, divided by the median RNA content of normal control lymphocytes). The RNA index of AML (20.6 ± 4.0) was found to be significantly higher than ALL (11.2 ± 2.0). These results were confirmed by Maddox *et al.* (*132*), Barlogie *et al.* (*31*), and Maslack *et al.* (*136*). (A comparison of Figs 3 and 4 exemplifies this difference.) No clear correlation was found between RNA index and FAB subgroups (*5,136,216*).

The additional measurement of cellular RNA content also proved valuable in detecting changes in the $G_{0/1}$ compartment, which could not be appreciated by DNA measurements alone. Figure 3 shows that G_0, G_1, S and G_2M as well as DNA aneuploidy can be discriminated. The importance of RNA lies also in the identification of populations of quiescent G_0 cells with low RNA content, which are conceptually of great interest since they probably harbor those luekemic stem cells that survive the initial chemotherapy and lead to relapse. Studies of the transition between G_0 and the different G_1 compartments (G_{1A}, G_{1B}) are also aided by AO staining of RNA content, which is directly associated with proliferation.

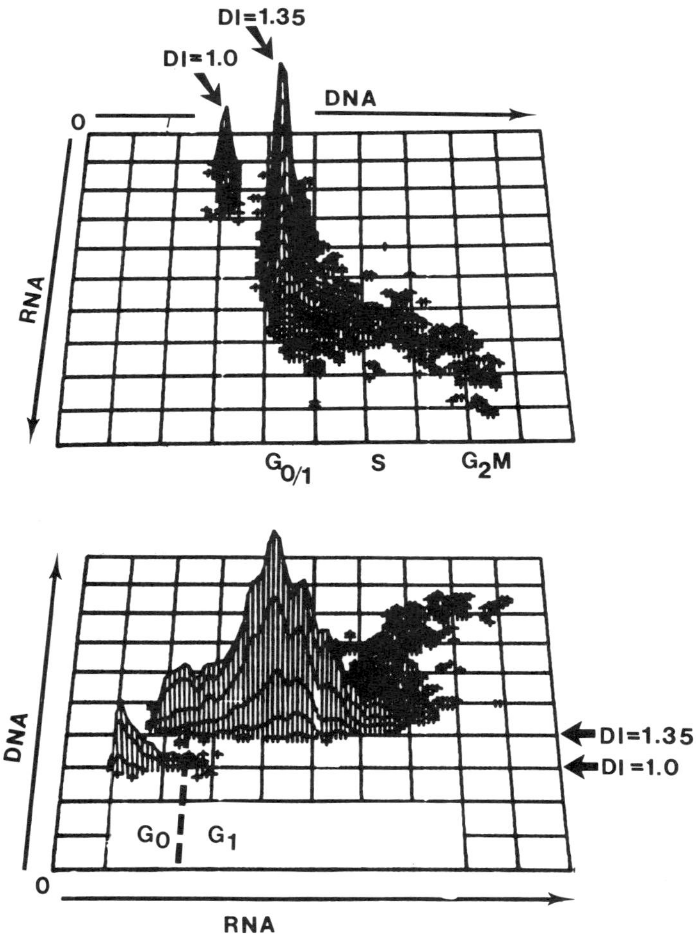

Fig. 3. DNA/RNA histograms of acute leukemia. Diploid (DNA index 1.0) cells can be distinguished from hyperdiploid cells and cell cycle stages can also be discriminated for the aneuploid cells (from (*4*), with permission).

Acute leukemia with high pretreatment RNA index (cells with low RNA content have low and cells with high RNA content high proliferative activity) is associated with high response rate and long remission duration (*5*). None of 16 patients with low RNA content (RI $G_1 < 16$) at diagnosis survived 12 months, but 35 out of 81 patients with RNA index $G_1 > 16$ did (*19*). This difference was also apparent in the response to induction therapy, and RNA index remained an independent variable in Cox multivariate analysis for survival ($p = 0.026$). The prognostic importance of RNA content was only apparent when specific indices for "G_0" (RI G_0) and "G_1" (RI G_1) were used. The prognostic importance of the RNA index was confirmed in Maddox' analysis (*132*) in which AML patients with RNA index $G_{0/1} > 20$ had longer CR durations ($p = 0.04$) than patients with RNA index < 20, but another series from M. D. Anderson Hospital which did not discriminate RNA index subgroups did not confirm the prognostic significance of RNA content (*216*).

Ffrench *et al.* (*76*) studied protein content in acute myeloid leukemia by staining with FITC and propidium iodide. Cells with low protein content, assumed to be quiescent, were correlated with FAB groups in that the peak value of protein content histograms was significantly lower in myeloid (M1–M3) than in monocytic (M4,M5) leukemias. The fraction of cells with low protein content (LPC) was higher in the CR group than in non-responders ($p < 0.01$). In a second report, the LPC fraction was lower for patients older than 50 years who did not achieve CR and higher for patients over 50 years who achieved remission as well as for younger patients who failed to respond (*75*).

A modification of DNA FCM allows for the cellular quantitation of double stranded RNA (dsRNA). DNA is treated with DNAse and the remaining dsRNA is stained with propidium iodide (PI) (*111*). High numbers of cells with dsRNA were found in AML and the persistence of dsRNA excess in the marrow of patients with AML in remission predicted a short remission duration (7 vs 22 months, $p = 0.05$).

When RNA is digested with RNAse, and DNA is partially denatured *in situ* the metachromatic properties of AO permit selective staining of single-stranded (ss) and double stranded (ds) FNA (*60*). All phases of the cell cycle can be distinguished with this method and significant differences were found between AML and ALL (*8,9,13*): the chromatin of ALL G_1 cells is more sensitive to standardized denaturation than that of AML leading to higher α_t (ratio of denatured to total DNA) values for ALL than for AML. Besides their diagnostic utility these results may suggest basic kinetic differences between AML and ALL; the high α_t associated with ALL may indicate a more quiescent cell cycle state than the low α_t values found in AML.

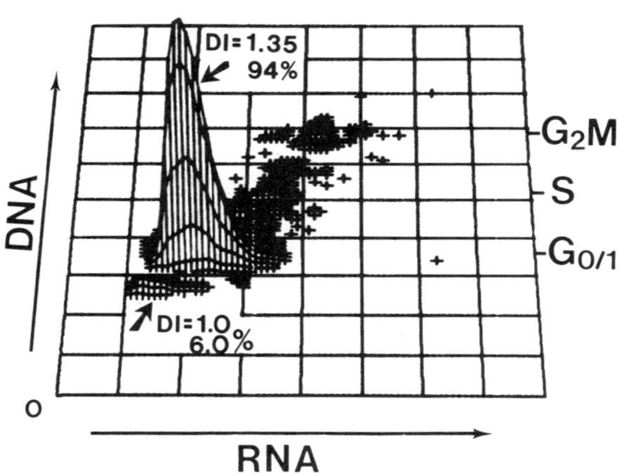

Fig. 4. DNA/RNA histogram from acute lymphoblastic leukemia. Note difference in RNA content of hyperdiploid G_1 cells in comparison to Fig. 3.

With BUdR and IUdR now available for *in vivo* studies of leukemia kinetics, questions related to growth fraction and recruitment in leukemia will be reassessed in the next few years. As noted, BUdR quenches not only Hoechst 33258 but also suppresses AO green fluorescence of stimulated lymphocytes: thus cycling cells in the presence of BUdR will have diminished AO green fluorescence and cells entering G_0 after having progressed through DNA synthesis are distinguishable from cells that remained quiescent during the period of BUdR exposure (*58*). Hopefully, the considerable gap in our knowledge of leukemia kinetics will be closed in the next few years.

Meanwhile, *in vitro* experiments using fresh leukemic myeloblasts are testing the hypothesis that recruitment of leukemic cells into the cell cycle will render them more susceptible to cell cycle specific drugs (*24,25,190,191*). A first clinical trial has already demonstrated that GM-CSF can recruit leukemic myeloblasts into the cell cycle and that good cliical responses are achievable (*40*). This concept requires additional verification.

C. Aneuploidy

Detection of DNA aneuploidy by FCM allows identification of malignant cells without the need for dividing cells, as is required for cytogenetic analysis. But it is also important to recognize the limitations of FCM in the detection of aneuploidy.

(1) The sensitivity is limited by the coefficient

variation (CV) of the measurement. Under optimum conditions some instruments can provide a CV of < 1% and distinguish male and female cells. Even under those ideal conditions balanced translocations with no gain or loss in DNA cannot be detected, by definition. Many leukemias have only small abnormalities in DNA content; monosomies of chromosomes 5 or 7, for example, will usually remain undiscovered by one-parametric DNA measurements. This abnormality is more easily detected by *in situ* hybridization using chromosome-specific DNA sequences as markers (*153*). Nevertheless, amongst 44 patients with AML exhibiting balanced translocations and inversion of chromosome 16, 36% had hyperdiploid DNA stemlines and 20% of patients with diploid karyotype displayed an abnormal DNA index (*216*).

(2) FCM measures the binding of certain dyes to DNA and all DNA may not be stained (*61*); accessibility of DNA is affected by chromatin proteins and varies in different cell types and with the degree of differentiation. Appropriate controls are, therefore, required from the same (normal) tissue, individual or species as outlined in the "Convention of Nomenclature for DNA Cytometry" (*101*). The degree of DNA aneuploidy in AML is usually small (DNA index (DI) ≤ 1.15) compared with ALL ($DI = 1.2 \pm 0.04$); and DNA aneuploidy is also less frequent in AML (27%) (*216*), 21% (*5*) and adult ALL (26%) (*216*), 20% (*5*), than in pediatric ALL (39%) (*102,129,168,189*). Once an aneuploid DI is determined it can serve as an excellent leukemia cell marker for determining the expression of differentiation antigens, and of specialized proteins including oncogene products. It is this ability of DNA FCM to positively identify the leukemic cell that makes it invaluable in leukemia research, where other methods are to able to positively identify individual leukemia cells in large cell populations.

Multiple DNA stemlines are not infrequent and allow the investigation of polyclonality and lineage fidelity. Redner *et al.* (*163*) observed a patient with four DNA stemlines, all of which were leukemic as evidenced by their expression of the common leukocyte antigen (CALLA) and pre-B cell markers. If a specific marker is expressed, DNA aneuploidy identifies these cells unequivocally as leukemic. Of particular clinical value is the detection of small numbers of aneuploid cells after induction therapy in patients in clinical complete remission. These patients have incomplete responses to chemotherapy and relapse early (*22*). In the spinal fluid of the central nervous system small numbers of hyperdiploid cells can be detected by FCM and establish CNS involvement as there is no background of proliferating normal cells as in bone marrow samples. Central nervous system relapse was detected by FCM up to 6 months prior to clinical diagnosis (*162,166*) and the doubling time of leukemic lymphoblasts in the CNS could be calculated from cell counts and percentage of aneuploid cells as 20 days in one case.

DNA aneuploidy is not prognostic in AML, which is in marked contrast to the significant contribution of cytogenetics to the prognostication of AML.

D. Myeloid Differentiation Markers

Differentiation was initially studied by enzymatic histochemical reactions on cells deposited on slides. They still are useful in identifying myeloid cells by myeloperoxidase (POX) or Sudan Black (SB) positivity, monocytic cells by esterase reactions (naphthyl AS-C acetate esterase inhibited by fluoride) and erythroid cells by periodic acid Schiff (PAS) positivity as seen prominently in erythroleukemia (FAB: M6). One of the commercial FCM instruments measures light scatter, peroxidase and esterase content and provides WBC differentials of good quality (*72*). In the rare and underdiagnosed acute megakaryoblastic leukemia (*35*) (FAB: M7), platelet peroxidase (PPO) detected by electron microscopy and MoAb to antiplatelet glycoprotein IIb/IIIa are the best diagnostic tests (*113*).

Morphological criteria alone are insufficient and frequently misclassify M7 as ALL.

The development of MoAbs reactive with lineage- and differentiation-associated cell surface antigens has provided specific reagents for the discrimination and subdivision of leukemic phenotypes. They were initially used in the classification of ALL. The dissection of the myelomonocytic differentiation pathway has developed more slowly, but a large number of myeloid-specific MoAbs are now available. Four international workshops on human leukocyte differentiation antigens were conducted to assess the interrelationships between MoAbs and to determine their reactivities (see Chapter 3.4).

As a result, groups of MoAbs that recognize the same or similar antigens have been identified and designated as cluster groups of differentiation (CD). Some groups of antibodies appear to identify the same or similar antigens by immunofluorescence; however, when not confirmed, or when different molecules appear to be recognized, the CD is designated as a provisional working group (CDw). The expression of several myeloid antigens corresponds to the different stages of myeloid differentiation as assessed morphologically (*91*). Cell sorting can be used to determine antigen expression by multipoint stem cells. For instance, CFU-GEMM express MY10 and MY9, CFU-GM (d14) are positive for MY10, MY9, MY7, p67 and CD15 and CFU-GM (d7) react with MY9, MY7, p67, CD15, VIM2 and CDw17. BFU-E have antigens for MY10, MY9 and p67, CFU-E are positive for p67 and FCU-Meg are positive for the MY10, p67 and VIM 2 antigens.

In leukemia diagnosis MoAbs are most useful in discriminating ALL and AML, though none detect leukemia-specific determinants. Majdic *et al.* (*134*) studied 208 leukemias and found that 91% of AML reacted with MoAb VIM2 while all non-myeloid leukemias were negative. Chan *et al.* (*46*) found that 86% of AML and no ALL were positive with MY9. Mirro and Behm (*141*) summarize results from several large series as in Table I.

Table I

REACTIVITY OF MONOCLONAL ANTIBODIES WITH AML SUBTYPES

	FAB category			
Antibody	M1 + M2	M3	M4 + M5	Overall
MY9	77%	76%	75%	76%
MY7	71%	73%	63%	67%
MY8	33%	32%	70%	52%
MY-1	32%	33%	54%	46%
Mol	38%	40%	84%	46%
MY4/Mo2	17%	2%	54%	35%

The prognostic importance of MoAb in AML is not entirely clear. Griffin *et al.* (*90*) reported that MY4 and MY7 predicted for low complete remission (CR) rates ($MY4^+$:53%, $MY4^-$: 69%, $p = 0.03$; $MY7^+$: 55%, $MY7^-$: 73% $p = 0.01$). The paired combination of antigen expression enhanced the prognostic significance: CR rate in $MY4^-MY7^-$ was 87% and in all other cases was 54% ($p = 0.001$). Expression of HLA-DR, MY8 and Mol was associated with decreased remission duration ($p < 0.05$) and expression of MY8 was associated with decreased survival ($p = 0.03$). Interesting and possibly limiting the prognostic value of MoAb phenotyping are the observed differences between the phenotypes of the total leukemic population and that of the corresponding clonogenic cells (*134*).

Differentiation of myeloid cells is also accompanied by changes in chromatin structure which result in changes in the ability of certain fluorochromes to intercalate into DNA (*61*), in differences in the sensitivity of chromatin to denaturation (*62*), and in a decrease in cellular RNA content (*22*).

The classic definition of acute leukemia as the result of a block in differentiation pathways deserves amendment: recent studies using the molecular method of restriction fragment length polymorphism (RFLP) have unequivocally documented clonal abnormalities in granulocytes of patients with AML (*74*), suggesting terminal differentiation of a small number of leukemic cells. *In vitro* experiments have demonstrated that the induction of differentiation of DNA aneuploid myeloid leukemic blasts resulted in macrophage-like cells with persistent DNA aneuploidy (*48*). Similar conclusions can be drawn from studies using premature chromosome condensation (*105*) and *in situ* hybridization using chromosome-specific DNA probes (*87*). The clonal origin of cells which are either characterized by specific MoAbs and sorted in a cell sorter, or identified by their nuclear morphology (e.g. granulocytes), can be ascertained by subsequent *in situ* hybridization (*23*). This will undoubtedly result in more precise phenotyping and lineage determination of leukemias since the leukemic nature of cell populations will be recognized without doubt. An example are T cells in patients with myelodysplastic syndromes and monosomy of chromosome 7 (Fig. 5): the majority of T cells were also found to be monosomic. This abnormality of cytokine-producing cells may have important conceptual implications for the pathogenesis of leukemias and myelodysplastic syndromes.

IV. ACUTE LYMPHOBLASTIC LEUKEMIA

A. Classification

The acute lymphoblastic leukemias (ALL) are defined by morphological criteria and are characterized by an accumulation of immature blast cells in the bone marrow. The French–American–British (FAB) classification discriminates three different subgroups: L1 is characterized by small cells with homogeneous nuclear chromatin, regular nuclear shape,

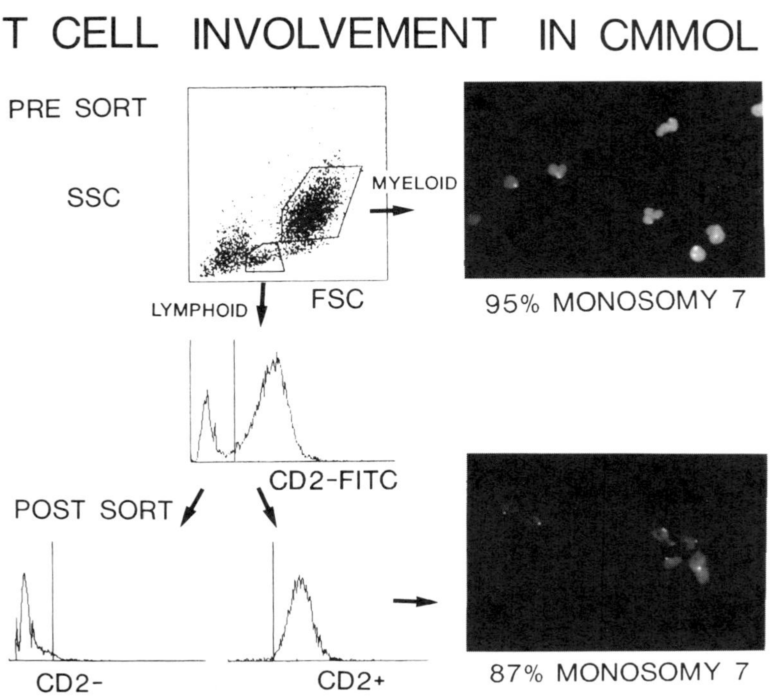

Fig. 5. Combined FACS and FISH analysis of myelodysplastic syndromes. Cells from a patient with chronic myelomonocytic leukemia were sorted based on light scatter (upper left panel) and positivity with CD2 (lower left panel). The sorted cells were remeasured for purity and analyzed by FISH using a centrometric DNA probe for chromosome number 7. Results indicate that both myeloid and CD2 positive cells are clonally abnormal.

invisible of small nucleoli, and scanty cytoplasm; L2 is characterized by larger, heterogeneous cells with more variable nuclear chromatin, irregular nuclear shape with common indentations, one or more nucleoli, and often moderately abundant cytoplasm; L3 consists of large and homogeneous cells with finely stippled chromatin, regular nuclear shape, one or more prominent nucleoli and abundant cytoplasm with deep basophilia and frequent vacuolization. Cytochemical stains are essentially negative except for PAS, which unfortunately can also be positive in acute non-lymphocytic leukemia (ANLL), especially in erythroleukemia. In children, the presence of L2 or L3 ALL, or a significant number of L2 cells in otherwise L1 disease, has been described as conferring a poor prognosis. In adult ALL, morphology has no prognostic significance except for L3 which is associated with poor prognosis. White blood cell count, karyotype, immunophenotype as well as the time to achieve remission are now universally accepted prognostic factors in both children and adults (*47*).

B. Cell Cycle Kinetics of ALL

Cell kinetics of ALL has been studied extensively. The reported mean DNA synthesis time is 18.3 hours and no systematic differences have been seen between T-ALL and common ALL (*66*). DNA synthesis time is significantly longer in ALL than in normal lymphoblasts from various tissues (thymus 7.9 hours, spleen 10.5 hours, lymph nodes 12.7 hours, bone marrow 13.5 hours) (*66*). ALL therefore appears to be the only type of leukemia in which a prolonged DNA synthesis time has been documented in comparison to normal cells. DNA synthesis time and labeling index of hyperdiploid ALL are increased over those of diploid cases, and this observation is consistent with equal DNA synthesis rates in both groups. LI in hyperdiploid ALL was found to be higher than in diploid cases: S phase in 65 patients with diploid DNA stemline was 4.7% and 9.5% in 44 cases with aneuploid stemline, respectively (*189*). Because the speed of DNA synthesis is constant for both groups, T_S is prolonged in hyperdiploid ALL, which requires more DNA to be synthesized before cells enter the following mitotic division. Peripheral blood lymphoblasts have a lower LI or fraction of cells in S phase as compared to bone marrow blasts from the same patients (*189,209*). Using the percent labeled mitosis method, Wagner *et al.* (*213*) found a T_C of 42–47 h, T_S of 42–47 h, T_{G2} of 6 h and T_M of 1 h. This estimate of T_{G2} leaves very little time for G_1. Clarkson *et al.* (*51*) found intermitotic times between 2 and 12 days. The grain count halving time of interphase cells was twice that of mitotic cells, indicating that many daughter cells entered a quiescent state after their parents had divided. The TTP index for ALL bone marrow was 28.3% and 5.4% for peripheral blood, respectively (*122*). This is in concordance with studies of chromatin structure by AO, indicating a more quiescent state for ALL as compared to AML (*3*), and with the observation that cellular RNA content is significantly lower in ALL than in AML (*145,216*). RNA index of G_0/G_1 cells in ALL was 11.2 ± 2.0, only slightly higher than that of normal control lymphocytes (RI = 10.0), and no difference was seen between L1 or L2 subtypes (*189*). A higher S phase was found for L2 lymphoblasts (12.4%) as compared to L1 (5.8%) ($p = 0.01$) (*189*). As in AML, better cell kinetic data can be obtained from marrow biopsy material since it avoids contamination with low-proliferating peripheral blood cells. In pediatric ALL, bone marrow aspirate S phase was 5.5 ± 0.3% as compared to 8.0 ± 0.5% in biopsies (Haimi, Redner, Andreeff, unpublished data). RNA index $G_{0/1}$ was 11.6 ± 0.4 in bone marrow aspirates and 10.6 ± 0.2 in biopsies. Based on these RNA measurements, it appears that most of the $G_{0/1}$ cells are quiescent ("G_0"). A definitive assessment of the growth fraction in ALL will be possible when studies of continuous BUdR infusions in patients are combined with second labels of cell proliferation, such as cell cycle specific proteins. BUdR incorporation *in vitro* into circulating leukemic blasts in lymphoblastic leukemia was higher than incorporation into bone marrow blasts in spite of the much smaller number of S phase cells in the blood, indicative of a higher DNA synthesis rate of circulating leukemic blasts (*4*).

Most studies have failed to demonstrate a significant effect of pretreatment S phase fraction on response in ALL. In one study, patients with marrow LI over 6% or blood LI over 4% were found to have significantly shorter remission durations than patients with lower LI (*70*). In pediatric and adult ALL at Memorial Hospital, no significance of pretreatment S phase was apparent (*5*). Pretreatment RNA content of cells in G_0/G_1 however, appears to be of prognostic importance with regard to achievement of remission, achievement of early remission and remission duration. Extensive multivariate analysis in a larger series of adult ALL showed that a low G_0 RNA index was the fourth most important unfavorable prognostic characteristic ($p = 0.011$) (after L3 or AUL morphology, white blood cell count (WBC) > 10 000/mm^3 and weight loss) for achieving complete remission. For the analysis of remission duration, a high RNA index (> 14) was the third most important factor ($p = 0.017$). In pediatric lymphoblastic leukemia, like in adult ALL the RNA index $G_{0/1}$ was significantly lower (11.66 ± 0.44) than in myeloblastic leukemia (14.24 ± 0.51) (Redner and Andreeff, unpublished data). Early response to treatment, the single most important factor for remission duration, was influenced by four main variables identified in stepwise logistic regression analysis: WBC ($p = 0.0221$), age ($p = 0.0158$), platelet count ($p = 0.0006$) and RNA index G_0 ($p = 0.04$). The lower the RNA content, the more quiescent the cell population, the slower was the cytoreduction induced by chemotherapy. Consequently, recruitment into the cell cycle prior to and during chemotherapy should make cells more susceptible to some of the drugs used in combination chemotherapy as already demonstrated for AML (see above). In adult ALL, once remission was achieved, low pretreatment RNA Index (<12.7) was associated with longer remission duration ($p = 0.017$) (*5*). Recent cell kinetic analysis sheds some light on a biological consequence of high WBC in pediatric ALL (*94*): WBC was found to be inversely correlated with the $G_{0/1}$ RNA index but not with S phase of marrow cells, indicating a more quiescent state of leukemic marrow cells in patients with high WBC at diagnosis ($r = 0.51$, $p = 0.02$). For these reasons, RNA index appears to be a more sensitive indicator of cell kinetic activity than standard S phase measurements and will be useful in evaluating the effects of cytokines on leukemic cell populations.

A new indicator of kinetic activity appears to be the nuclear protein detected by the monoclonal antibody Ki67 (*83*). Ki67 is not detectable in non-proliferating peripheral blood G_0 lymphocytes, but is expressed and can be quantitated by flow cytometry in G_1, S, and G_2M cells (Fig. 6). A systematic evaluation of Ki67 in leukemia has not yet been published, but expression of Ki67 in non-Hodgkin's lymphoma was found to be of prognostic value in one study. The "proliferating cell nuclear antigen" (PCNA) "cyclin" has likewise not yet been systematically investigated in leukemia. Recent findings suggest that PCNA is an auxiliary protein of DNA polymerase delta. Kurki *et al.* (*121*) developed a flow cytometric method to identify cells expressing PCNA and showed a pattern similar to BUdR. In our own experience, PCNA is expressed in cells in G_1, prior to the initiation of DNA synthesis and may precede the expression of Ki67 in normal lymphocytes stimulated to proliferate (*173*). In leukemic cells, no clear correlation was found between c-myc, c-myb, c-fos, or Ki-67 and cell proliferation (*84*) or between levels of Ki-67, c-myc, PCNA and cell proliferation (*173*). An example is shown in Fig. 6, where diploid cells are all Ki-67 positive and PCNA negative, while hyperdiploid cells are Ki-67 positive and positive or negative for PCNA. Clearly, the functional role of these "proliferation markers" remains to be understood.

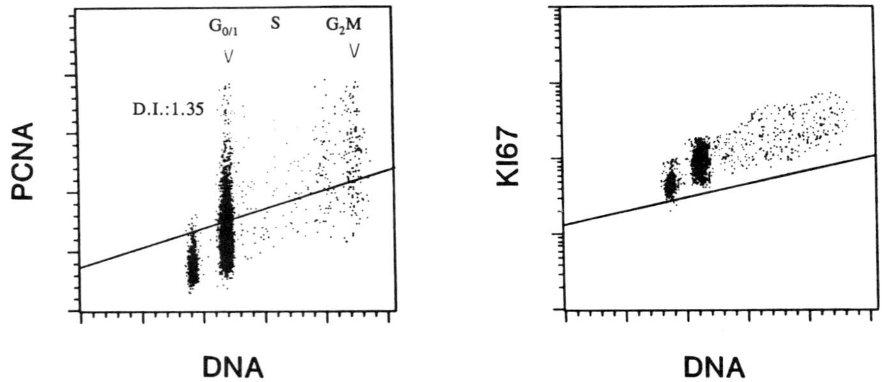

Fig. 6. DNA/PCNA and DNA/Ki67 double staining of hyperdiploid ALL cells (for explanation, see text).

C. Clonal Abnormalities

Approximately 40% of pediatric ALL (*102,129,168,189*) and 30% of adult ALL (*5,216*) have DNA aneuploid stemlines detectable by standard DNA or DNA/RNA flow cytometry with CVs of 2–3%. The frequency of DNA aneuploidy is related to FAB morphology; in pediatric ALL, 34% of L1 and 71% of L2 ALL were aneuploid (*189*). It is likely that the frequency of detectable DNA aneuploidies will increase with more precise measurements, because many karyotypic abnormalities are undetectable with low or "standard" resolution DNA measurements. As demonstrated by Look (*129,130*), DNA ploidy can be directly correlated with differentiation antigen expression, e.g. with the common ALL antigen (CALLA) which was found to be expressed on diploid, hypotetraploid and tetraploid cells in the same patient. Redner *et al.* (*163*) reported a case with four DNA stemlines and differences in the expression of antigens for each stemline.

In addition to being a marker for malignancy, DNA ploidy may be a prognostic factor in ALL. Look *et al.* reported in children treated at St Jude's Hospital that DNA index $\geq 1.15 (\geq 53$ chromosomes) was associated with better prognosis than $\mathrm{DI} < 1.16$ (*129,130*). The markedly hyperdiploid group generally had B cell precursor phenotype with expression of CALLA and HLA-DR. Combining two prognostic variables, children with hyperdiploid leukemia, $\mathrm{DI} \geq 1.16$, and leukocyte count $\leq 25 \times 10^9/\mathrm{l}$ were found to have the lowest probability of relapse. The near-diploid group included many chromosomal translocations (t(9;22), t(4;22), t(8;14), etc.) which are known to be unfavorable, based on the analysis of the "Third International Workshop on Chromosomes in Leukemia" (*1*) and results from the Pediatric Oncology Group (*56*). These results are not uniformly accepted. A series of Memorial Hospital patients did not show prognostic importance of DI, which may reflect that prognostic models are frequently valid only in the context of the therapeutic regimen on which they are based. Improving therapeutic efficacy will successively eliminate prognostic factors until all prognostic factors become insignificant, that is when all patients are cured. Nevertheless, DNA aneuploidy may be an important prognostic factor in the natural history of the disease, but this hypothesis is beyond verification. Supporting evidence can be found in the German BFM trials of pediatric ALL, which have the best therapeutic results reported to date: an initial study (ALL protocol 79/81) showed a tendency for longer remissions for patients with DNA aneuploidy ($p = 0.053$f) (*102*) but the follow-up study (ALL protocol 81/83) did not identify significant differences between DNA diploid and DNA aneuploid patients (*100*).

A patient group of particular importance are those with markedly hypodiploid (DI < 0.8) or near-haploid stemlines. Many cases have a secondary DNA stemline, with twice the DNA content of the primary stemline (*157,164,165*). The median survival from diagnosis of hypodiploid ALL was only 6.0 ± 3.5 months, and survival from initial diagnosis was 16.1 ± 2.9 months in our series (some patients were identified as hypodiploid at relapse). In one such case high expression of the multidrug resistance associated P-glycoprotein (gp180) was found and the bone marrow cells were deeply quiescent and responded only minimally to chemotherapy (*165*).

Many structural and numerical cytogenic abnormalities that are not evident by conventional DNA flow cytometry may be detectable by fluorescence *in situ* hybridization (FISH) using chromosome-specific DNA probes (*153*). In cases with known abnormal DNA content, this technique adds information on whether specific types of chromosomal abnormalities are present. The method was adapted for flow cytometry by Gray, Pinkel, Trask *et al.* and has been successfully used in the detection of minimal numbers of blast cells in sex-mismatched bone marrow transplantation (*18,186*) and also for identifying t(4;11) and the Philadelphia chromosome with t(9;22) in leukemia cells (*152,204*).

D. Lymphoid Differentiation Markers

ALL cells are "frozen" at early stages of differentiation, but most cases can be identified as pre-B or T cell ALL, with very few expressing a true "null" cell phenotype. The most useful marker is the common ALL antigen, CALLA, detectable by several antibodies. Smets *et al.* (*180*) and Look *et al.* (*129*) both reported a close association between DNA hyperdiploidy and CALLA expression that remains to be fully understood.

Foon and Todd (*78*) have reviewed the phenotying of ALL with panels of monoclonal antibodies directed against cell surface and cytoplasmic antigens expressed during B (*125,127*) and T cell differentiation and their schedules may serve as guidelines for an antigen-defined classification of ALL (*4*).

The rare subgroup of L-3 ALL is almost invariably classified as B cell, though exceptions have been found (*116*). A special entity are the adult T cell leukemias (ATL) described in Japan and the Caribbean (*137*). They are virus-induced (HTLVI) and essentially nonresponsive to chemotherapy. They are also characterized by a high number of interleukin-2 receptors. Contrary to initial reports, T-cell phenotype in adult ALL (excluding ATL) is associated with a better prognosis than the pre-B phenotype (*82,106*). Of interest is the absence of detectable IL-2 receptors in common and pre-B ALL, which can be induced *in vitro* by the lectin phytohemagglutinin or by phorbol ester (*206*), resulting in proliferation and colony formation. Our own studies have confirmed the inducibility of IL-2 receptors on aneuploid leukemic lymphoblasts by IL-2, TPA (phorbol 12-myristate 13-acetate) and OKT3. Recruitment into G_1, but not into S phase, was observed in 80% of samples studied (*14*). This approach may be useful in rendering leukemic cells more sensitive to cell cycle-specific drugs as discussed (*191*).

The typical pre-B ALL are not completely frozen in early stages of B cell differentiation. Korsmeyer *et al.* (*114,115*) and Cossman *et al.* were able to differentiate these cells with TPA *in vitro* and Koziner *et al.* (*117*), using the concept of "clonal excess" (CE), found evidence of kappa light chain expression in six patients and lambda CE in two patients in a series of 35 patients studied. They also found SIg lambda expression in the REH cell line, which is derived from a "common" ALL patient, and rearrangement of the lambda gene by southern blotting. Though CE, indicative of clonal predominance of restricted B cells, cannot be found in normal blood and marrow, additional markers to identify leukemic cells would increase confidence in the existence of fully differentiated leukemic B cells. Double staining for DNA and kappa or lambda light chain expression revealed CE in 16 out of 18 aneuploid ALL studied (*26*). It was estimated that approximately 5–8% of cells expressed restricted light chains. FACS sorted, fully differentiated, kappa light chain positive leukemic cells may also have a more "differentiated" morphology and thus not be recognized as leukemic by routine examination (*4*). The initiation of a program of cellular differentiation in ALL would open new therapeutic options.

V. MULTILINEAGE LEUKEMIA

With increasingly sophisticated techniques to identify hematopoietic lineages affected by leukemic transformation, a surprisingly complex picture is evolving. A considerable number (22%) of acute non-lymphoctyic leukemias were found to exhibit deoxy-nucleotidyl-transferase (TdT) activity, as measured in a biochemical assay and TdT^+ patients had lower remission incidence (48% vs 68% for TdT^- patients) and experienced shorter remission durations ($p = 0.003$) (*146*). TdT was thought to be a marker specific for the lymphoid lineage in leukemia and two alternative explanations were postulated: (1) myeloid leukemia samples contain a small number of cells with lymphoid phenotype that express TdT, or (2) immature cells in "AML" are TdT positive but also have myeloid characteristics. The term "biphenotypic" leukemia has been used for these "hybrid" leukemias and the expression of "aberrant" markers was interpreted as "lineage infidelity" (i.e. the cell is committed to a specific lineage and expresses aberrant markers) or "lineage promiscuity" (the cell has not been committed to a specific lineage and expresses markers of two or more differentiation pathways) (*39,139,144,149, 156,181*).

In some cases, lymphoid and myeloid markers were expressed on different cells, in others on the same cells indicating malignant transformation of a pluripotent stem cell. In certain human myeloid leukemic cell lines, a rearrangement of the immunoglobulin heavy chain genes that typically occurs during the early stages of B cell differentiation was detected (*149*). In some cases, a lineage switch was identified when the patient relapsed (*187,215*) usually from lymphoid to myeloid. Two alternative explanations are possible: (1) chemotherapy eradicates the dominant clone present at diagnosis, permitting expression of a secondary clone with a different phenotype, or (2) drug-induced changes in the original clone may either amplify or suppress differentiation programs resulting in a phenotypic shift. Indeed, lineage infidelity could be induced *in vitro* by exposing MOLT-3 cells to a single dose of 5-azacytidine (*182*).

Additional complexity is introduced when antigen expression is examined with regard to different DNA stemlines in cases with aneuploidy or polyploidy. Expression of TdT or CALLA on diploid cells in cases of hyperdiploid ALL indicates that at least some of the DNA diploid cells are leukemic as well (*127,163*). Large variations in the amount of antigen expressed can be found on cells with similar DNA ploidy (*163*). Clonal evolution may be the mechanism responsible for many, if not all, cases with multiple DNA stemlines.

Some cases express myeloid and lymphoid markers with identical or different aneuploid DNA stemlines. An example is shown in Fig. 7, which has diploid (DI = 1) and hyperdiploid (DI = 1.35) cells. Myelomonocytic antigens (MY7, MY4, OKM), HLA-DR (Ia) and TdT are expressed on both diploid and aneuploid cell populations. A particularly interesting case expressed TdT on diploid, near-triploid and tetraploid cells, but only the tetraploid cells also expressed the myeloid-specific M2 antigen. *In vitro* treatment with retinoic acid induced the expression of the M2 antigen on diploid cells as well. Cytogenetic analysis revealed four clones: 46XX, 92XXXX, 92XXXX (2q-), and 65-73 chromosomes (near-triploid) all with 2q-. It is possible that the deletion of 2q after cells became tetraploid introduced clonal instability

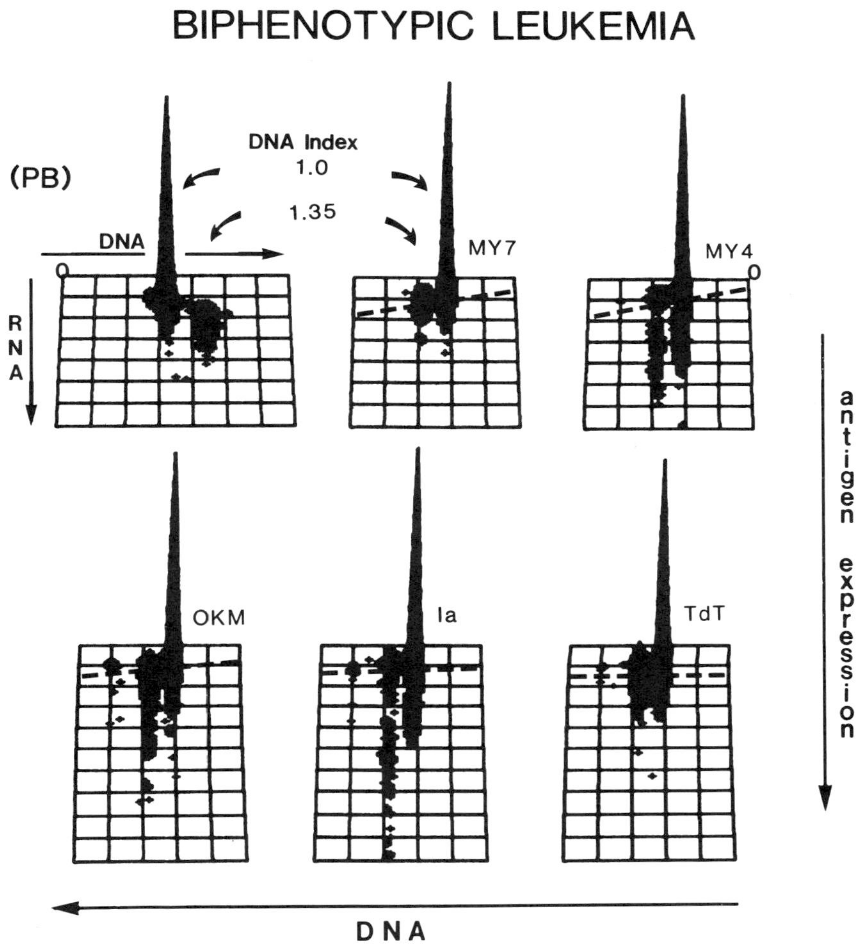

Fig. 7. Biphenotypic leukemia: sample consists of cells with DNA index 1.0 and D.I. 1.35 (hyperdiploid). Myeloid (My7, My4, OKM) and lymphoid (TdT) markers are expressed on both populations. Ia is also highly expressed on both cell populations. Histograms represent acridine orange-stained cells (DNA/RNA) (upper left panel), and DNA/antibody-stained cells (remaining panels).

and that all DNA stemlines and karyotypically defined clones originated from the same cell (*146*). This and other (*16*) examples demonstrate the complexity of clonal and phenotypic evolution in luekemia. Multilineage leukemias with heterogeneous phenotypes (lineage infidelity or promiscuity (*88*)) and clonal variability may be appropriate targets of differentiation therapy in the future. At present they are a formidable challenge to the hematologist and molecular biologist.

VI. MYELODYSPLASTIC SYNDROMES

The myelodysplastic syndromes (MDSs) comprise a number of hematopoietic malignancies with poor prognosis. Five different entities are defined by the French–American–British classification (*36*): (1) refractory anemia (RA), (2) refractory anemia with ring sideroblasts (RARS), (3) refractory anemia with excess of blasts (RAEB), (4) refractory anemia with excess blasts in transformation (RAEBIT), and (5) chronic myelomonocytic leukemia (CMMoL). The FAB subtype is highly predictive of survival with median survivals of 64 months for RA, 71 months for RARS, 7 months for RAEB, 5 months for RAEBIT and 8 months for CMMoL (*79*). MDSs involve two or more hematopoietic cell lineages, indicating involvement of a more undifferentiated stem cell than in many acute leukemias (*158*). Chromosomal abnormalities in MDSs include monosomies of chromosomes 5 and 7, and deletions of the long arms of chromosomes 5, 7 and 20 (*63,124*). A number of genes crucial for hematopoietic growth and differentiation are localized on 5q, including the genes for interleukin-3 (5q23-q31), granulocyte–macrophage colony-stimulating factor (5q23-q32, macrophage-stimulating factor (5q23-q31) and the c-*fms* proto-oncogene which encodes the receptor of M-CSF (5q23-34). Because of the cellular heterogeneity, few flow cytometric studies of MDSs have been reported. In a study of 24 patients by Montecucco *et al.* (*142*), only two patients had aneuploid DNA stemlines, proliferation was higher in RARS and RA than in RAEB and CMMoL and, regardless of diagnosis, low S phase predicted short survival. Maiolo *et al.* found no cell kinetic differences between MDS and AML (*133*). Pedersen-Bjergaard *et al.* (*150*) analyzed patients with secondary preleukemia by flow cytometry and cytogenetics and demonstrated DNA hypodiploidy in the granulocytes of four out of nine karyotypically hyploid (monosomy 7) samples, indicative of the leukemic origin of these granulocytes. Peters *et al.* (*151*) described both hyperdiploidy and hypodiploidy in patients with RA, and Clark *et al.* (*49*) reported DNA aneuploidy in 34 out of 70 patients with MDS. In their study, hypodiploidy was noted in RA and RAEB and was associated with poor prognosis ($p = 0.001$). It was found to be a better prognostic factor than blast percentage. Fluorescence *in situ* hybridization techniques, combined with MoAb detection of incorporated BUdR and FACS sorting may be best suited to investigate lineage involvement, proliferation and clonal abnormalities in MDS. An example of T cell monoclonality in CMMoL is shown in Fig. 5.

VII. CHRONIC LYMPHOCYTIC LEUKEMIA

The chronic lymphocytic leukemias are characterized by a uniform population of lymphoid cells with distinct morphology. They are mostly (95%) of B cell origin with rearrangement of immunoglobulin heavy and light chain genes. Clonality of CLL has been demonstrated by expression of a single Ig light chain on the cell surface. Typical is the relatively weak fluorescence of surface Ig, which can be used to discriminate CLL from lymphocytic lymphomas and from prolymphocytic leukemia. Ig isotype analysis indicates that most CLL display a single heavy chain class (*77*). Cytoplasmic Ig heavy chains have also been reported. B-CLL display receptors for mouse erythrocytes, IgG and complement. Antigens expressed include Ia, Ba1, B1, B2, and B4, and surprisingly, a 65 kDa glycoprotein (T1) thought to be restricted to T lymphocytes (CD5, detected by the T101 and Leu 1 antibodies). A study from M. D. Anderson found two-color analysis of CD5 and surface light chain expression useful for monitoring patients with CLL treated with fludarabine (*169*). The lineage infidelity discussed for the acute leukemias is also evident in CLL: rearrangement of the T receptor has been reported in 10% of B-CLL and in 25% of pre-B ALL. This again emphasizes that Ig and receptor rearrangements alone are not adequate for lineage assignment.

Prolymphocytic leukemia (PLL) differs from B-CLL both morphologically and clinically (*79*). Like B-CLL, PLL express Ia, B1, B2 and B4, but surface Ig expression is more intense and these cells usually lack the T1 receptor. There is high binding of the FMC-7 antibody (*77*), which can be considered diagnostic for PLL.

Hairy cell leukemia is characterized by distinct mononuclear cells with "hairy" cytoplasmic projections. Histochemically, cells are positive for tartrate-resistant acid phosphates. The surface markers expressed and Ig gene arrangements confirm the monoclonal B cell nature of this disease. Antigens detected include Ia, B1, FMC-1 FMC-7, (BA1) and, surprisingly, the plasma cell antigen PCA-1. In addition, the Tac antigen (IL-2 receptor) is expressed in most cases (for review, see *77*).

The cell kinetics of CLL has been reviewed elsewhere (*3*). Characteristically, S phase is very low, almost undetectable in peripheral blood and marrow cells (0.2%), but has been found to be as high as 60% in CLL lymph nodes (*202*). In studies of continuous infusion of ^{3H}-thymidine over 7 days, about 95% of small peripheral blood lymphocytes were long lived, with turnover times greater than a year. Hence, CLL is considered to represent an accumulative and not a proliferative type of growth (*65*). RNA content determined by DNA/RNA flow cytometry is characteristically low (RNA index = 8), comparable to that of normal B cells, and DNA aneuploidy is usually not detectable by standard resolution DNA measurements (CV 2–3%) (*8*). Of interest may be an unexpectedly large variability of α_t in AO measurements of chromatin structure and cell cycle position (Andreeff, unpublished data): α_t values vary over a large range from "deep G_0" to "G_1". Supporting the notion of considerable proliferative activity in some CLL or in subpopulations of CLL cells is the observation of a high expression of the cell cycle-related proto-oncogene p53 in the blood of patients with CLL (*186*).

VIII. CHRONIC MYELOGENOUS LEUKEMIA

Chronic myelogenous leukemia (CML) is usually easy to diagnose by virtue of increased production of granulocytes. CML often involves multiple cell lineages, and is cytogenetically characterized by the "Philadelphia" translocalization (9;22) (*183*). This results in rearrangement of the abl proto-oncogene (*96*) and formation of a new hybrid gene ("bcr-abl") encoding the CML-specific protein p210 with demonstrated tyorsine kinase activity (*57*). The cellular defect leading to CML was described by Strife and Clarkson as "discordant maturation" (*188*). Griffin *et al.* (*89*) isolated progenitor cells from the peripheral blood of CML patients and obtained cells of which 47% formed myeloid colonies or clusters in culture. The kinetics of bone marrow cell production has been analyzed by Doermer *et al.* (*64*).

AO flow cytometry with measurement of green pulse width (GPW) can identify CML by a characteristic pattern with very high GPW: Ficoll-Hypaque separation of mononuclear cells fails to remove all granulocytes, which can be identified by their unique GPW pattern. This pattern, in conjunction with high S phase of blood cells, identifies CML in AO flow cytometry (*4*).

Discrepancies between flow cytometric S phase and autoradiographic labeling index determinations were observed by Raza (*68*). Holdrinet *et al* (*109*) analyzed CML cells and found no consistent differences in cell kinetics when compared with normal hematopoiesis, but observed DNA aneuploidy in 4 out of 11 patients undergoing blastic transformation. Following a 3–4 year chronic phase, CML eventually evolves into a blastic phase, with myeloblastic, lymphoblastic or undifferentiated features. As in the case of acute leukemias, myeloblastic transformation of CML is usually characterized by high RNA index (*43*), and lymphoblastic transformation by low RNA index (*7,8*). Ligler *et al.* (*126*) attempted to predict the transformed phenotype by studying cell surface markers, but early detection of transformation has not yet been achieved with consistency by any method. Some patients have mixed lineage blastic transformation (*93*).

With improving therapeutic options which now include bone marrow transplantation, the need for predicting transformation at an early stage has increased and flow cytometric techniques could be helpful. The detection of minimal disease in CML at present can be achieved using the PCR technique, which can identify less than one CML cell per 100 000 normal cells (*112*). However, such low levels of leukemic cells are of unknown clinical importance. A FISH technique hybridizing regions of the bcr and abl genes was successfully applied to identify CML and may allow better quantitation of residual disease and of lineage involvement (*205*).

IX. MONITORING OF LEUKEMIA/MININAL RESIDUAL DISEASE

Clinical monitoring of leukemia is conventionally done by serial morphological evaluations of bone marrow and blood samples. The clinician's aim of achieving complete remission (CR) is thought to be reached when the leukemic cell mass is reduced to less than 5% of marrow blasts and when subsequent normalization of peripheral blood counts indicates repopulation of the marrow with normal progenitors, which are able to proliferate and differentiate. These clinical criteria for CR are insufficient, since most patients with ALL or AML will relapse with the same leukemic clone. Flow cytometry can improve the detection of minimal disease.

A. Cell Kinetic Monitoring

The introduction of flow cytometry made cell kinetic analysis much faster and easier and a number of reports appeared in the 1970s, starting with the work of Buechner, Barlogie, Hiddemann, Hillen and others. (*29,30,32,33,42,43,45,103,104*). In particular, effects of the cell cycle-specific drug ara-C (*42*) on leukemic cells *in vivo* were investigated (*2,80*). Low doses of ara-C result in an increase of S phase cells due to a block of DNA synthesis, whereas high ara-C doses kill cells in S, and DNA histograms show a depletion of S phase cells. In a study of AML treated with a daunomycin/ara-C-containing regimen, responding and non-responding patients showed different effects of induction therapy on the kinetics of the leukemic cells. In patients achieving CR, S phase decreased to 1.33 ± 0.99%, and to 2.65 ± 2.3% in patients failing therapy ($p = 0.04$) (*22*). Another study found a decreasing RNA index (mean 1.4) in responders, but not in non-responders (mean 2.5) (*132*). In the same study, S phase recovery was different between patient groups: S phase was 16.9% for responders and 12.0% for non-responders ($p < 0.001$) between days 18 and 22 of treatment: persistent leukemic cells probably prevent recovery of normal hematopoiesis. Future studies of BUdR incorporation will better define kinetics as a monitoring tool.

B. Cell Kill Kinetics

The quantitation of cytoreduction in terms of cell number per volume of bone marrow was made possible by two techniques, developed by Hiddemann *et al.* (*99*) and Blumenreich *et al.* (*41*). Hiddemann's method requires simultaneous bone marrow aspirate biopsy and blood samples and corrects for the admixture of blood cells in the marrow aspirate. Cell kinetic differences identify the *number* of contaminating blood cells, and the ratio of red cell hematocrits in blood and aspirate corrects for the *volume* of contaminated blood. The total cell number found in acute leukemia marrows was determined to be 10^5–10^6 blast cells/mm^3, which amounts to a total of $1–5 \times 10^{12}$ bone marrow blasts per patient. Cell kill with daunomycin, ara-C 6-thioguanine in 5 days was between 1.1 and 3.3 $\log^{10}$, and was significantly higher in responding patients than in non-responders ($p < 0.01$). Patients treated with high dose thymidine had up to 3.6 $\log^{10}$ cell kill. The method also allows the determination of cell kill for the different cell lineages present in the marrow (*8,179*). The efficacy of different drug schedules can be objectively quantitated (*97*). Blumenreich's method is technically easier and requires the complete disaggregation and precise counting of cells in bone marrow biopsy samples. In a recent study, it was used to assess ALL induction therapy in children. Continuous infusion of daunomycin induced a 1 $\log^{10}$ higher cell kill than the same total dose given i.v. push (*94*). This important information cannot be obtained with other methods, since almost all children achieve CR. It is expected (*82*), but remains to be proven, that rapid initial cell kill results in prolonged remission duration.

C. Monitoring of Minimal Residual Disease by DNA Aneuploidy

The high rate of DNA aneuploidy in ALL (40% in children) makes DNA flow cytometry an excellent technique for monitoring therapy-induced cytoreduction. The sensitivity of detecting small numbers of DNA aneuploid luekemic cells is only 1–3%, but in hypocellular marrows after cytotoxic therapy, and in regenerating marrows with high proliferative activity of normal hematopoietic cells, monitoring of DNA aneuploid cells is clearly superior to morphological assessment. Patients with persistent aneuploid cells invariably relapse within 3 months. Long-term follow-up of 93 children in remission whose ALL cells were characterized by DNA aneuploidy identified only 3 out of 15 relapsed patients prior to clinical relapse, with a lead time of 1–6 months (*22*). Despite the usefulness of DNA aneuploidy monitoring during induction therapy, more sensitive methods are clearly needed for long-term follow-up studies. Fluorescence *in situ* hybridization (FISH) using chromosome-specific DNA markers has already been shown to detect as few as 2 in 2000 abnormal cells, when slides were counted manually (*87*). Flow cytometry may increase the sensitivity of FISH by one or two orders of magnitude and may help in detecting recurrent disease early enough for successful clinical intervention.

D. Immunological Monitoring of Minimal Residual Disease

Since no leukemia-specific immunological markers have been identified, the use of flow cytometry in this context appears limited to the early stages of therapy, when reduction and/or persistence of the abnormal phenotype can give clues regarding residual disease. In mixing experiments using CD1-positive malignant T lymphoblasts and thymocytes, the detection limit was 1–2%. Likewise, measurements of TdT activity and CALLA were not useful in predicting ALL relapse (W. D. Ludwig, personal communication), since both markers are present in normal marrow (*110*) and blood cells. When cell populations were gated by light scatter properties and irrelevant cells were tagged with phycoerythrin-labeled antibodies, the sensitivity for detecting CALLA-positive lymphoblasts was one per 100 000 in the peripheral blood (*174*). Enumeration of $CALLA^+TdT^+$ cells by double-staining in children with ALL on maintenance therapy showed that levels $> 0.03\%$ were indicative of imminent relapse. The technique of kappa–lambda clonal excess (*28,217*) could be useful in B cell lymphomas and B-ALL, but has not been used extensively in leukemia monitoring.

In myeloid leukemia, the rat AML model BNML was investigated by Martens, Hagenbeek and Visser and the high expression of a myeloid antigen allowed detection of small numbers of leukemic cells (*135,212*). It was estimated that 1 per 10^4 to 1 per 10^5 cells could be identified. Wormann *et al.* (*219*) reported coexpression of early and late antigens in AML and used these aberrant phenotypes as tools for monitoring residual disease.

Other techniques to detect minimal residual disease in leukemia include the detection of chromosomal abnormalities by premature chromosome condensation, which identifies cells with high or low proliferative potential (PPI, or cells in late G_1). Fourteen of 16 patients with high PPI relapsed after 23 weeks on average, and 19 out of 44 patients with low

PPI relapsed after 68 weeks (*105*). Patients with specific translocations, including those with t(9;22), t(8;14) and t(4;11), will be excellent candidates for monitoring with the DNA polymerase chain reaction technique (PCR) (*112*), which can detect less than one abnormal cell per 100 000. A combination of PCR, FISH and flow cytometry is conceivable and may provide the quantum jump in sensitivity urgently needed for the detection of truly minimal, quantifiable residual leukemia.

X. DRUG EFFECTS AND DRUG RESISTANCE IN LEUKEMIA

Drug effects on leukemic cells have been studied extensively by *in vitro* and *in vivo*. We will briefly discuss the effects of the three drugs most frequently used in leukemia.

A. Cytosine-arabinoside (Ara-C)

Ara-C is the classic blocker of DNA synthesis. Early studies by Buechner *et al.* (*32,42,44*), Barlogie (*30*), and Yataganas (*220*) confirmed the expected cell cycle specificity of ara-C. Using the AO techniques, recruitment of cells into the cell cycle could be demonstrated in patients receiving therapy with high-dose ara-C (*21*). Of potential clinical usefulness is Waldmann's observation that cells resistant to ara-C can be identified based on their persistent ability to incorporate BUdR. The sensitivity of the method is remarkable and approaches 10^{-3} (*4,214*).

B. Methotrexate (MTX)

MTX has also been studied extensively with regard to its cell cycle effects (S phase block), its uptake and metabolism. The uptake of fluorescinated MTX was studied in MTX-sensitive and resistant leukemic cells (*154,170*). A fluorescent MTX analog was used as a dihydrofolate reductase (DHFR) probe by Rosowsky *et al.* (*171*) and can be utilized in a flow cytometry assay to detect methotrexate resistance based on DHFR overproduction. Taylor and Tattersall (*194*) investigated cell cycle effects using DNA/RNA FCM. FCM is potentially advantageous as it can identify small numbers of MTX-resistant cells that are undetectable in pharmacological assays.

C. Anthracyclines

The three main anthracyclines, daunomycin, adriamycin and idarubicin, fluoresce upon excitation with a 488 nm argon ion laser and thus allow measurement of their intracellular concentration, as pioneered by Krishan *et al.* (*118*). Drug uptake and modulation of uptake were extensively studied *in vitro* (*83,170*). A comparison of anthracycline concentrations in human hematopoietic cells measured by flow cytometry and high pressure liquid chromatography showed good correlation (*185*). Comparative studies of daunomycin and adriamycin uptake in sensitive and resistant leukemia cells were also conducted (*193*) and effects of calmodulin inhibitors were studied (*81*). These studies of leukemia cell lines showed effects of calcium channel blockers on anthracycline uptake, but a study by Ross *et al.* (*172*) failed to demonstrate an effect of verapamil on the uptake of daunomycin in cells from clinically resistant AML.

D. Multidrug Resistance (MDR)

Multidrug resistance describes the phenomenon of cells selected for resistance to a single cytotoxic agent developing resistance to a group of unrelated agents which include colchecin, daunomycin, doxorubicin, actinomycin D, vinca alkaloids and epipodophyllitoxins. The gene conferring multidrug resistance (MDR 1) and its product, a P-glycoprotein, have been identified (*203*) and antibodies to detect the MDR phenotype are available. One such antibody (HYB241) (*140*) detects a 180 kDa cell surface P-glycoprotein and has been used in the study of primary human leukemia cells (*95*). Relapsed leukemias expressed twice the amount of P-glycoprotein as newly diagnosed leukemias, and patients failing anthracycline/vinca-containing treatment regimens expressed much higher levels of (gp180) than responding patients (*20*). Another antibody, MRK16, was found to identify only a small number of cases with detectable P-glycoprotein in leukemia (*208*) and the clinical importance of MDR expression is still unresolved (*177,178*).

Flow cytometric studies of drug uptake, retention and metabolism and of drug effects on cell cycle progression and survival *in vitro* combined with investigations of drug resistance may be useful in designing better therapeutic regimen. One such example is the ability of tamoxifen to increase intracellular daunorubicin fluorescence in cell lines expressing MDR (*38*). Combined measurement of P-glycoprotein and DNA content is shown in Fig. 8. The multidrug-resistant leukemia cell line HL-60/RV expresses high levels of P-glycoprotein, but these levels are constant as cells progress from G_1 to S and G_2M, indicating cell cycle independence of MDR gene regulation in these cells. The simultaneous measurement of P-glycoprotein and cellular daunorubicin (DNR) content illustrates the working of the MDR pump: the lymphoblastic leukemic cell line CEM-S has low levels of P-glycoprotein and high levels of DNR, and the multidrug-resistant sub-line CEM-VBL has low levels of DNR and high levels of P-glycoprotein (Fig. 9). Cellular pharmacology by flow cytometry is still in its infancy and deserves greater attention by clinical researchers. As cellular heterogeneity is increasingly appreciated by pharmacologists, flow cytometry will become an indispensible tool in drug research.

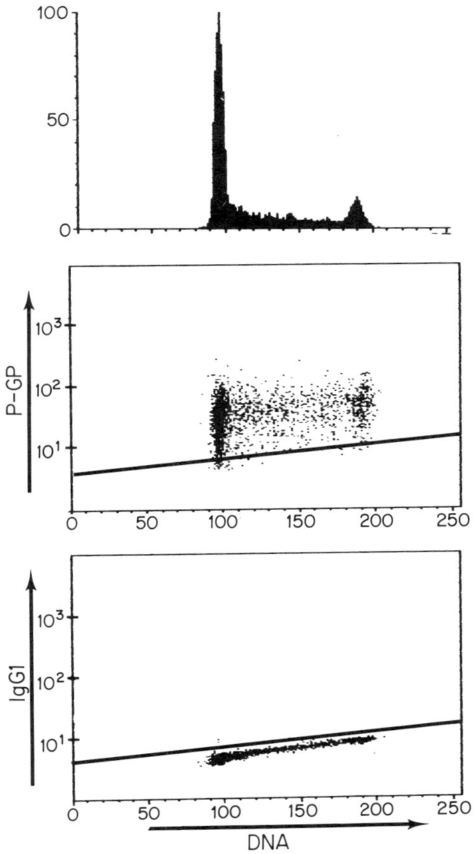

Fig. 8. Biparametric measurements of DNA content (propidium iodide) and P-glycoprotein, the product of the mdR-1 gene. Antibody: HYB241 (Hybritech) (middle panel), IgG_1 control (bottom panel). MDR is expressed and not cell cycle regulated (multidrug resistant HL-60 cells).

XI. OUTLOOK

Flow cytometry has immensely contributed to the study of leukemia, both in laboratory and clinical investigations. Many cell kinetic, clonality and immunologic studies are directly related to this technique and the routine clinical application of FCM attests to its usefulness. As more tools are being developed, FCM will further contribute to leukemia research. The breathtaking developments in molecular biology and genetics have had perhaps even greater impact on our understanding of the molecular basis of hematopoietic tumors, and of the mechanisms regulating differentiation and proliferation. It will be vital to blend molecular approaches and FCM technology in the future. FCM is still unsurpassed in its ability to

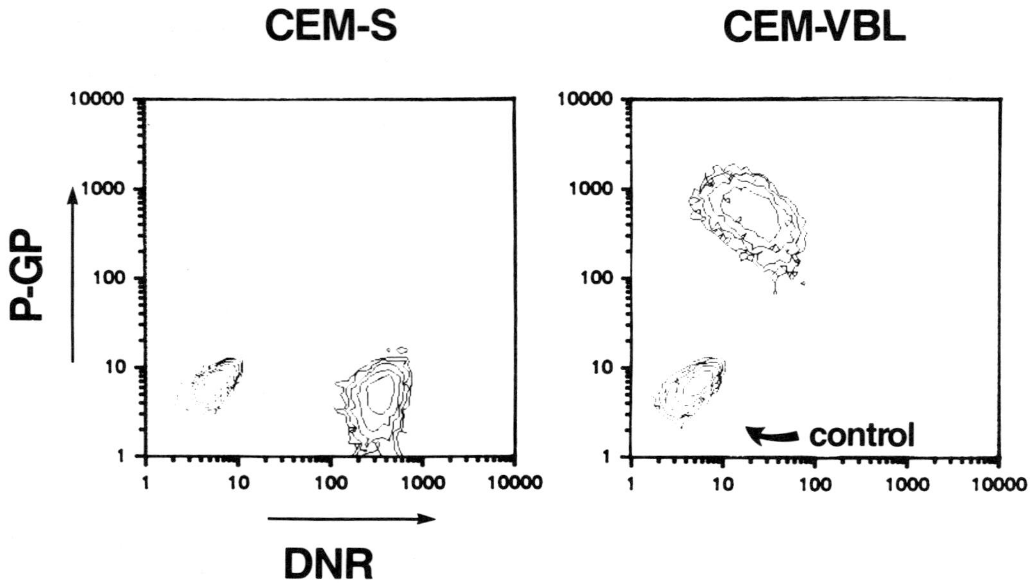

Fig. 9. Biparametric measurements of P-glycoprotein (antibody HYB241 green fluorescence) and daunorubicin (DNR, red fluorescence). CEM-S cells are drug sensitive, lack P-glycoprotein and have high levels of DNR. CEM-VBL cells are drug resistant, have low levels of DNR and high levels of P-glycoprotein (isotypic control cell displayed in lower left corner).

characterize individual cells, and molecular techniques such as PCR and FISH could serve as a link to the molecular universe. As flow cytometry and cell sorting technologies have matured, development has shifted to new probes. In the context of modern biology, flow cytometry will continue to play a major role in the elucidation of mechanisms regulating normal and malignant cells.

ACKNOWLEDGEMENTS

The author wishes to thank Bayard Clarkson, Zbigniew Darzynkiewicz and Myron Melamed for many years of friendship, support and stimulating exchanges. Wolfgang Hiddemann, Carlos Suarez, Joseph Haimi, Susanna Hegewisch-Becker, Agostino Tafuri, Gabriel Ruiz, and Arlene Redner have all contributed to the work done at the Leukemia Cell Biology Laboratory at Memorial Sloan-Kettering Cancer Center. Jim Steinmetz, Barry Eagel, Suzanne Swarthout, Edith Espiritu, Gordon Assing, Jan Bressler, James Squires, Loren Godfrey, Walter Verbeek, Carsten Bokemeyer, Barbara Rehermann, Richard Chen and Maria Josephina Pereira Leahy have worked hard to translate concepts into results. Nguyen Van and Susan Escudier are continuing this tradition at the M. D. Anderson Cancer Center, and Jane Schoppe and Vicki Dennison provided excellent secretarial assistance in the preparation of the manuscript.

REFERENCES

1. (1983). Third International Workshop on Chromosomes in Leukemia: Chromosomal abnormalities and their clinical significance in acute lymphoblastic leukemia. *Cancer Res.* **43**, 868–873.

2. Aardal, N. P., Talstad, I., and Laerum, O. D. (1979). Sequential flow cytomeric analysis of cellular DNA-content in peripheral blood during treatment for acute leukaemia. *Scand. J. Haematol.* **22**, 25–32.

3. Andreeff, M. (1986). Cell kinetics of leukemia. *Sem. Hematol.* **23**, 300–314.

4. Andreeff, M. (1990). Flow cytometry of leukemia. *In* "Flow Cytometry and Cell Sorting" M. R. Melamed and M. L. Mendelsohn (eds), pp. 697–724. A. R. Liss, New York.

5. Andreeff, M., Assing, G., and Cirrincione, C. (1986). Prognostic value of DNA/RNA flow cytometry in myeloblastic and lymphoblastic leukemia in adults: RNA content and S-phase predict remission duration and survival in multivariate analysis. *Ann. NY Acad. Sci.* **468**, 387–406.

6. Andreeff, M., Beck, J., Darzynkiewicz, Z. D., Traganos, F., Gupta, S., Melamed, M. R., and Good, R. A. (1978). RNA content of human lymphocyte subpopulations. *Proc. Natl. Acad. Sci. USA* **75**, 1938–1942.

7. Andreeff, M., Beck, J. D., Darzynkiewicz, Z., Arlin, Z., Melamed, M. R., and Clarkson, B. (1979). DNA, RNA and chromatin structure assayed *in situ* by flow cytometry: A means for the classification of human leukemias. *Verh. Dtsch. KrebsGes.* **2**, 348.

8. Andreeff, M., and Darzynkiewicz, Z. (1981). Multiparameter flow cytometry part II. Application in hematology. *MSKCC Clin. Bull.* **3**, 120–130.

9. Andreeff, M., and Darzynkiewicz, Z. (1982). Chromatin structure in leukemia: Classification of subtypes and cell kinetics during induction therapy. *Tissue Kinet.* 679.

10. Andreeff, M., Darzynkiewicz, Z., Melamed, M. R., Gee, T. and Clarkson, B. (1978). Nucleic acid content and chromatin structure of leukemic cells *in situ* as studied by flow cytometry. *Proc. Am. Assoc. Cancer Res.* **19**, 786.

11. Andreeff, M., Darzynkiewicz, Z, Sharpless, T. K., Clarkson, B. D., and Melamed, M. R. (1980). Discrimination of human leukemia subtypes by flow cytometric analysis of cellular DNA and RNA. *Blood* **55**, 282–293.

12. Andreeff, M., Darzynkiewicz, Z., Sharpless, T. K., and Clarkson, B. (1977). Cytofluorometric DNA and RNA measurements in the classification of human leukemias. *Blood* (Suppl. 1) **1**, 181.

13. Andreeff, M., Darzynkiewicz, Z., Steinmetz, J., Melamed, M. R., and Clarkson, B. D. (1981). Chromatin structure and nucleic acid content determined *in situ* by flow cytometry (FC) in acute luekemia. *Blood* **58**, 127a.

14. Andreeff, M., Espiritu, E., Bressler, J., and Welte, K. (1991). Induction of Interleukin-2 receptor and recruitment of aneuploid lymphoblastic leukemia by Interleukin-2. Unpublished.

15. Andreeff, M., Gaynor, J., Chapman, D., Little, C., Gee, T., and Clarkson, B. D. (1987). Prognostic factors in acute lymphoblastic leukemia in adults: the Memorial Hospital experience. *In* "Haematology and Blood Transfusions" (T. Buechner, G. Schellong, W. Hiddemann, D. Urbanitz, and W. Ritter, eds), pp. 111–124. Springer-Verlag, Berlin.

16. Andreeff, M., Gee, T., Mertelsmann, R., McKenzie, S., Steinmetz, J., Chaganti, R., Koziner, B., and Clarkson, B. (1980). Biclonal lymphoblastic and myeloblastic acute leukemia. *Proc. Am. Assoc. Cancer Res.* **21**, 313.

17. Andreeff, M., and Haag, D. (1975) Fluorochromierung mit Acridinorange in der Impulscytophotometric. *In* "Impulscytophotometrie", (M. Andreef, ed), pp. 31–37. Springer-Verlag, New York.

18. Andreeff, M., Haimi, J., Redner, A., Steinhertz, P., O'Reilly, R., Gray, J., and Pinkel, D. (1991). Monitoring of minimal residual disease in leukemia by fluorescence *in situ* hybridization using chromosome-specific DNA Probes and nucleic acid flow cytometry. Unpublished.

19. Andreeff, M., Hansen, H., Cirrincione, C., Filippa, D., and Thaler, H. (1986). Prognostic value of DNA/RNA flow cytometry in B-cell non-Hodgkin's lymphoma: development of laboratory model and correlation with four taxonomic systems. *Ann. NY Acad. Sci.* **468**, 368–386.

20. Andreeff, M., Hegewisch-Becker, S., Rehermann, B., Godfrey, L., and Maslak, P. (1990). Flow cytometric determination of p-glycoprotein in acute leukemia. *Proc. Am. Assoc. Cancer Res.* **31**, 1134 (Abstract).

21. Andreeff, M., Kempin, S., Arlin, Z., Mertelsmann, R., Espiritu, E., and Gee, T. (1983). High dose cytosine arabinoside (HDARAC) in acute luekemia (AL). Correlation of clinical response and cell kinetics. *Proc. Am. Assoc. Cancer Res.* **24**, 166.

22. Andreeff, M., Redner, A., Thongsprasert, S., *et al.* (1985). Multiparameter flow cytometry for determination of ploidy, proliferation and differentiation in acute leukemia: Treatment effects and prognostic value. *In* "Tumor Aneuploidy" (T. Buechner, ed.), pp. 81–105. Springer-Verlag, Berlin.

23. Andreeff, M., Stone, R., Young, C., Tong, W., Ervin, T., Arthur, K., Salvia, B., Tafuri, A., Kolitz, J., Rifkind, R., Kufe, D., and Marks, P. (1990). Treatment of myelodysplastic syndromes (MDS) and acute myeloid leukemia (AML) with hexamethylene bisacetamide (HMBA). *Blood* **76**, (Suppl. 1) 251a (Abstract).

24. Andreeff, M., Tafuri, A., and Hegewisch-Becker, S. (1990). Colony stimulating factors (rhG-CSF, rhGM-CSF, rhIL-3 and BCGF) recruit myeloblastic and lymphoblastic leukemia cells and enhance the cytotoxic effects of cytosine-arabinoside. *In* "Acute Leukemias II", (T. Buechner, G. Schellong, W. Hiddemann, and J. Ritter, eds), pp. 747–762. Springer-Verlag, Berlin.

25. Andreeff, M., Tafuri, A., and Hegewisch-Becker, S.

(1990). Stimulating factors (rhG-CSF, rhGM-CSF, rhIL-3 and BCFG) recruit myeloblastic and lymphoblastic leukemic cells and enhance the cytotoxic effects of cytosine-arabinoside. *In* "Acute Leukemias II", (T. Buechner, G. Schellong, W. Hiddemann, and J. Ritter, eds), pp. 747–762. Springer-Verlag, Berlin.

26. Andreeff, M., Wong, G., Koziner, B., Chaganti, R. S. K., Al-Katib, A., Espiritu, E., Espiritu, B., and Hermann, T. K. (1984). Evidence for complete B cell differentiation of a subpopulation of DNA aneuploid leukemic cells in "non B-non T" acute lymphoblastic leukemia (ALL). *Blood* **64**, 650.

27. Atzpodien, J., Gulati, S. C., Kwon, J.H., Wacheter, M., Fried, J., and Clarkson, B. D. (1987). Human bone marrow CFU-GM and BFU-E localized by light scatter well sorting. *Exp. Cell Biol.* **55**, 265–270.

28. Ault, K. A. (1979). Detection of small numbers of monoclonal B lymphocytes in the blood of patients with lymphoma. *N. Engl. J. Med.* **300**, 1401–1405.

29. Bakkeren, J. A. J. M., Vaan, G. A. M. de, and Hillen, H. F. P. (1979). Interrelationship of immunologic characteristics, proliferation pattern, and prednisone sensitivity in acute lymphoblastic leukemia of childhood. *Blood* **53**, 883–891.

30. Barlogie, B., Buechner, T., Asseburg, U., and Kamanabroo, D. (1974). Zellkinetischer effekt von Cytosin-arabinosid im Klinishchen Test. *Med. Welt.* **25**, 1532.

31. Barlogie, B., Maddox, A. M., Johnston, D. A., Raber, M. N., Drewinko, B., Keating, M. J., and Freireich, E. J. (1983). Quantitative cytology in leukemia research. *Blood Cells* **1**, 35–55.

32. Barlogie, B., Roessner, A., Asseburg, U., Kamanabroo, D., Hiddemann, W., and Buechner, T. (1974). Untersuchungen zur Teilsynchronisation von Zellen der experimentellen Rattenleukaemie L5222. *Verh. Dtsch. Ges. Inn. Med.* 80.

33. Barlogie, B., Spitzer, G., Hart, J. S., Johnston, D. A., Buechner, T., Schumann, J. and Drewinko, B. (1976). DNA Histogram Analysis of Human Hemopoietic Cells. *Blood* **48**, 245–257.

34. Bennett, J. M., Catovsky, D., Daniel, M.-T., Flandrin, G., Galton, D. A. G., Gralnick H. R., and Sultan, C. (1976). Proposals for the classification of the acute leukaemias. *Br. J. Haematol* **33**, 451.

35. Bennett, J. M., Catovsky, D., Daniel, M. T., Flandrin, G., Galton, A. G., Gralnick, H. R., and Sultan, C. (1985). Criteria for the diagnosis of acute leukemia of megakaryocytic lineage (MT). *Ann. Intern. Med.* **103**, 460–462.

36. Bennett, J. M., Catovsky, D., Daniel, M. T., Flandrin, G., Galton, D. A. G., Gralnick, H. R., and Sultan, C. (1982). Proposals for the classification of the myelodysplastic syndromes. *Br. J. Haematol.* **51**, 189–199.

37. Bennett, J. M., Catovsky, D. M. T., Flandrin, G., Galton, D. A. G., Gralnick, H. R. and Sultan, C. (1985). Proposed revised criteria for the classification of acute myeloid leukemia. *Ann. Intern. Med.* **103**, 626–629.

38. Berman, E., Adams, M., Duigou-Osterndorf, R., Godfrey, L., Clarkson, B., and Andreeff, M. (1991). Effect of tamoxifen on cell lines displaying the multidrug-resistant phenotype. *Blood* **77**, 818–825.

39. Bettelheim, P., Paietta, E., Majdic, O., Gadner, H., Schwarzmeier, J., and Knapp, W. (1982). Expression of a myeloid marker on TdT-positive acute lymphocytic leukemic cells: evidence by double-fluorescence staining. *Blood* **60**, 1392–1396.

40. Bettelheim, P., Valent, P., Andreeff, M., Tafuri, A., Haimi, J., Gorischek, C., Muhm, M., Sillaber, Ch., Haas, O., Vieder, L., Maurer, D., Schulz, G., Speiser, W., Geissler, K., Kier, P. Hinterberger, W., and Lechner, K. (1991). Recombinant human granulocyte-macrophage colony-stimulating factor in combination with standard induction chemotherapy in de nova acute myeloid leukemia. *Blood* **77**, 700–711.

41. Blumenreich, M. A., Woodcock, T. M., Andreeff, M., Hiddemann, W., Chou, T.-C., Vale, K., O'Hehir, M., and Clarkson, B. D. (1984). Effect of very high-dose thymidine infusions on leukemia and lymphoma patients. *Cancer Res*, **44**, 2203–2207.

42. Buechner, T., Barlogie, B., Hiddemann, W., Kamanabroo, D. and Gohde, W. (1974). Accumulation of S-phase cells in the bone marrow of patients with acute leukemia by cytosin-arabinosid. *Blut* **28**, 299.

43. Buechner, T., Dittrich, W., and Gohde, W. (1971). Die Impulscytophotometrie in de Haematologischen Cytologie. *Klin. Wochenschr.* **49**, 1090–1092.

44. Buechner, T., Gohde, W., Schuman, J., and Barlogie, B. (1976). In vivo kinetic response of human leukemia cells to antileukemic drugs as seen in the DNA histogram. *Excerp. Med. Int. Cong. Top. Hematol. 265–270.*

45. Buechner, T., Hiddemann, W., Schneider, R., Kamanabroo, D., and Gohde, W. (1973). Untersuchungen uber zellkinetische Effekte der Leukaemiebehandlung in der Klinik anhand des DNS-Histogramms mittels Impulscytophotometrie. *Med. Welt.* **24**, 1616–1617.

46. Chan, L. C., Pegram, S. M. and Greaves, M. F. (1985). Contribution of immunophenotype to the classification and differential diagnosis of acute leukaemia. *Lancet* **1** 475–480.

47. Chang, L. J.-A. and McCulloch, E. A. (1981). Dose-dependent effects of a tumor promotor on blast cell progenitors in human myeloblastic leukemia. *Blood* **57**, 361–367.

48. Chiao, J. W., Andreeff, M., Freitag, W. B., and Arlin, Z. (1982). Induction of in vitro proliferation

and maturation of human aneuploid myelogenous leukemia cells *J. Exp. Med.* **54**, 1397.

49. Clark, R., Peters, S., Hoy, T., Smith, S., Whittaker, K., and Jacobs, A. (1986). Prognostic importance of hypodiploid hemopoietic precursors in myelodysplastic syndromes. *New Engl. J. Med.* **314**, 1472–1473.

50. Clarkson, B., Berman, E., Little, C. *et al.* (1990). Update on clinical trials of chemotherapy and bone marrow transplantation in acute myelogenous leukemia at Memorial Sloan-Kettering Cancer Center (MSKCC) 1966 to 1989. *In* "ULCA Symposia on Molecular and Cellular Biology" (R. P. Gale, ed.). Wiley-Liss, New York.

51. Clarkson, B., Gee, T., Arlin, Z. *et al.* (1984). Current status of treatment of acute leukemia in adults: an overview. *In* "Therapie der akuten leukaemien" (T. Buechner, D. Urbanitz, and J. Van de Loo, eds), Springer-Verlag, Berlin. p. 1–31.

52. Clarkson, B., Strife, A., Fried, J., and Gulati, S. (1986). Cytokinetics and cancer treatment. *In* "Proc. General Motors Cancer Research Foundation" (E. J. Freireich and T. Frei, eds), pp. 131–190. J. P. Lippincott, Philadelphia.

53. Cossman, J., Neckers, L. M., Arnold, A., and Korsmeyer, S. J. (1982). Induction of differentiation in a case of common acute lymphoblastic leukemia. *New Engl. J. Med.* **307**, 1251–1254.

54. Coulter, W. H. (1956). High speed automatic blood cell counter and cell size analyzer. *Proc. Natl. Electron. Conf.* **12**, 1034–1042.

55. Crissman, H. A., Oishi, N., and Steinkamp, J. A. (1988). Analysis of DNA replication patterns in BrdU-labeled cell populations and conditions for correlated measurement of RNA and FITC-labeled components. *Cytometry* **233**, 18.

56. Crist, J. W., Pullen, J., Boyett, J., Falletta, J., van Eys, J., Borowitz, M., Jackson, J., Dowell, B., Russell, C., Quddus, F., Ragab, A., and Vietti, T. (1988). Acute lymphoid leukemia in adolescents: Clinical and biologic features predict a poor prognosis. A Pediatric Oncology Group Study. *J. Clin. Oncol.* **6**, 34–43.

57. Daley, G. Q., McLaughlin, J., Witte, O. N. and Baltimore, D. (1987). The CML-specific p210 bcr/abl protein, unlike v-abl, does not transform NIH/3T3 fibroblasts. *Science* **237**, 532–536.

58. Darzynkiewicz, Z., Andreeff, M., Traganos, F., Sharpless, T., and Melamed, M. R. (1978). Discrimination of cycling and noncycling lymphocytes by BUdR-suppressed acridine orange fluorescence in a flow cytometric system. *Exp. Cell Res.* **115**, 31–35.

59. Darzynkiewicz, Z. and Kapuscinski, J. (1990). Acridine orange: a versatile probe of nucleic acids and other cell constituents. *In* "Flow Cytometry and Sorting" (M. R. Melamed, T. Lindmo, and M. L. Mendelsohn, eds), pp. 291–314. Wiley-Liss, New York.

60. Darzynkiewicz, Z., Traganos, F., Andreeff, M., Sharpless, T., and Melamed, M. R. (1979). Different sensitivity of chromatin to acid denaturation in quiescent and cycling cells as revealed by flow cytometry. *J. Histochem. Cytochem.* **27**, 478–485.

61. Darzynkiewicz, Z., Traganos, F., Kapuscinski, J., Straino-Coico, L. and Melamed, M. (1988). Accessibility of DNA *in situ* to various fluorochromes: relationship to chromatin changes during erythroid differentiation of friend leukemia cells. *Cytometry* **5**, 355–363.

62. Darzynkiewicz, Z., Traganos, F., Sharpless, T., Friend, C., and Melamed, M. R. (1976). Nuclear chromatin changes during erythroid differentiation of friend virus induced leukemic cells. *Exp. Cell Res.* **99**, 301–309.

63. Dewald, G. W., Davis, M. P., Pierre, R. V., O'Fallon, J. R., and Hoagland, H. C. (1985). Clinical characteristics and prognosis of 50 patients with a myeloproliferative syndrome and deletion of part of the long arm of chromosome 5. *Blood* **66**, 189–197.

64. Doermer, P., Lau, B., and Wilmanns, W. (1980). Kinetics of bone marrow cell production in human acute and chronic myeloid leukemias. *Leuk. Res.* **4**, 231–237.

65. Doermer, P., Theml, H., and Lau, B. (1983). Chronic lymphocytic leukemia: a proliferative or accumulative disorder. *Leuk. Res.* **7**, 1–10.

66. Doermer, P., Ucci, G., Lau, B., Hass, R., and Janka, G. E. (1984). *In vivo* production of childhood acute lymphoblastic leukemia cells in relation to ploidy and immunological subtype. *Leuk. Res.* **8**, 587–595.

67. Dolbeare, F., Graftner, H., Pallacini, M. G., and Gray, J. W. (1983). Flow cytometric measurement of total DNA content and incorporated bromodeoxyuridine. *Proc. Natl. Acad. Sci. USA* **80**, 5573.

68. Dormer, P., Ucci, G., Hershko, C. H. and Wilmanns, W. (1988). Cell kinetics in leukemia and preleukemia. *In* "Acute Leukemias" (Th. Buechner, G. Shellong, eds). Berlin. Springer Verlag.

69. Dosik, G. M., Barlogie, B., Cohde, W., Johnston, D., Tekell, J. L., and Drewinko, B. (1980). Flow cytometry of DNA content in human bone marrow: a critical reappraisal. *Blood* **55**, 734–740.

70. Dow, L. W., Chang, L. A., Tsiatis, A. A., Melvin, S. L., and Bowman, W. P. (1982). Relationship of pretreatment lymphoblast proliferative activity and prognosis in 97 children with acute lymphoblastic leukemia. *Blood* **59**, 1197–1202.

71. Drapkin, R. L., Andreeff, M., Koziner, B., Strife, A., Wisniewski, D., Darzynkiewicz, Z., Realmed, R. M., and Clarkson, B. (1979). Subpopulations of

human peripheral blood cells: analysis of granulocytic progenitor cells by flow cytometry and immunologic markers. *Am. J. Hematol.* **7**, 163–172.

72. Drewinko, B., Bollinger, P., Brailas, C., Moyle, S., Wyatt, J., Simson, E., Johnston, D., and Trujillo, J. M. (1987). Flow cytochemical patterns of white blood cells in human haemotopoietic malignancies. I. Acute leukaemias. *Br. J. Haematol.* **66**, 27–36.
73. Estey, E., Smith, T. L., Keating, M. J., McCredie, K. B., Gehan, E. A., and Freireich, E. J. (1989). Prediction of survival during induction therapy in patients with newly diagnosed acute myeloblastic leukemia. *Leukemia* **3**, 257–263.
74. Fearon, E. R., Burke, P. J., Schiffer, C. A., Zehnbauer, B. A., and Vogelstein, B. (1986). Differentiation of Leukemia Cells to Polymorphonuclear Leukocytes in Patients with Acute Nonlymphocytic Leukemia. *New Engl. J. Med.* **315**, 15–58.
75. Ffrench, M., Bryon, P. A., Fiere, D., Vu-Van, H., Guyotat, D., Extra, J. M., and Viala, J. J. (1986). Cell cycle prognostic value in adult acute myeloid leukemia. The choice of the best variables. *Leuk. Res.* **10**, 51–57.
76. French, M., Byron, P. A., Fiere, D., Vu-Van, H., Gentilhomme, O., Adeleine, P., and Viala, J. J. (1985). Cell-cycle, protein content and nuclear size in acute myeloid leukemia. *Cytometry* **6**, 47–53.
77. Foon, K. A., Gale, R. P., and Todd, R. F. (1986). Recent advances in the immunologic classification of leukemia. *Sem. Hematol.* **23**, 257–283.
78. Foon, K. A., and Todd, R. F. (1986). Immunologic classification of leukemia and lymphoma. *Blood* **68**, 1–31.
79. Foucar, K., Langdon II, R. M., Armitage, J. O., Olson, D. B., and Carroll, T. (1985). Myelodysplastic syndromes. *Cancer* **56**, 553–561.
80. Fried, J., Arlin, Z., and Clarkson, B. (1976). Flow microfluorometry of bone marrow cells from a patient with acute lymphocytic leukemia during chemotherapy. *In* "Proceedings of the Second International Symposium on Pulse Cytophotometry" (J. Schumann, W. Gohde, and T. Buchner, eds). European Press, Ghent, Belgium.
81. Ganapathi, R., Grabowski, D., Turinic, R., and Valenzuela, R. (1984). Correlation between potency of calmodulin inhibitors and effects on cellular levels and cytotoxic activity of doxorubicin (adriamycin) in resistant p388 mouse leukemia cells. *Eur. J. Cancer Clin. Oncol.* **20**, 799–806.
82. Gaynor, J., Chapman, D., Little, C., McKenzie, S., Miller, W., Andreeff, M. A. Z., Berman, E. K. S., and Gee, T. C. B. (1988). A cause-specific hazard rate analysis of prognostic factors among 199 adults with acute lymphoblastic leukemia: The Memorial Hospital Experience Since 1969. *J. Clin. Oncol.* **6**, 1014–1030.
83. Gerdes, J., Lemke, H., Baisch, H., Wacker, H.-H., Schwab, U., and Stein, H. (1984). Cell cycle analysis of a cell proliferation-associated human nuclear antigen defined by the monoclonal antibody Ki-67. *J. Immunol.* **133**, 1710–1715.
84. Gore, S. D., Albornoz, L., Kastan, M. B., Burke, P. J., and Civin, C. I. (1990). C-MYC, C-MYB, C-FOS, and Ki-67 levels do not correlate with proliferative status (PS) of acute myelogenous leukemia (AML) blasts. *Blood* **76** (Suppl. 1), 276a (Abstract).
85. Gratzner, H. G. (1982). Monoclonal antibody to 5-bromo and 5-iododeoxyuridine. A new reagent for detection of DNA replication. *Science* **218**, 474.
86. Gratzner, H. G., and Leif, R. C. (1981). An immunofluorescence method for monitoring DNA synthesis by flow cytometry. *Cytometry* **1**, 385–389.
87. Gray, J. W., Pinkel, D., Trask, B. *et al.* (1989). Application of flow karyotyping and fluorescence "in situ" hybridization to detection of chromosome aberrations. *In* "Prognostic Cytometry and Cytopathology of Prostate Cancer" (J. P. Karry, D. S. Coffey, and W. Gardner, eds), pp. 304–313. Elsevier, New York.
88. Greaves, M. F., Chan, L. C., Furley, A. J. W., Watt, S. M., and Molgaard, H. V. (1986). Lineage promiscuity in hematopoietic differentiation and leukemia. *Blood* **1**, 1–11.
89. Griffin, J. D., Beveridge, R. P., Schlossman, S. F. (1982). Isolation of myeloid progenitor cells from peripheral blood of chronic myelogenous leukemia patients. *Blood* **60**, 30–37.
90. Griffin, J. D., Davis, R., Nelson, D. A., Davey, F. R., Mayer, R. J., Schiffer, C., McIntyre, O. R., and Bloomfield, C. D. (1986). Use of surface marker analysis to predict outcome of adult acute myeloblastic leukemia. *Blood* **68**, 1232–1241.
91. Griffin, J. D., Larcom, P., and Schlossman, S. F. (1983). Use of surface markers to identify a subset of acute myelomonocytic leukemia cells with progenitor cell properties. *Blood* **62**, 1300–1303.
92. Griffin, J. D., Ritz, J., Nadler, L. M., and Schlossman, S. F. (1981). Expression of myeloid differentiation antigens in normal and malignant myeloid cells. *J. Clin. Invest.* **68**, 932–941.
93. Ha, K., Freedman, M. H., Hrincu, A., Petsche, D., Poon, A., and Gelfand, E. W. (1985). Separation of lymphoid and myeloid blasts in the mixee blast crisis of chronic myelogenous leukemia: no evidence for Ig gene rearrangement in CALLA-positive blasts. *Blood* **667**, 1404–1408.
94. Haimi, J., Redner, A., Steinherz, P. and Andreeff, M. (1988). Quantitative cytoreduction and cell kinetics during induction therapy in acute lymphoblastic leukemia (ALL). *Blood* **72**, 202a (Abstract).
95. Hegewisch, S., Andreeff, M., and Bertino, J. (1988). Correlation of p-glycoprotein (pgp180) expression and response to therapy in human leukemia. *Blood* **72**, 204a (Abstract).
96. Heisterkamp, N., Stam, K., Groffen, J., Klein, A.,

and Grosveld, G. (1985). Structural organization of the bcr gene and its role in the Ph' translocation. *Nature* **316**, 758–761.

97. Hiddemann, W., Buechner, T., Andreeff, M., Arlin, Z., Woermann, B., and Clarkson, B. D. (1983). Vergleich der Therapieeffekivitaet von zwei Induktionasprotokollen bei akuter myeloischer Leukaemie (AML) mittels exakter quantifizierung der Knochenmarkzellularitaet. *Onkologie* **6**, 179–183.

98. Hiddemann, W., Buechner, T., Andreeff, M., Woermann, B., Melamed, M., and Clarkson, B. D. (1982). Bone Marrow Biopsy instead of Marrow Juice for Cell Kinetic Analysis Comparison of Bone Marrow Biopsy and Aspiration Material. *Leuk. Res.* **6**, 601–612.

99. Hiddemann, W., Clarkson, B. D., Buechner, T., Melamed, M. R., and Andreeff, M. (1982). Bone marrow cell count per mm^3 bone marrow: a new parameter for quantitating therapy induced cytoreduction in acute leukemia. *Blood* **59**, 216–225.

100. Hiddemann, W., Ritter, J., Wormann, B., Budde, M., Creutzig, U., Schellong, G., Buchner, Th., and Riehm, H. (1985). DNS-Aneuploidie bei Kindern mit akiten Leukamien: I. Inzidenz und klinische Bedeuting im Rahmen der BFM-Studien. *Klin. Padiatr.* **197**, 215–220.

101. Hiddemann, W., Schumann, J., Andreeff, M., Barlogie, B., Herman, C. J., Leif, R. C., Mayall, B. H., and Murphy, R. F. (1984). Convention on nomenclature for DNA cytometry. *Cytometry* **5**, 445–446.

102. Hiddemann, W., Woermann, B., Ritter, J., *et al.* (1986). Frequency and clinical significance of DNA aneuploidy in acute leukemia. *In* "Annals of the New York Academy of Sciences. Clinical Cytometry" (M. Andreeff, ed.). pp. 227–240. New York Academy of Sciences, New York.

103. Hillen, H., Haanen, C., and Wessels, J. (1975). Pulse-cytophotometry as a monitor for treatment in acute leukemia. *In* "First International Symposium of Pulse Cytophotometry" (C. A. M. Haanen, H. F. P. Hillen, and J. M. C. Wessels, eds), pp. 315–332. European Press Mediken, Ghent, Belgium.

104. Hillen, H. F. P., Burghouts, J., and Haanen, C. (1978). The application of pulse-cytophotometry for the early detection of the relapse in acute leukemia *In* "Pulse Cytometry, Third International Symposium" (D. Lutz, ed). European Press, Ghent.

105. Hittelman, W. N., Menegaz, S. D., McCredie, K. B., and Keating, M. F. (1984). Premature chromosome condensation studies in human leukemia: 5 prediction of early relapse. *Blood* **64**, 1067–1073.

106. Hoelzer, D., and Gale, R. P. (1987). Acute lymphoblastic leukemia in adults: recent progress, future directions. *Semin. Hematol.* **24**, 27–39 (Abstract).

107. Holdrinet, R. S. G., Egmond, J. V., Wessels, J. M. C., and Haanen, C. (1980). A method for quantification of peripheral blood admixture in bone marrow aspirates. *Exp. Hematol.* **8**, 103–107.

108. Holdrinet, R. S. G., Pennings, A., Drenthe-Schonk, A. M., Egmond, J., Wessels, J. M. C., and Haanen, C. (1983). Flow cytometric determination of the S-phase compartment in adult acute leukemia. *Acta Haematol.* **70**, 369–378.

109. Holdrinet, R. S. G., Pennings, A., Egmond, J., Wessels, J. M. C. and Haanen, C. (1983). DNA-flow cytometry of blood and bone marrow in chronic myelogenous leukemia. *Acta Haematol.* **69**, 98–105.

110. Janossy, G., Bollum, F. J., Bradstock, K. F. and Ashley, J. (1980). Cellular phenotypes of normal and leukemic hemopoietic cells determined by analysis with selected antibody combinations. *Blood* **56**, 430.

111. Kantarjian, H. M., Barlogie, B., Pershouse, M., Swartzendruber, D., Keating, M. J., McCredie, K. B., and Freireich, E. J. (1985). Preferential expression of double-stranded ribonucleic acid in tumor versus normal cells: biological and clinical implications. *Blood* **66**, 39–46.

112. Kawasaki, E. S., Clark, S. S., Coyne, M. Y., Smith, S. D., Champlin, R., White, O. N., and McCormick, F. P. (1988). Diagnosis of chronic myeloid and acute lymphocytic leukemias by detection of leukemia-specific mRNA sequences amplified *in vitro*. *Proc. Natl. Acad. Sci. USA* **312**, 711–716.

113. Koike, T. (1984). Megakaryoblastic leukemia: the characterization and identification of megakaryoblasts. *Blood* **64**, 683–692.

114. Korsmeyer, S. J., Arnold, A., Bakhshi, A., Ravetch, J. V., Siebenlist, U., Hieter, P. A., Sharrow, S. O., and LeBien, T. W. (1983). Immunoglobulin gene rearrangement and cell surface antigen expression in acute lymphocytic leukemias of T cell and B cell precursor origins. *J. Clin. Invest.* **71**, 301–313.

115. Korsmeyer, S. J., Hieter, P. A., Ravetch, J. V., Poplack, C. G., Waldmann, T. A., and Leder, P. (1981). Developmental hierarchy of immunoglobulin gene rearrangements in human leukemic pre-B-cells. *Proc. Natl. Acad. Sci. USA* **78**, 7096–7100.

116. Koziner, B., Mertelsmann, R., Andreeff, M., Arlin, Z., Hansen, H., deHarven, E., McKenzie, S., and Gee, T. (1980). Heterogeneity of cell lineages in L3 leukemias. *Blood* **55**, 694–698.

117. Koziner, B., Stavnezer, J., Al-Katib, A. *et al.* (1986). Surface immunoglobulin light chain expression in pre-B cell leukemia. *In* "Clinical Cytometry" (M. Andreeff, ed.), pp. 211–226. Anals of New York Academy of Science, New York.

118. Krishan, A. (1986). Flow cytometric monitoring of anthracycline transport in tumor cells. *Ann. NY Acad. Sci.* **468**, 80–84.

119. Kubbies, M., Poot, M., Schindler, D., Seyschab, H., Hoehn, H., and Rabinovitch, P. (1988). Flow

cytometric analysis of continuous BrdUrd-labeling with Hoechst 33258/ethidium bromide for cell cycle kinetics and clinical diagnosis. *Cytometry* **2**, 18.

120. Kuo, W. L., Dolbere, F., Vanderlaan, M., Thomas, C. B., and Gray, J. B. (1988). Simultaneous cytometric analysis of bromo- and iodo-deoxyuridine labeled BN myelocytic leukemia. *Cytometry* **2**, 90 (Abstract).

121. Kurki, P., Ogata, K., and Tan, E. M. (1988). Monoclonal antibodies to proliferating cell nuclear antigen (PCNA/cyclin) as probes for proliferating cells by immunofluorescence microscopy and flow cytometry. *J. Immunol. Methods* **109**, 49–59.

122. Lange, W. G. (1977). Nuclear labelling of leukaemic blast cells with triated thymidine triphosphate in 35 patients with acute leukaemia. *Br. J. Haematol.* **37**, 475–482.

123. Lansdorp, P. M., Sutherland, H. J., and Eaves, C. J. (1990). Selective expression of CD45 isoforms on functional subpopulations of CD34+ hemopoietic cells from human bone marrow. *J. Exp. Med.* **172**, 363–366.

124. LeBeau, M. M., Albain, K. S., Larson, R. A., Vardiman, J. W., Davis, E. M., Blough, R. R., Golomb, H. M., and Rowley, J. D. (1986). Clinical and cytogentic correlations in 63 patients with therapy-related myelodysplastic syndromes and acute nonlymphocytic leukemia: further evidence for characteristic abnormalities of chromosomes No. 5 and 7. *J. Clin. Oncol.* **4**, 325–345.

125. Liendo, C., Danieu, L., Al-Katib, A., and Koziner, A. (1985). Phenotypic analysis by flow cytometry of surface immunoglobulin light chains and B and T cell antigens in lymph nodes involved with non-Hodgkin's lymphoma. *Am. J. Med.* **79**, 445–454 (Abstract).

126. Ligler, G. S., Brodsky, I., Schlam, M. L., and Fuscaldo, K. E. (1985). Cytogenetics and cell surface marker analysis in CML-1. Prediction of phenotype of acute phase transformation. *Leuk. Res.* **9**, 1093–1098.

127. Loken, M. R., Shah, V. O., Dattilio, K. L., and Civin, C. I. (1987). Flow cytometric analysis of human bone marrow. II. Normal B lymphocyte development. *Blood* **70**, 1316–1324.

128. Loken, M. R., Shah, V. O., Dattilio, K. L., and Civin, C. I. (1987). Flow cytometric analysis of human bone marrow: I. Normal erythroid development. *Blood* **69**, 255–263.

129. Look, A. T., Melvin, S. L., Williams, D. L., Brodeur, G. M., Dahl, G. V., Kalwinsky, D. K., Murphy, S. B., and Mauer, A. M. (1982). Aneuploidy and percentage of S-phase cells determined by flow cytometry correlate with cell phenotype in childhood acute leukemia. *Blood* **63**, 959–967.

130. Look, T. (1985). The emerging genetics of acute lymphoblastic leukemia: Clinical and biologic implications. *Sem. Oncol.* **12**, 92–104.

131. Lu, L. I., Walker, D., Broxmeyer, H. E., Hoffman, R., Hu, W., and Walker, E. (1987). Characterization of adult human marrow hematopoietic progenitors highly enriched by two-color cell sorting with my10 and major histocompatibility class II monoclonal antibodies. *J. Immunol.* **139**, 1823–1829.

132. Maddox, A. M., Johnston, D. A., Barlogie, B., Youness, E., Keating, M., and Freireich, E. J. (1985). DNA-RNA measurements in patients with acute leukemia undergoing remission induction therapy *J. Clin. Oncology* **3**, 799–807.

133. Maiolo, A. T., Foa, P., Mozzana, R., Chiorboli, O., Ciani, A., Maisto, A., and Starace, G. (1982). Flow cytometric analysis of cellular DNA in human acute nonlymphatic leukemias and dysmyelopoietic syndromes. *Cytometry* **4**, 265–267.

134. Majdic, O., Liszka, K., Lutz, D., and Knapp, W. (1981). Myeloid differentiation antigen defined by a monoclonal antibody. *Blood* **58**, 1127–1133.

135. Martens, A. C. M., and Hagenbeek, A. (1985). Detection of minimal disease in acute leukemia using flow cytometry: studies in a rat model for human acute leukemia. *Cytometry* **6**, 342–347.

136. Maslak, P., Espiritu, E., Godfrey, L., and Andreeff, M. (1990). Clinical application of DNA/RNA flow cytometry in adult leukemia. *Cytometry* **4**, 97 (Abstract).

137. Matsuoka, M., Hattori, T., Chosa, T., Tsuda, H., Kuwata, S., Yoshida, M., Uchiyama, T., and Takatsuki, K. (1986). T3 surface molecules on adult T-cell leukemia cells are modulated *in vivo*. *Blood* **67**, 1070–1076.

138. Melamed, M. R., Adams, L. R., Zimring, A., Murnick, J. G., and Mayer, K. (1972). Preliminary evaluation of acridine orange as a vital stain for automated differential leukocyte counts. *Am. J. Clin. Pathol.* **57**, 95–102.

139. Mertelsmann, R., Koziner, B., Ralph, P., Filippa, D., McKenzie, S., Arlin, Z. A., Gee, T. S., and Moore, M. A. S. (1978). Evidence for distinct lymphocytic and monocytic populations in a patient with terminal transferase-positive acute leukemia. *Blood* **51**, 1051–1056.

140. Meyers, M. B., Rittmann-Grauer, L., O'Brien, J. P., and Safa, S. R. (1989). Characterization of monoclonal antibodies recognizing an Mr 180,000 p-glycoprotein: differential expression of the Mr 180,000 and Mr 170,000 p-glycoproteins in multidrug-resistant human tumor cells. *Cancer Res.* **49**, 3209–3214.

141. Mirro, J. J. R. and Behm, F. G. (1987). Immunophenotype in the characterization and diagnosis of acute leukemia. Myeloid markers. *In* "The Acute Leukemias" (S. Stass, ed.), pp. 271–298. Marcel Dekker. New York.

142. Montecucco, C., Riccardi, A., Traversi, E., Giodano, P., Mazzini, G. and Ascari, E. (1983).

Proliferative activity of bone marrow cells in primary dysmyelopoietic (preleukemic) syndromes. *Cancer* **52**, 1190–1195.

143. Mullaney, P. F., Van Dilla, M. A., and Coulter, J. R. *et al.* (1969). Cell sizing: a light scattering photometer for rapid volume determination. *Rev. Sci. Instrum.* **40**, 1029–1034.

144. Neame, P. (1985). Simultaneous or sequential expression of lymphoid and myeloid phenotypes in acute leukaemia. *Blood* **65**, 142–148.

145. Nitta, M., Kato, Y., Strife, A., Wachter, M., Jerrold, F., Perez, A., Jhanwar, S., and Clarkson, B. (1985). Incidence of involvement of the B and T lymphcyte lineages in chronic myelogenous leukemia. *Blood* **66**, 1053–1061.

146. Paietta, E., Andreeff, M., Papenhausen, P., Gucalp, R., and Wiernik, P. H. (1989). Distinct antigen expression related to DNA ploidy in a case of biphenotypic leukemia. *Leukemia* **3**, 76–78.

147. Paietta, E., Mittermayer, K., and Schwarzmeier, J. (1980). Proliferation kinetics and cyclinc AMP as prognostic factors in adult acute leukemia. *Cancer* **46**, 102–108.

148. Pallavicini, M. G., Summers, L. J., Giroud, F. J., Dean, P. N., and Gray, J. W. (1985). Multivariate analysis and list mode processing of murine hemopoietic subpopulations for cytokinetic studies. *Cytometry* **6**, 539–549.

149. Palumbo, A., Minoxada, J., Erikson, J., Croce, C. M., and Rovera, G. (1984). Lineage infidelilty of a human myelogenous leukemia cell line. *Blood* **64**, 1059–1063.

150. Pedersen-Bjergaard, J., Vindelov, L., Philip, P., Ruutu, P., Elmgreen, J., Repo, H., Christensen, I. J., and Killmann, S.-A. (1982). Varying involvement of peripheral granulocytes in the clonal abnormality-7 in bone marrow cells in preleukemia secondary to treatment of other malignant tumors: cytogentic results compared with results of flow cytometric DNA analysis and neutrophil chemota. *Blood* **60**, 172–179.

151. Peters, S. W., Clark, R. E., Hoy, T. G., and Jacobs, A. (1986). DNA content and cell cycle analysis of bone marrow cells in myelodysplastic syndromes (MDS). *Br. J. Haematol.* **62**, 239–245.

152. Pinkel, D., Landegent, J., Collins, C., Fuscoe, J., Segraves, R., Lucas, J., and Gray, J. (1988). Fluorescence *in situ* hybridization with human chromosome-specific libraries: Detection of trisomy 21 and translocation of chromosome 4. *Proc. Natl. Acad. Sci. USA* **85**, 9138–9142.

153. Pinkel, D., Straume, T., and Gray, J. W. (1986). Cytogenetic analysis using quantitative, high-sensitivity fluorescence hybridization. *Proc. Natl. Acad. Sci. USA* **83**, 2934–2938.

154. Poppitt, D. G., McGown, A. T., and Fox B. W. (1984). The use of a fluorescent methotrexate probe to monitor the effects of three vinca alkaloids on a mixed population of parental L1210 and gene-amplified methotrexate-resistant cells by flow cytometry. *Cancer Chemother. Pharmacol.* **13**, 54–57.

155. Preisler, H. D., Epstein, J., and Barcos, M., *et al.* (1984). Prediction of response of acute nonlymphocytic leukaemia to therapy with "high dose" cytosine arabinsoide. *Br. J. Haematol.* **58**, 19–32.

156. Prentice, A. G., Smith, A. G., and Bradstock, K. F. (1980). Mixed lymphoblastic myelomonoblastic leukemia in treated Hodgkin's disease. *Blood* **56**, 129–133.

157. Pui, C. H., Crist, W. M., and Look, A. T. (1990). Biology and clinical significance of cytogenetic abnormalities in childhood acute lymphoblastic leukemia. *Blood* **76**, 1449–1463.

158. Raskind, W. H., Tirumali, N., Jacobson, R., Singer, J., and Fialkow, P. J. (1984). Evidence for a multistep pathogenesis of a myelodysplastic syndrome. *Blood* **63**, 1318–1323.

159. Raza, A., Maheshwari, Y., and Preisler, H. D. (1987). Differences in cell cycle characteristics among patients with acute nonlymphocytic leukemia. *Blood* **69**, 1647–1653.

160. Raza, A., Preisler, H. D., Day, R., Yasin, Z., White, M., Lykins, J., Kukla, C., Barcos, M., Bennett, J., Browman, G., Goldberg, J., Grunwald, H., Larson, R., Vardiman, J., and Vogler, R. (1990). Direct relationship between remission duration in acute myeloid leukemia and cell cycle kinetics: a leukemia intergroup study. *Blood* **76**, 2191–2197.

161. Raza, A., Spiridonidis, C., Kalust, U., Mayers, G., Bankert, R., and Preisler, H. D. (1985). Double labeling of S-phase murine cells with bromodeoxyuiridine and a second DNA-specific probe. *Cancer Res.* **45**, 2283–2287.

162. Redner, A., Andreeff, M., Miller, D. R., Steinherz, P., and Melamed, M. R. (1984). Recognition of CNS Leukemia by Flow Cytometry. *Cytometry* **5**, 614–618.

163. Redner, A., Grabowski, E., Al-Katib, A., and Andreeff, M. (1986). Common acute lymphoblastic leukemia characterized by four different DNA stemlines with heterogeneity in RNA content, antigen expression and sensitivity to chemotherapy. *Leuk. Res.* **10**, 671–676.

164. Redner, A., Haimi, J., Steinherz, P., and Andreeff, M. (1989). Near haploid leukemia: correlation between kinetic quiescence and failure to respond to chemotherapy. *Proc. Am. Assoc. Cancer Res.* **30**, 226 (Abstract).

165. Redner, A., Hegewisch, S., Haimi, J., Steinherz, P., Jhanwar, S., and Andreeff, M. (1990). A study of multidrug resistance and cell kinetics in a child with near-haploid acute lymphoblastic leukemia. *Leuk. Res.* **14**, 771–777.

166. Redner, A., Melamed, M. R., and Andreeff, M.

(1986). Detection of central nervous system relapse in acute leukemia by multiparameter flow cytometry of DNA, RNA and CALLA. *In* "Annals of the New York Academy of Sciences. Clinical Cytometry" (M. Andreeff, ed.), pp. 241–255. New York Academy of Science, New York.

167. Ricciardi, A., Danova, M., Montecucco, G., Ucci, G., Cassano, E., Giordano, M., Maxxini, G., and Giordano, P. (1986). Adult acute non-lymphoblastic leukaemia: reliability and prognostic significance of pretreatment bone marrow S-phase size determined by flow cytofluorometry. *Scand. J. Haematol.* **36**, 11–17.

168. Ritter, J., Hiddemann, W., Woermann, B., Buechner, T., and Schellong, G. (1985). DNA aneuploidy in children with acute leukemia: II. Correlation with the phenotype of blasts, clinical picture and course of disease. *Klin. Padiatr.* **3**, 221–224.

169. Robertson, L. E., Huh, Y., Hirsch-Ginsberg, C., Kantarjian, H., O'Brien, S., Koller, C., McCredie, K. B., and Keating, M. J. (1990). Clinical, immunophenotypic, and molecular analysis of the completeness of response in chronic lymphocytic leukemia after fludarabine. *Blood* **76** (Suppl. 1) 314a–314a (Abstract).

170. Rosowsky, A., Wright, J. E., Cucchi, C. A., Boeheim, K., and Frei, E. (1986). Transport of a fluorescent antifolate by methotrexate-sensitive and methotrexate-resistant human leukemia lymphoblasts. *Biochem. Pharmacol.* **35**, 356–360.

171. Rosowsky, A., Wright, J. E., Shapiro, H., Beardsley, P., and Lazarus, H. (1986). A new fluorescent dihydrofolate reductase probe for studies of methotrexate resistance. *J. Biol. Chem.* **35**, 356–360.

172. Ross, D. D., Joneckis, C. C., and Schiffer, C. A. (1986). Effects of Verapamil on in-vitro intracellular accumulation and retention of daunorubicin in blast cells from patients with acute nonlymphocytic leukemia. *Blood* **68**, 83–88.

173. Ruiz, J. G., Tafuri, A., Chen, R., Godfrey, L., Lister, C. D., and Andreeff, M. (1991). Comparative analysis of proliferating related proteins (C-MYC, PCNA, Ki67) in stimulated lymphocytes and human acute leukemia. Unpublished.

174. Ryan, D. H., Mitchell, S. J., Hennessy, L. A., Bauer, K. D., Horan, P. K., and Cohen, H. J. (1984). Improved detection of rare CALLA-positive cells in peripheral blood using multiparameter flow cytometry. *J. Immunol. Methods* **74**, 115–128.

175. Salzman, G. C., Crowell, J. M. and Martin, J. C., *et al.* (1975). Cell identification by laser light scattering: identification and separation of unstained leukocytes. *Acta Cytol.* **19**, 374–377.

176. Sasaki, K., Murakami, T., Ogino, T., Takahashi, M., and Kawasaki, S. (1986). Flow cytometric estimation of cell cycle parameters using a monoclonal antibody to bromodeoxyuridine. *Cytometry* **7**, 391–395.

177. Sato, H., Gottesman, M. M., Goldstein, L. J., Pastan, I., Block, A. M., Sandberg, A. A., and Preisler, H. D. (1990). Expression of the multidrug resistance gene in myeloid leukemia. *Leuk. Res.* **14**, 11–22.

178. Sato, H., Preisler, H., Day, R., Raza, A., Larson, R., Browman, G., Goldberg, J., Vogler, R., Grunwald, H., Gottlieb, A., Bennett, J., Gottesman, M., and Pastan, I. (1990). MDR-1 transcript levels as an indication of resistant disease in acute myelogenous leukemia. *Br. J. Haematol.* **75**, 340–345.

179. Shank, B., Andreeff, M. and Li, D. (1983). Cell survival kinetics in peripheral blood and bone marrow during total body irradiation for marrow transplantation. *Int. J. Radiat. Oncol. Biol. Phys.* **9**, 1613–1623.

180. Smets, L. A., Slater, R. M., Behrendt, H., Vant-Veer, M. D., and Homan-Block, J. (1985). Phenotypic and karyotypic properties of hyperdiploid acute lymphoblastic leukaemia of childhood. *Br. J. Haematol.* **61**, 113–123.

181. Smith, L. J., Curtis, J. E., Messner, H. A., Senn, J. S., Furthmayr, H., and McCulloch, E. A. (1983). Lineage infidelity in acute leukemia. *Blood* **61**, 1138–1145.

182. Smith, L. J., and McCulloch, E. A. (1984). Lineage infidelity following exposure of T lymphoblasts (MOLT-3 cells) to 5-azacytidine. *Blood* **63**, 1324–1330.

183. Sokal, J., and Gomez, G. A. (1986). The Philadelphia chromosome and Philadelphia chromosome mosaicism in chronic granulocytic leukemia. *J. Clin. Oncol.* **4**, 104–111.

184. Spangrude, G. J., Heimfeld, S., and Weissman, I. L. (1988). Purification and characterization of mouse hematopoietic stem cells. *Science* **241**, 58–62.

185. Speth, P. A. J., Linssen, P. C. M., Boszeman, J. B. M., Wessels, H. M. C., and Haanen, C. (1985). Quantitation of anthracyclines in human hematopoietic cell subpopulations by flow cytometry correlated with high pressure liquid chromatography. *Cytometry* **6**, 143–150.

186. Squires, J., Redner, A. and Andreeff, M. (1988). P53 expression in human leukemia: restriction to lymphoid cell lineage and potential prognostic importance in acute lymphoblastic leukemia. *Proc. Am. Assoc. Cancer Res.* **29**, 454 (Abstract).

187. Stass, S., Mirro, J., Melvin, S., Pui, C.-H., Murphy, S. B., and Williams, D. (1984). Lineage switch in acute leukemia. *Blood* **64**, 701–706.

188. Strife, A., and Clarkson, A. (1988). Biology of chronic myelogenous leukemia: Is discordant maturation the primary defect? *Sem. Hematol.* **25**, 1–19.

189. Suarez, C., Miller, D. R., Steinherz, P. G., Melamed, M. M., and Andreeff, M. (1985). DNA and RNA determination in 111 cases of childhood acute lymphoblastic leukaemia (ALL) by flow cytometry: correlation of FAB classification with DNA stemline and proliferation. *Br. J. Haematol.* **60**, 677–686.

190. Tafuri, A. and Andreeff, M. (1990). Kinetic rationale and limitations of combined cytokine-chemotherapy treatment of acute myeloblastic leukemia: effects of G-CSF, GM-CSF and IL-3 on recruitment and cytoxicity. *Leukemia* **4**, 826–834.

191. Tafuri, A. and Andreeff, M. (1990). Cell kinetic modulation of hematologic malignancies by cytokines: recruitment therapy to overcome cytokinetic resistance. *In* "Hematological Growth Factors in Clinical Applications" (R. Mertelsmann, and F. Herrman, eds.), pp. 381–399. Marcel Dekker, New York.

192. Tanke, H. J., Niewenhuis, I. A. B., Koper, G. J. M. *et al.* (1983). Flow cytometry of reticulocytes applied to clinical hematology. *Blood* **61**, 1091–1097.

193. Tapiero, H., Fourcade, A., Vaigot, P., and Farhi, J. J. (1982). Comparative uptake of adriamycin and daunorubicin in sensitive and resistant friend leukemia cells measured by flow cytometry. *Cytometry* **2**, 298–302.

194. Taylor, I. W., and Tattersall, M. H. (1981). Methotrexate cytotoxicity in culture human leukemic cells studied by flow cytometry. *Cancer Res.* **41**, 1549–1558.

195. Terry, N. H. A., White, R. A., Meistrich, M. L., and Calkins, D. P. (1991). Evaluation of flow cytometric methods for determining population potential doubling times using cultured cells. *Cytometry* **12**, 234–241.

196. Terstappen, L. W. M. M., Hollander, Z., Meiners, H., and Loken, M. R. (1990). Quantitative comparison of myeloid antigens on five lineages of mature peripheral blood cells. *J. Leuk. Biol.* **48**, 138–148.

197. Terstappen, L. W. M. M., Huang, S., Safford, M., Lansdorp, P. M., and Loken, M. R. (1991). Sequential generations of hematopoitic colonies derived from single nonlineage-committed CD34 + CD38 − progenitor cells. *Blood* **77**, 1–10.

198. Terstappen, L. W. M. M., Johnsen, S., Segers-Nolten, M. J., and Loken, M. R. (1990). Identification and characterization of plasma cells in normal human bone marrow by high-resolution flow cytometry. *Blood* **76**, 1739–1747.

199. Terstappen, L. W. M. M., and Loken, M. R. (1990). Myeloid cell differentiation in normal bone marrow and acute myeloid leukemia assessed by multidimensional flow cytometry. *Anal. Cell. Pathol.* **2**, 229–240.

200. Terstappen, L. W. M. M., Mickaels, R. A., Dost, R., and Loken, M. R. (1990). Increased light scattering resolution facilities multidimensional flow cytometry analysis. *Cytometry* **11**, 506–512.

201. Terstappen, L. W. M. M., Safford, M., and Loken, M. R. (1990). Flow cytometric analysis of human bone marrow III. Neutrophil maturation. *Leukemia* **4**, 657–663.

202. Theml, H., Trepel, G., and Schick, P. *et al.* (1973). Kinetics of lymphocytes in chronic lymphocytic leukemia: Studies using continuous 3H-thymidine infusion in two patients. *Blood* **42**, 623–636.

203. Thiebaut, F., Tsuruo, T., Hamada, H., Gottesman, M. M., Pastan, I., and Willingham, M. C. (1987). Cellular localization of the multidrug-resistance gene product P-glycoprotein in normal human tissues. *Proc. Natl. Acad. Sci. USA* **84**, 7735–7738.

204. Tkachuk, D. C., Westbrook, C. A., Andreeff, M., Donlon, T. A., Clearly, M. L., Homge, M., Redner, A., Gray, J., and Pinkel, D. (1990). Direct detection of fusion gene in leukemic cells by two-color fluorescence by *in situ* hybridization. *Science* **250**, 559–562.

205. Tkachuk, D. C., Westbrook, C. A., Andreeff, M., Donlon, T. A., Clearly, M. L., Survanaravan, K., Homge, M., Redner, A., Gray, J., and Pinkel, D. (1990). Detection of bcr-abl fusion in chronic myelogenous leukemia by *in situ* hybridization. *Science* **250**, 559–561 (Abstract).

206. Touw, I., Delwel, R., Bolhuis, R., van Zanen, G., and Loewenberg, B. (1985). Common and pre-B acute lymphoblastic leukemia cells express interleukin 2 receptors and interleukin 2 stimulates *in vitro* colony formation. *Blood* **66**, 556.

207. Traganos, F., Darzynkiewicz, Z., Sharpless, T. and Melamed, M. R. (1977). Simultaneous staining of ribonucleic and deoxyribonucleic acids in unfixed cells using acridine orange in a flow cytofluorometric system. *J. Histochem. Cytochem.* **25**, 46–56.

208. Tsuruo, T., Sugimoto, Y., and Hamada, H. *et al.* (1987). Detection of multidrug resistance markers, p-glycoprotein in MDR1 mRNA, in human leukemia cells. *J. Cancer Res.* **78**, 1415–1419.

209. Ucci, G., Riccardi, A., Doermer, P., *et al.* (1986). Rate of time of DNA synthesis of human leukemic blasts in bone marrow and peripheral blood. *Cell Tissue Kinet* **19**, 429–435.

210. Uckun, F. M., Jaszcz, W., Ambrus, J. L., Fauci, A. S., Gajl-Peczalska, K., Song, C. W., Wick, M. R., Myers, D. E., Waddick, K., and Ledbetter, J. A. (1988). Detailed studies on expression and function of CD19 surface determinant by using B43 monoclonal antibody and the clinical potential of anti-CD19 immunotoxins. *Blood* **71**, 13–29.

211. Van-Dilla, M. A., Trujillo, T. T., Mullaney, P. F., and Coulter, J. T. (1969). Cell Microfluorometry: a Method for Rapid Fluorescent Measurements. *Science* **163**, 1213–1214.

212. Visser, J. W., Martens, A. C., and Hagenbeek, A. (1986). Detection of minimal residual disease in acute leukemia by flow cytometry. *Ann. NY Acad. Sci.* **468**, 268–275.

213. Wagner, H. P., and Hirt, A. (1980). Sequential surface marker and cytokinetic studies on individual cells from children with acute lymphoid leukemia. *Blood* **56**, 648–652.

214. Waldman, F. M., Dolbeare, F., and Gray, J. W. (1985). Detection of cytosine arabinoside resistant cells at low frequency using the bromodeoxyuridine/DN assay. *Cytometry* **6**, 657–662.

215. Walker, L. M. S., and Sandler, R. M. (1978). Acute myeloid leukaemia with monosomy-7 follows acute lymphoblastic leukaemia. *Br. J. Haematol.* **38**, 359.

216. Walters, R. S., Johnston, D. A., Dixon, D. O., *et al.* (1987). Nucleic acid flow cytometry: an aid to diagnosis and prognosis in acute leukemia in adults. *In* "The Acute Leukemias" (S. A. Stass, ed.). pp. 203–245. Marcel Dekker, Inc., New York.

217. Weinberg, D. S., Pinkus, G. S., and Ault, K. A. (1984). Cytofluorometric detection of B cell clonal excess: a new approach to the diagnosis of B cell lymphoma. *Blood* **63**, 1080–1087 (Abstract).

218. White, R. A., and Meistrich, M. L. (1986). A comment on "a method to measure the duration of DNA synthesis and the potential doubling time from a single sample". *Cytometry* **7**, 486–490.

219. Wormann, B., Konemann, S., Safford, M., Schreiber, K., Zurlutter, K., Loken, M. R., Buechner, T., Hiddemann, W., and Terstappen, L. W. M. M. (1990). Immunophenotypic heterogeneity of leukemic cells in AML at diagnosis and after chemotherapy. *Blood* **76**, (Suppl. 1) 337a (Abstract).

220. Yataganas, X., Strige, A., Perez, A., and Clarkson, B. (1974). Microfluorimetric Evaluation of Cell Kill Kinetics with 1-B-D-Arabinofuranosylcytosine. *Cancer Res.* **34**, 2795–2806.

Selected Definitions

BFU-E	Burst forming unit for erythropoiesis. This is an early progenitor cell in the bone marrow which is capable of forming a large family of erythropoietic cells, as evidenced by the formation of large colonies ("bursts) in semisolid culture.
BrdU	Bromodeoxiuridine. An analog of thymidine, one of the four nucleotides which form DNA. The analog will be incorporated into DNA if available during DNA synthesis.
CD antigens	A group of surface antigens in hemopoietic cells which can be defined by corresponding monoclonal antibodies. According to the current nomenclature, they are numbered from 1 upwards.
CFU-E	A colony forming unit for erythropoiesis. A late progenitor cell for the formation of erythropoiesis, as evidenced by formation of small erythropoietic colonies in semisolid culture medium.
CFU-GEMM	A colony forming unit for granulocytes, erythrocytes, macrophages and megakaryocytes which forms corresponding colonies in semisolid medium; considered to be a stem cell.
CFU-Meg	A colony forming unit for megakaryocytes. A progenitor cell capable of forming megacytic colonies in semisolid medium.
CFU-Mix	The same as CFU-GEMM.
CFU-S	Colony forming unit in spleen. A stem cell capable of forming all types of hemopoietic cells as colonies in the spleen following transfusion of bone marrow cells to lethally irradiated mice.
Circadian variations	Variations which occur with a periodicity of 24 ± 3 hours.
CSF	Colony stimulating factors. Glycoproteins which are capable of stimulating progenitor cells for the production of hemopoietic colonies in semisolid medium. There are altogether four of these: G-CSF, stimulating production of granulocytes; GM-CSF, stimulating granulocyte and macrophage formation; M-CSF (CSF_1) stimulating formation of macrophages; and interleukin-3 (IL-3) stimulating the formation of all types of hemopoietic cells. Several other stimulators of polypeptide nature have recently been described but are not yet classified as CSFs.
Interleukins	Polypeptide growth factors which stimulate hemopoiesis and regulate leukocyte function. So far 12 different interleukins have been described, IL1–12.
Ki-67	A nuclear antigen of protein nature which is expressed in proliferating cells.
Lymphoblasts	Cells capable of forming lymphocytes, i.e. proliferating early lymphocytes.
Metamyelocytes	Maturing myelopoietic cells which have lost the ability to divide, i.e. the last stage before a mature granulocyte.
Myeloperoxidase	An enzyme (peroxidase) present in myelopoietic cells, i.e. mainly in granulocytes.
Myelopoiesis	The formation of granulocytes and monocytes, or cells belonging to the class of cells which are maturing to granulocytes and monocytes through a defined number of cell divisions.

Opsonins	Proteins (mainly immunoglobulins and complement factors) which act as cofactors for phagocytosis by binding to the targets of phagocytic cells, e.g. bacteria.
Normoblasts	Cells maturing to erythrocytes.
Phagocytes	Cells capable of taking up foreign material, either of exogenous or endogenous origin in the body.
Promyelocytes	An early myelopoietic cell, descendant of myeloblasts and ancestor of myelocytes.
Respiratory burst	The increase of oxidative metabolism in phagocytic cells caused by phagocytosis.
Reticulocyte	An erythropoietic cell which has lost nucleus but still contains some nuclear fragments, i.e. a newly formed erythrocyte.
Stem cell	An early hemopoietic cell capable of forming all types of hemopoietic cells according to the actual needs.

Index